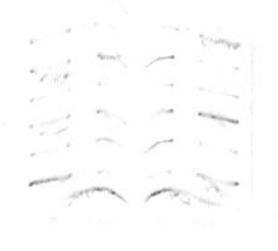

PROTEIN STRUCTURE PREDICTION

BIONFORMATIC APPROACH

IUL Biotechnology Series

Igor F. Tsigelny, *Series Editor*

1. ***Protein Crystallization: Techniques, Strategies, and Tips.***
   A Laboratory Manual.
   Terese M. Bergfors, ed.
2. ***Pharmacophore Perception, Development, and Use in Drug Design.***
   Osman Güner, ed.
3. ***Protein Structure Prediction: Bioinformatic Approach.***
   Igor F. Tsigelny, ed.
4. ***Methods and Results in Crystallization Membrane Proteins.***
   So Iwata, ed.

PROTEIN STRUCTURE PREDICTION

BIOINFORMATIC APPROACH

edited by

IGOR F. TSIGELNY

INTERNATIONAL UNIVERSITY LINE
La Jolla, California

Library of Congress Cataloging-in-Publication Data

Protein Structure Prediction: Bioinformatic Approach / edited by Igor F. Tsigelny

p. ; cm. -- (IUL biotechnology series ; 3)
Includes bibliographical references and index.
ISBN 0-9636817-7-X (hardcover)
1. Proteins--Structure--Computer simulation. 2. Bioinformatics.
[DNLM: 1. Protein Conformation. 2. Computational Biology. QU 55 P96875 2002] I. Tsigelny, Igor F. 1950- II. Series.

QP551 .P697685 2002
547'.750442--dc21

2002002352
CIP

Cover illustration: Plate 13.2 from Chapter 13: *Protein Structure Prediction on the Basis of Combinatorial Peptide Library* by Igor Tsigelny, Yuriy Sharikov, Vladimir Kotlovyi, Michael J. Kelner, and Lynn F. Ten Eyck.

Post Office Box 2525,
La Jolla, CA 92038-2525, USA

Library of Congress Catalog Card Number 2002002352

Printed in the United States of America

10 9 8 7 6 5 4 3 2 1

ISBN 0-9636817-7-X $129.95 Hardcover

Contents

Preface

Prediction of protein structure is very important today. Whereas more than 17,000 protein structures are stored in PDB, more than 110,000 proteins are stored only in SWISSPROT. The ratio of solved crystal structures to a number of discovered proteins to about 0.15, and I do not see any improvement of this value in the future. At the same time development of genomics has brought an overwhelming amount of DNA sequencing information, which can be and already is used for constructing the hypothetical proteins.

This situation shows the great importance of protein structure prediction. The field is growing very rapidly. A simple analysis of publications shows that the number of articles having the words 'protein structure prediction' has almost doubled since 1995.

So many really great ideas are used as a basis for the current prediction systems. Some of them will evolve into the next generation of the prediction software but some of them, even very promising, will be lost and rediscovered in the future. Here we tried to include the variety of methods representing the most interesting concepts of current protein structure prediction.

This compendium of ideas makes this book an invaluable source for scientists developing prediction methods. In many cases authors describe successful prediction methods and programs that make this book an invaluable source of information for numerous users of prediction software.

The first chapter describes the protein structure prediction program PROSPECT that produces a globally optimal threading alignment for a typical threading energy function, and allows users to easily incorporate experimental data as constraints into the threading process. PROSPECT also provides a confidence assessment of a threading result based on a neural network. The second chapter presents a protein fold-recognition method that selects the best fold model for a given protein sequence from a library of structural hidden Markov models (HMMs). The HMMs are built from protein structures following their modular decomposition into the secondary structure elements and representing those elements by a pre-designed set of submodels. The third chapter describes a method to fold proteins into simplified three-dimensional structures constructed from small fragments cut out of a representative set of known three-dimensional structures. The three-dimensional protein structures and fragments are represented in a simplified form as a sequence of angle pairs, one angle pair per residue. Chapter 4 describes the application of HMM constructed on the basis of structural alignments for protein structure prediction. An example system HMM-SPECTR is given with the description of different types of HMMs based on structural alignments. Chapter 5 reviews the different methods of extraction of information from multiple sequence alignments and illustrates how to use them as a primary source of information. The chapter describes the application of rarely used features such as sequence conservation, variations between sub-families, correlation between the patterns of mutation of pairs of positions, and the distribution of apolar residues for structure prediction. Chapter 6 illustrates how knowledge of protein three-dimensional structure can be used to identify homologues of known structure, generate sequence-structure alignments and assist model building. It describes the programs: HOMSTRAD, a database of structure-based alignments for protein families of known structure, JOY, a program to annotate local environments in structure-based alignments and FUGUE, a program to perform sequence-structure homology recognition. Chapter 7 proposes a different concept of sequence homology. This concept is derived from a periodicity analysis of the physicochemical properties of the residues constituting proteins primary structures. The analysis is performed using a front-end processing technique in automatic speech recognition by means of which the cepstrum (measure of the periodic wiggliness of a frequency response) is computed that leads to a spectral envelope that depicts the subtle periodicity in physicochemical characteristics of the sequence.

Chapter 8 describes the building block protein folding model. Via a building block assigning algorithm, sequence comparisons and weighting scheme, building blocks are assigned to a target protein sequence. The problem of the 'building block' is very important for both protein folding modeling and protein structure prediction. Authors of several chapters in this book propose different 'building blocks' for discretization of the prediction process. In most cases they do not discuss the physical properties of these blocks, paying attention only to the information coding. The approach of the authors of the chapter 8 clearly defining the physical and informational properties of the building blocks looks very promising.

Chapter 9 describes a new fold recognition method called FROST (Fold Recognition Oriented Search Tool). It includes 1D and 3D comparison and a database of representative three-dimensional structures. The chapter uses information theory concepts for embedding of a number of sequence and structure parameters in one scoring function. This approach makes this chapter very elegant and useful for the developers of protein structure prediction systems. Chapter 10 continues the discussion of how to combine different levels of resolution and representation of a protein and the rationalizations of score functions for protein structure prediction. The statistical mechanical parameters are used together with purely empirical and even ad hoc parameters.

Chapter 11 describes one of the most effective HMM system for biological applications—SAM. The chapter shows an approach to fold recognition that relies on HMMs for both selecting the template and for aligning the target to the template. The technique has been used successfully in three of the Critical Assessment of Structure Prediction (CASP) experiments.

Chapter 12 discusses the important link between genomic information and protein structure. The chapter describes the clues that could be used to help infer the evolutionary relationship via structural similarity and improve the ability to predict the biochemical function. The first such clue is a positional conservation along the genome, i.e., nearby genes tend to be structurally related more often than expected by chance alone. The second such clue is present in expression data: genes that are correlated in expression are more apt to share a common fold than two randomly chosen genes.

Chapter 13 proposes a comprehensive system for computer based drug design. The chapter describes the program HMM-ELONGATOR, which predicts putative protein targets based on a set of peptides shown to bind a drug molecule from combinatorial libraries.

Chapter 14 on the basis of three examples shows how the use of fold recognition helped biologists in planning and devising experiments and in generating verifiable hypotheses. This chapter describes the meta-predictor approach for protein structure prediction. Chapter 15 describes in details the Structure Prediction Meta Server that collects prediction models from many high quality services and translates them into standard formats enabling convenient analysis of the results. The Meta Server offers an infrastructure for the creation of automated jury algorithms, which analyze the set of results for the user and calculate the reliability score for a consensus prediction. Chapter 16 describes a new method for fold recognition, Pcons that utilizes the "consensus analysis." This chapter shows the advantages of Pcons based on the large scale benchmarking.

Chapter 17 starts the part of the book devoted to the concepts of structural alignment. It is obvious that proper structural alignment of proteins is the cornerstone of the majority of prediction methods. This chapter introduces several new views of protein fold space which will help to further understand protein evolution and interpret structural similarities. Differences between the manual (SCOP) and automated (CE) approaches to the structural classification problem are described. Chapter 18 discusses the design principles of a structure alignment system that can be used for structure prediction assessments. This system is based on a hierarchical representation of a protein shape. Such a representation makes the system suitable for effective alignments of structures with low similarity. Chapter 19 describes the Monte-Carlo approach to the construction of multiple structural alignments. Chapter 20 describes the specific example, where an alignment of eukaryotic protein kinases generated using the combinatorial extension algorithm (CE) is compared with a manually derived alignment. Implications for CE are discussed, as well as implications for automated structural alignment in general.

Overwhelmed by current errands, proposals, and papers, we mostly do not think in global terms of our place in building of knowledge, building of science. Nevertheless it is going on and in one way or another we build the structure of scientific knowledge. If the scientific articles are the 'bricks' in this building, books are the cornerstones.

I would like to thank all the authors for devoting their time to the writing of the chapters. I hope this book will be useful to professionals and students in the field.

Igor F. Tsigelny
La Jolla

Contributors

Nick Alexandrov
nicka@ceres-inc.com

Jadwiga Bienkowska
Jadwiga.Bienkowska@serono.com

Philip E. Bourne
bourne@sdsc.edu

David de Juan
dajs@gredos.cnb.uam.es

Carlos Adriel Del Carpio Muñoz
carlos@translell.eco.tut.ac.jp

Damien Devos
devos@cnb.uam.es

Arne Elofsson
arne@sbc.su.se

Huisheng Fang
fang@sbc.su.se

Jose M. Fernández
jmfernandez@gredos.cnb.uam.es

Daniel Fisher
dfischer@cs.bgu.ac.il

Jean-François Gibrat
gibrat@versailles.inra.fr

Osvaldo Graña
osvaldog@gredos.cnb.uam.es

Chittibabu Guda
babu@sdsc.edu

Nurit Haspel
nurith@post.tau.ac.il

Hongxian He
hxian@darwin.bu.edu

Thomas Huber
huber@maths.uq.edu.au

Kevin Karplus
karplus@soe.ucsc.edu

Michael J. Kelner
mkelner@ucsd.edu

Vladimir Kotlovyi
vlad@sdsc.edu

Roland Lüthy
rluethy@amgen.com

Jesper Lundström
jesper@sbc.su.se

Antoine Marin
marin@versailles.inra.fr

Kenji Mizuguchi
kenji@cryst.bioc.cam.ac.uk

Ricardo Núñez Miguel
ricardo@cryst.bioc.cam.ac.uk

Ruth Nussinov
ruthn@ncifcrf.gov

Osvaldo Olmea
olmea@gredos.cnb.uam.es

Florencio Pazos
pazos@cnb.uam.es

Joël Pothier
jompo@abi.snv.jussieu.fr

Robert Rogers Jr.
rogers@darwin.bu.edu

Leszek Rychlewski
leszek@bioinfo.pl

Eric D. Scheeff
scheeff@sdsc.edu

Yuriy Sharikov
sharikov@sdsc.edu

Jiye Shi
jiye@cryst.bioc.cam.ac.uk

Ilya N. Shindyalov
shindyal@sdsc.edu

Naomi Siew
nomsiew@cs.bgu.ac.il

Lynn F. Ten Eyck
teneyckl@sdsc.edu

Andrew E. Torda
Andrew.Torda@anu.edu.au

Chung-Jung Tsai
tsai@ncifcrf.gov

Igor Tsigelny
itsigeln@ucsd.edu

Alfonso Valencia
valencia@cnb.uam.es

Wayne Volkmuth
wvolkmuth@ceres-inc.com

Christer von Wowern
christervw@chello.se

Björn Wallin
bjorn@sbc.su.se

Haim Wolfson
wolfson@post.tau.ac.il

Dong Xu
xud@ornl.gov

Ying Xu
xyn@ornl.gov

Atsushi Yoshimori
yosimori@translell.eco.tut.ac.jp

Lihua Yu
Lihua.Yu@asrtazeneca.com

Jun Zhu
junz@amgen.com

Karel Zimmermann
karel@versailles.inra.fr

PART I

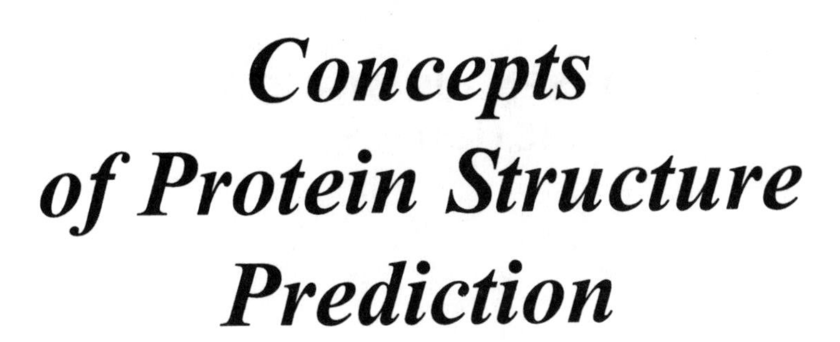

A. PREDICTION METHODS AND SYSTEMS

B. CONSENSUS STRUCTURE PREDICTION

Prediction Methods and Systems

- **Studying Proteins Using PROSPECT**
- **Bayesian Protein Fold Recognition**
- **Simplified Structure Models and Bayesian Blocks**
- **HMM Based Structural Homology Detection**
- **Sequence Information in Protein Structure Prediction**
- **Protein Fold Recognition and Comparative Modeling**
- **Folding Pattern Recognition in Proteins Using Spectral Analysis**
- **Building Blocks and Structure Prediction**
- **Protein Threading Statistics**
- **Protein Structure Prediction by Threading**
- **Using SAM HMMs**
- **Folds, Genomes, and Expression**
- **Protein Structure Prediction Using Library Screening**

1

Computational Studies of Protein Structure and Function Using Threading Program PROSPECT

Dong Xu and Ying Xu

Abstract

In this chapter, we will describe our computer program PROSPECT and show how to use it effectively in protein structure and function analysis. PROSPECT employs threading method to predict protein structures, i.e., given a query protein sequence of unknown structure, PROSPECT searches the templates to find the native-like fold for the query. Compared with other threading programs, PROSPECT has two unique features: (1) it guarantees to find a globally optimal threading alignment for a typical threading energy function; and (2) it allows users to easily incorporate experimental data as constraints into the threading process, and guarantees to find optimal alignment under the specified constraints. PROSPECT also provides a confidence assessment of a threading result based on a neural network. We have successfully applied PROSPECT in the community-wide experiments on the Critical Assessment of Techniques for Protein Structure Prediction (CASP). In collaboration with experimentalists, we have also used PROSPECT to study some biologically interesting proteins, including vitronectin, DNA-activated protein kinase, and PTR3 protein. Based on these applications, we have developed an effective procedure for making structural predictions and structure-based functional inference.

1

Computational Studies of Protein Structure and Function Using Threading Program PROSPECT

Dong Xu and Ying Xu

Oak Ridge National Laboratory, Oak Ridge, Tennessee

1.1. Introduction

One of the key goals in the post-genome era is to characterize the gene products (proteins) obtained from large-scale genome sequencing efforts. Generally the first step to annotate a new protein is to carry out a sequence comparison using tools such as BLAST/PSI-BLAST.[1] However, for a significant portion of new proteins, sequence comparison cannot establish any relationship between them and the proteins with known cellular roles or molecular functions. These unknown proteins may fall outside the limit of sequence-based techniques to detect homology. Characterizing them represents an urgent need and a grand challenge. One of the methods for characterizing an unknown protein is to determine or predict its three-dimensional (3D) structure, which often provides the basis for functional studies and drug designs at the molecular level. Protein structures have traditionally been solved by X-ray crystallography or NMR methods. However, due to high cost and low throughput, these methods can only solve a small fraction of the new proteins being discovered. In addition, many proteins have very low expression level or low solubility, so it could be very difficult to solve their structures experimentally. On the other hand, existing prediction methods for protein structures have matured to a level that useful information can be extracted for functional inferences. In many cases the predicted protein 3D structures are highly useful for guiding experimentalists in designing new experiments for further investigations of protein functions. It is expected that protein structure prediction will play an important role for many years to come.

Existing protein structure prediction methods fall into two main classes: (a) *ab initio* methods[2,3] give predictions based on physicochemical principles directly, and (b) comparative modeling methods give predictions based on identified structural relationship with known structures. Comparative modeling methods, if applicable, are generally quicker and more accurate than *ab initio* methods. Statistics from the PDB Web site (http://www.rcsb.org/pdb/holdings.html) show that about 90% of the proteins submitted to the PDB database[4] during 1997–2000 share similar folds to structures already in PDB. This suggests that many of these protein structures are potentially solvable by comparative modeling methods. It by no means indicates that comparative modeling may apply to 90% of proteins because of the biased sampling from the protein space. In particular, membrane proteins are clearly under-represented in PDB. Nevertheless, it is generally believed that 60–70% of new proteins are potentially solvable using comparative modeling methods.[5] As more protein structures are solved, comparative modeling methods will become more applicable, and they are becoming more popular than *ab initio* methods.

Comparative modeling consists of two approaches: (1) sequence-sequence comparison,[1,5] and (2) sequence-structure comparison (threading).[7,8] Threading makes a structural fold prediction from the amino-acid sequence by recognizing a structural template that represents the native-like fold of a query protein in a database of experimentally determined structures. Fold recognition is typically achieved by finding an optimal alignment of residues of the query with residue positions of each structural template in the database, and by identifying those sequence-structure alignments that are statistically significant. For each statistically significant alignment, the residues of the query sequence are predicted to have the coordinates of the aligned backbone positions in the template structure. Since protein threading uses structural information in addition to the sequence-based information for alignment, it is often more effective than sequence-based methods for identifying native-like folds. This is particularly true for a significant portion of proteins in a genome that do not have detectable sequence similarity to any known protein sequences (so-called "orphan genes"), since no profile of multiple-sequence alignment[1] can be generated for sensitive profile-based methods in sequence comparison.

In this chapter we will focus on the threading tool that we have developed, PROSPECT (PROtein Structure Prediction and Evaluation Computer Toolkit).[9] Compared to other threading programs, PROSPECT has two unique features. First, it guarantees to find the globally optimal alignment between a query sequence and a template structure for an energy function consisting of a pairwise contact energy, and does so efficiently.[10] Second, it allows the user to incorporate experimental data as "soft" restraints (allowing violation) or "hard" constraints (without violation) in the threading process, and it guarantees to find the globally optimal threading under the specified restraints/constraints.[11] The advantage of having a rigorous threading algorithm has been demonstrated in the CASP (Community Wide Experiment on the Critical Assessment of Techniques for Protein Structure Prediction) contests.[12,13] The capability of allowing the user to add partial experimental data as threading inputs has greatly enhanced the applicability and flexibility of the PROSPECT system, as we will show in the following. PROSPECT can basically run on any platform with a C compiler. It has been compiled and tested on PC (with Linux or Windows operating systems), Silicon Graphics, Sun, DEC, and HP. It has also been implemented on an IBM/SP3 supercomputer at Oak Ridge National Laboratory, with a Web interface at http://compbio.ornl.gov/structure/prospect_server/. A detailed description of the PROSPECT package can be found in reference.[9] The availability and the manual of PROSPECT can be found at the Web site http://compbio.ornl.gov/structure/prospect/. This chapter mainly focuses on its application from the user's point of view. Most of what will be addressed also applies to other threading methods.

Threading can generate backbone structures with accuracies better than 4 Å RMSD (root mean square deviation) in many cases.[14] However, in some cases, the native-like fold of the query protein may not be recognized correctly as the top hit of the threading templates, and the alignment between the query and the template may not be accurate even when the native-like fold is identified correctly. Unlike homology modeling, which deals with structure modeling when a protein has a close homolog among solved structures, threading attempts to solve a much more general and difficult problem, including structure modeling of proteins that have no detectable sequence similarities to any proteins with known structures. Compared to the prediction accuracies by homology modeling, threading clearly has a significant room for improvement, particularly in its energy function, confidence assessment of threading results, and integration between threading and other methods. While application of threading tools is far from a rigorous science yet, it is crucial for users to learn how to effectively apply a threading tool. Just like many experimental tools, user expertise is important in obtaining high-quality results. For example, the quality of protein structures determined by different groups can be quite different, even when they use the same type of NMR machines. In threading, human interpretation of threading results often affects prediction accuracy greatly. The fourth CASP (CASP4)[15] shows that the prediction accuracies of fully automated prediction servers (even the combined best predictions by all the servers) are clearly lower than computer-assisted human predictions. It is even more complicated to use the predicted structural information to analyze protein function. In this chapter we will address how to use PROSPECT effectively to help elucidate protein structures and to infer functions. We will explain how to take advantage of the strengths of PROSPECT and how to deal with the current limitations of the program.

The rest of this chapter is divided into five sections. Section 1.2 outlines the methods used in the PROSPECT program. Section 1.3 describes the protocol of using PROSPECT in protein structure and function analysis. Section 1.4 shows the performance of PROSPECT. Section 1.5 gives several examples of using PROSPECT in conjunction with experimental studies. Section 1.6 provides a summary.

1.2. Method of PROSPECT

PROSPECT has four components:

(1) a database of solved 3D protein structures used as threading templates;

(2) a knowledge-based energy function for measuring the fitness between a query sequence and a structural template;

(3) a "divide-and-conquer" threading algorithm to search for the lowest energy among the possible alignments for a given query-template pair; and

(4) a criterion for estimating the confidence level of the predicted structure based on a neural network

To use PROSPECT effectively, it is important to understand the design and implementation of the four components.

1.2.1. Threading Templates

PROSPECT uses both protein chains from the FSSP non-redundant set and protein domains from the DALI non-redundant domain library[17] as templates. The FSSP template dataset is used as the default in PROSPECT since it is more frequently updated. Sometimes a query protein and a template can structurally match each other partially rather than globally, while our threading algorithm is for a global alignment. To solve this problem, the user may want to use the DALI domain library as the threading templates, to avoid an alignment between structurally unalignable regions. Another method to address this issue is to cut the query protein into several segments and use each segment to do threading. We will discuss this method in detail in the next section. An advantage of using a non-redundant set, like FSSP (currently with less than 2700 protein chains), instead of the whole PDB (currently with more than 25,000 chains) as the template dataset is to save computing time. The drawback, however, is that a protein chain excluded from a template database may fit a query protein better than its representative in the library. To resolve this, we will add a new feature to PROSPECT so that the user can retrieve, from PDB, all protein chains represented by a template in a non-redundant set. The user can either carry out threading alignments with the protein chains to find the best fit or use them in a multiple-template homology modeling.[18]

Each template is derived from a chain (for FSSP) or a domain segment (for DALI domain library) in a PDB file. The following information is used to define a residue in a template:

(1) its amino acid type;

(2) its secondary structure type (three states: α-helix, β-strand, or loop) assigned by the DSSP package;[19]

(3) its solvent accessibility (three states: exposed to solvent, in the interior of the protein, or in between) defined as the percentage of exposed solvent accessible surface area of a residue's side chain according to DSSP; and

(4) its C_β atom coordinates.

The residues in a template are divided into core (α-helix or β-strand) residues and loop residues. Core residues are expected to have more accurate predictions than loop residues, since the former tend to be conserved among structures of the same fold, while the latter may not be.[20] PROSPECT has adopted two widely accepted assumptions:[20] (1) no gap can occur within a core secondary structure in a threading alignment; and (2) any pairwise interaction involving a loop residue is ignored. These assumptions substantially reduce the computational complexity.

1.2.2. Energy Function

PROSPECT uses four knowledge-based energy terms:

(1) A mutation energy term is used to describe the compatibility of substituting one amino acid type by another. PROSPECT uses PAM250,[21] which has been shown to be one of the best substitution matrices available for threading.[22]

(2) A singleton energy term is used to represent a residue's preference for its local secondary structures and its preference for being in certain solvent environment.

(3) A pairwise energy term between spatially close residues is used to describe the contact preference between residues. We only consider the residue pairs that are separated by at least three amino acids in the protein sequence, since the pairs separated by one or two amino acids in the protein sequence represent local interactions, which are less important in determining an overall fold in threading. The cutoff distance for a pairwise interaction is 7.0 Å between the C_β atoms. This accounts for most of the important inter-residue interactions.[8]

(4) The alignment gap penalty is a linear function of the gap size, which assigns a penalty for opening a gap and a smaller penalty for each extension thereafter.

The parameters for the singleton and pairwise energies were derived from a non-redundant set of known protein structures. The weights between different energy terms are optimized based on structure-structure alignments of protein pairs in the same folds.

PROSPECT has additional energy terms for incorporating experimental data or other predictions as restraints/constraints in the threading process:

(1) disulfide bonds between specified residues;

(2) active sites involving a specified set of residues;

(3) long-range NOE restraints from NMR experiments;

(4) secondary structures predicted by programs like PHD[23] or determined from chemical shift data of NMR experiments; and

(5) position-dependent profile based on a multiple-sequence alignment using SAM.[24]

These restraints/constraints are turned off in default, and the user can choose the weights of the restraint terms in the threading energy function. We highly recommend users to include the secondary structure prediction by PHD, which can help improve the accuracies of both fold recognition and threading alignment. A default optimized weight for the secondary structure term is also available in PROSPECT, and the confidence assessment, which we will discuss in Section 1.2.4, is based on threading with the secondary structure term.

1.2.3 Threading Algorithm

The goal of a threading algorithm is to find an alignment between a query sequence and a template structure, which optimizes the threading energy function as described in Section 1.2.2. PROSPECT employs a *divide-and-conquer* strategy to solve the optimal threading problem. The basic idea of this strategy is to first align part of the query sequence and part of the template optimally, and then merge the partial results to form an optimal global alignment, as shown in Fig. 1.1. For this purpose, we pre-process the template by repeatedly dividing (bi-partitioning) it into sub-structures until each sub-structure contains only one core secondary structure.

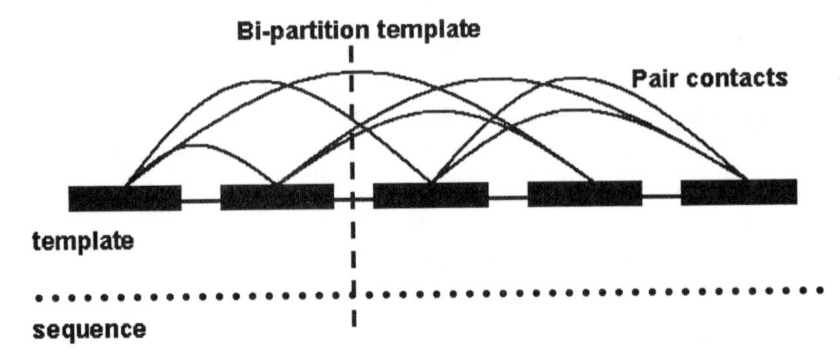

Figure 1.1. A schematic of sequence-structure alignment in PROSPECT. The dotted line at the bottom shows the query sequence. The connected boxes show the template, where each box represents a core secondary structure, and the lines between the boxes represent loop regions. An arc between two core secondary structures indicates that there exists at least one pairwise interaction between the two core secondary structures. The dashed line shows a partition of the template into two sub-structures, one with two core secondary structures and the other with three.

The divide-and-conquer algorithm works correctly on any bi-partition of the template. However, the way a template is partitioned affects the computing time, and we partition a template in such a way that the optimal computing efficiency is achieved.[9] The algorithm solves the entire optimal alignment problem by recursively solving a series of sub-alignment problems between sub-structures and sub-sequences, and then it combines the two sub-alignments that are consistent with the pairwise interactions between the two parts. PROSPECT starts from calculating the alignment score between a core secondary structure and each sequence position. Since we assume that there is no alignment gap within a core alignment, this score can be calculated by simply adding the mutation term, the singleton term, the pairwise interactions within the core, and the possible pairwise interactions between the core and the rest of the protein. PROSPECT then calculates the alignment scores for larger sub-structures consisting of multiple cores. This process continues for larger sub-structures until the top level of the partition tree is reached (i.e., the whole template is considered). When merging two sub-alignments, we check the consistency of the pairwise contacts between the two sub-sequences (from all possible pairwise contacts between two sub-sequences) and add the consistent pairwise interactions. In addition, the optimal loop alignment score in between the two sub-structures is included using the dynamic programming method.[25] It can be shown[9,10] that the final solution of this procedure gives a globally optimal threading score for the given energy function defined in Section 1.2.2.

The computing time of this algorithm is proportional to $mn + Mn^C N^{C/2}$, where m and n represent the lengths of the template and query protein sequences, respectively, M is the number of core secondary structures of the template, N is the maximum allowed difference between the lengths of two aligned loops, and C is the topological complexity of the template.[9] C is typically 3 or 4 for most templates. The actual threading time on a workstation is a few hours for one query sequence (with a typical size of 200–300 residues) against the whole FSSP library. The function of computing time shows how various parameters affect the running time. The computing time may increase substantially as a result of increasing the values of n, M, and C. From the user's perspective, it is very important to balance the threading accuracy and the computing time. If there is some evidence that a large query protein may fold into two compact domains, the user can submit the two domain sequences separately. Running two separate threading jobs would save a substantial amount of computing time by reducing n. If the user expects alignments without large gaps, he or she can decrease M, which has a default value of 20, to save some computing time. When no pairwise interaction is used (i.e., $C = 2$), the computing time can be saved by several folds. On the other hand, the user may lose some accuracy since pairwise interactions are important in threading. To balance the computing time and accuracy. The user can initially run PROSPECT without pairwise interactions for prescreening, and then select a small group of top hits (e.g., top 200 hits) to run threading with pairwise interaction. This is a good strategy for large query proteins.

1.2.4. Confidence Assessment of Threading Results

It is a well-known fact that existing threading programs do not have a reliable way to rank the sequence-fold alignments in terms of the degree of correctness of the alignments. One problem is that the scoring functions used by existing programs are generally not well normalized with respect to various factors of structural templates, which could bias the threading results towards or against some groups of structures. These factors may include (a) the lengths of the templates, (b) the amino acid compositions of the templates, and (c) the structural features of the templates. We have applied the neural network technique to provide a practical way for normalizing the threading scores by PROSPECT. A detailed description of our method can be found in reference.[27] The basic idea is to use a neural network to map a threading score to a real number in the range of [0, 1] in such a way that the closer the number is to zero, the lower the possibility that this particular sequence-structure alignment is a correct prediction. A number of parameters, which are potentially useful for the normalization, are used to help accomplish such a mapping. These parameters include information about the overall scoring distribution of each structural template, the lengths of the query sequence and the template protein, etc. The objective function, which defines what value each threading score is desired to map to, is defined as follows. We calculate the number of structurally alignable residues between each query-template as is defined by the SARF program.[26] A sequence-structure pair is considered to be a *true* pair if their FSSP indices[16] share the same first digit (i.e., they belong to the same fold family). We calculate the frequency of the true pairs given a particular number of alignable residues on a data set of 36,657 sequence-structure alignments, among which 1469 are true pairs. We train a neural network to accomplish the mapping between the parameters in the threading result, as described above, and the frequency, using 50% of the 36,657 pairs as the training set and the remaining 50% of the data points as testing set.[27] The mapped values (neural network score) now have a well-defined meaning. For example, if the mapped value is 0.5, we know that the conditional probability of having a correct fold recognition/alignment is about 60%, as plotted in Fig. 1.2. The ranking of templates according to neural network scores may be different from the one based on the raw scores. A template representing the native-like fold with a low ranking based on raw scores often moves up to top hits in ranking based on neural network scores. However, our neural network scheme is not perfect. Sometimes a template with the native-like fold among the top hits in raw scores may have a very low neural network score. We suggest that the user check the top hits in both raw scores and neural network scores. If a template ranks among the top 10 and it has a neural network score higher than 0.7, it has a high probability of representing the native-like fold. If no such case is found, the user should pay attention to top hits ranked by either raw scores or neural network scores.

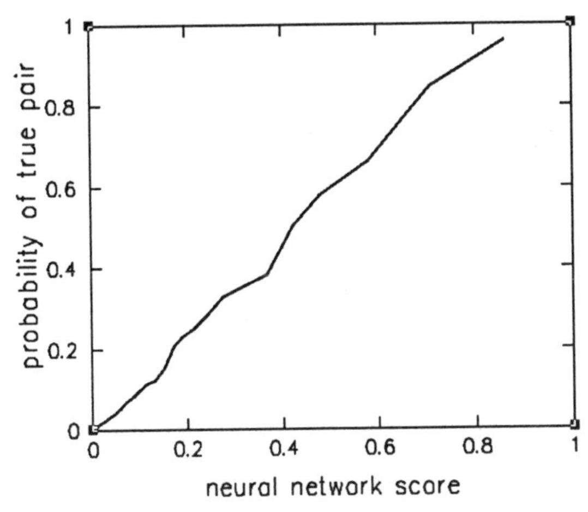

Figure 1.2. The conditional probability of being a true pair for a threading given a neural network score.

1.3. Protocols of Using PROSPECT

Although PROSPECT works well on a large class of proteins just by running the default options without manual inspection, one can improve the performance of structure prediction and obtain more functional information from the threading results by using additional bioinformatic tools and human evaluation. During the past three years since the inception of PROSPECT, we have developed an effective procedure for making structural predictions and structure-based functional inference through two CASP contests and a number of structure prediction exercises jointly with experimentalists. The procedure is outlined in Fig. 1.3. In this section, we will review the procedure in five stages: (1) pre-processing, (2) running PROSPECT, (3) human evaluation of threading results, (4) manual refinement of structural model, and (5) inference of function based on predicted structures.

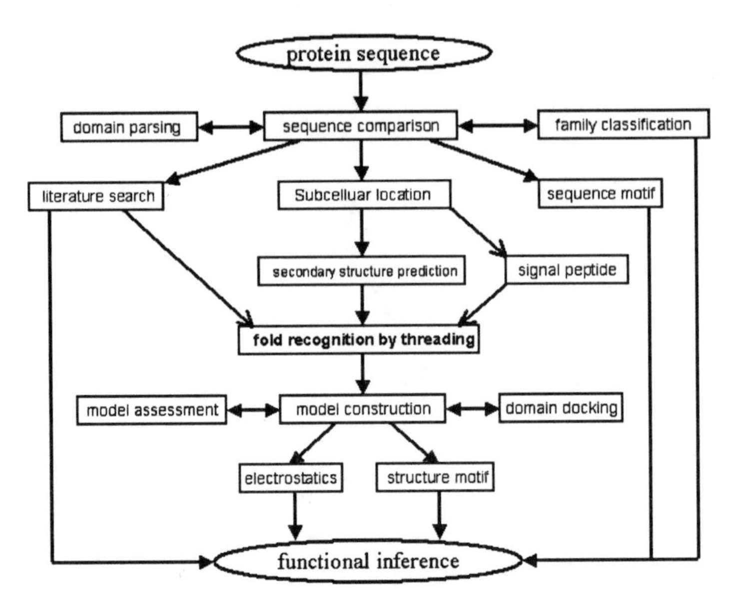

Figure 1.3. A summary of PROSPECT protocol.

1.3.1. Pre-Processing before Running PROSPECT

The goal of pre-processing is to extract information about a query protein from literature and from computational results using publicly available tools. The information collected can be used in three ways: (1) to input the information as restraints/constraints during threading and structure refinement, (2) to help determine native-like fold during human evaluation of the threading results, and (3) to assist functional inference based on predicted structures. Pre-processing involves the following aspects:

1. **Sequence-based search.** One can use sequence-based search tools, such as BLAST/PSI-BLAST[1] and SAM[6], to search possible hits, particularly in PDB and SWISSPROT (http://www.expasy.org/sprot/).[28] If a significant hit is found in PDB, threading may not be necessary. A structure can be built directly from homology modeling using tools like MODELLER.[18] A significant hit in SWISSPROT

can provide useful information such as the EC number of an enzyme and functional annotation in SWISSPROT. Hits in other databases, such as PIR,[29] may also help with the functional annotation of the protein.

2. **Protein classification between globular and membrane proteins.** Researchers can identify possible membrane proteins and their transmembrane regions using the transmembrane prediction tools such as SOSUI.[30]

   Since there are few templates available in PDB for membrane proteins, and the energy function used in threading is derived from globular proteins, threading methods generally do not work for membrane proteins. If a protein is predicted to be a transmembrane protein, using PROSPECT may not be the right approach. If a membrane protein has a soluble domain, PROSPECT can be used to predict the structure of this domain instead of the whole protein.

3. **Domain parsing.** Domain parsing can be carried out using tools such as PRODOM.[31] If the query protein or its homolog is found in SWISSPROT, information of possible domain boundaries may be found in the "Domain structure" entry in SWISSPROT. Identified domain sequences can then be used in threading and in identifying functional modules of the protein. As discussed above, it saves computing time substantially and typically increases the threading accuracy to run separate threading jobs using the contiguous sequences of the compact domains in a large protein.

4. **Signal peptide detection.** SignalP[32] can be used to detect signal peptides. Since a signal peptide is not present in the functional form of a folded protein, using a sequence with signal peptide may decrease the accuracy of threading. The predicted signal peptide should be removed from the input sequence before threading.

5. **Sequence-based function search.** A protein family classification based on Pfam[33] may provide useful function annotation. Motif searches by MOTIF (http://www.motif.genome.ad.jp), PROSITE,[34] PRINTS,[35] and BLOCKS[36] may identify potential active sites, which are often helpful to threading alignment (as shown in the following). A prediction of sub-cellular locations using PSORT[37] may assist functional inference. For example, if a protein is

predicted to be extracellular, it is not likely to be a DNA binding protein.

6. **Secondary structure prediction.** Several tools for secondary structure prediction, including PHD[23] and PsiPred,[38] can achieve about 80% accuracy in predicting α-helix, β-strand, and loop, as demonstrated in CASP4.[15] The output from them, e.g., PHD, can be used as restraints in threading. Although PROSPECT has its own energy function for the secondary structure preference of each amino acid type, it is not as effective as PHD in predicting secondary structures.
7. **Literature search.** A query protein can often be found in a database (e.g., SWISSPROT), which contains the literature references about the protein. The abstracts of these references and related publications in PubMed (http://www.ncbi.nlm.nih.gov/PubMed/) can provide valuable structural and functional information about the query protein. In addition, one can also use the keywords obtained from the annotations in protein databases to do the search in PubMed.

1.3.2. Running PROSPECT

Running PROSPECT includes three components: (1) fold recognition, (2) atomic model construction, and (3) quality assessment of atomic models. The procedures are straightforward and do not require any expertise in biology.

1. **Fold recognition.** Threading against each template in the database is carried out with predicted transmembrane and signal peptide regions removed from the query sequence. For a large query protein, particularly larger than 300 residues, sequences of individual domains can be run separately. If the user does not have any idea about the domain partition of the query sequence, he or she can use overlapping subsequences (such as 1-500, 400-900, 800-1300, 1200-1700, etc.) as presumed domains, to save computing time. Some of the information collected in Section 1.2.2 will be used as constraints during the threading process. Threading alignments generated by PROSPECT are ranked by neural network scores. Based on the neural network assessment,

the user can choose the best templates to build models. A confidence level is assigned for each model according to its neural network assessment score (a "high" confidence is when the score is larger than 0.7, a "medium" confidence is when the score is between 0.4 and 0.7, and a "low" confidence is when the score is less than 0.4).

2. **Atomic model construction.** Alignments with the selected templates in fold recognition are used as inputs for MODELLER,[18] which produces 3D atomic model that can be visualized. PROSPECT provides a command line to automatically generate all the input files needed for running MODELLER. Long unaligned regions (larger than 30 residues) of the query protein to the template are removed from the models, since their structures are unreliable. The user may generate multiple (e.g., 10) structures using MODELLER for each alignment.
3. **Quality assessment of atomic models.** The user can apply structure assessment tools WHATIF[39] and PROCHECK[40] to evaluate the packing and backbone conformations, the inside/outside occupancies of hydrophobic and hydrophilic residues, and stereochemical quality of a predicted structure. Based on this assessment, the user can pick the best of the multiple structures derived from an alignment. When there is no significant sequence similarity between the query and the template, only a backbone model will be provided since a detailed full atomic model is not reliable.

1.3.3. Human Evaluation

The confidence assessment in PROSPECT was designed to identify native-like folds automatically. However, it is far from perfect at this time. Human evaluation is often helpful to select a good template based on the collected structural and functional information. This is particularly true when the confidence level of a computer prediction is low. If some experimental data, such as the location of a disulfide bond, are available, they can be used to verify whether a template represents the native-like fold. If no such data exists, the user may confirm a native-like fold from a set of rules. These rules, which were used in our CASP predictions, require some knowledge of structural biology. Unlike the method used by Alexei Murzin,[15] our method does not require any specific knowledge of a particularly protein or protein family. On the contrary the rules are general to any globular proteins, and most of them could be automated. If no template can be found to represent the native-like fold with good confidence according to the rules and the neural network score is low (less than 0.4), the query may be assigned as a possible new fold. The identification of possible new folds may be useful for identifying novel fold targets in the structural genomics. The following is a summary of the rules for manual evaluation of threading results.

1. **Function comparison.** The most helpful information is often the function of the query protein. Many proteins with unknown structures have function annotation, either from experiments or from a sequence-based search. If a template is among the top ranking hits, either by neural network scores or raw scores, and has the same function as the query, we will have high confidence that the template represents the native-like fold of the query protein. For example, target T0095 of CASP4 is an alpha-catenin. If the user runs PROSPECT, the template 1dow-A (chain A in the PDB file1dow), which is also an alpha-catenin, will be returned as the highest-ranking hit. Even the sequence similarity is only 12.6%, the user can be certain that 1dow-A represents the native-like fold of T0095. Unfortunately, when we ran our CASP4 prediction for T0095, our template database did not have 1dow, since it was released after the CASP4 contest started. A good measurement of the same function for enzyme is the Enzyme Classification (EC) number. If a template with the same EC number of the query protein ranks among the top hits, the confidence for the template to represent the native-like fold is very high. The EC number of a template can be found at *Enzyme Structure Database* (http://www.biochem.ucl.ac.uk/bsm/enzymes/index.html).
2. **Secondary structure comparison.** The user can check the consistency between the predicted secondary structures by PHD and secondary structure assignments of the predicted structure by DSSP.[19] A poor agreement (less than 50%) between them usually indicates low confidence of the prediction. Based on this, many templates can be excluded from consideration for native-like folds.

3. **Motif matching.** Some motifs identified through sequence-based analysis, such as coiled coils and WD40 repeats, have long segments in sequence, and they can be predicted with low false positive rate. These motifs can be used to verify where a template represents the native-like fold of the query protein. For example, in vitronectin, as shown in Section 1.4, a sequence-based analysis found WD40 motifs. The template (1gen) that we identified through PROSPECT also contains WD40 repeats. This increase the confidence of our prediction.
4. **Consensus among different scores.** We found the templates of the native-like fold often have top ranking in both raw score and the neural network score. If a template meets the criterion of ranking top 50 using raw scores and top 10 using reliability assessment scores, it indicates that there is a good chance for the template to represent the native-like fold. Without any human evaluation on a threading result, such a criterion can often provide a best guess for the templates of the native-like fold. We made several successful CASP predictions by simply using this criterion. For example, we could not find any helpful information during the pre-processing for target T0127 (Magnesium chelatase) of CASP4. Hence, we simply chose the template (1a5t) that has the best consensus between the ranking of raw scores and the ranking of neural network scores. It turned out that we recognized the correct fold.
5. **Compactness of aligned region on template.** A good estimate for the compactness of the aligned portion in threading is the normalized radius of gyration. If the value is above 3.0, the aligned portion is not compact enough. This would indicate that the template probably does not represent the native-like fold, unless the template is a multi-domain protein. In addition, if a small query protein only aligns with a portion of a domain in the template, the prediction also has a very low confidence level. The user can easily check the aligned portion of all the top-hit templates by using a Web-interface of the PROSPECT output, which gives both alignments and 3D structures. With a Web browser plug-in (such as CHIME[41]) for viewing structures, the user can see the 3-D graphics of a template and where the aligned portion of the template is based on the color.

6. **Consensus among different templates.** A feature that we found for successful recognition of a native-like fold in many cases is that the top-hit templates of a query protein share the same or similar fold. The templates in a threading library may not share any significant sequence similarity, but they can be in the same fold. If multiple templates with the same fold are among the top hits, the confidence level for the templates to represent the native-like fold of the query protein is high. Here, we use $T0087_1$ (pyrophosphatase) of CASP4 as an example. By running PROSPECT and the reliability assessment of the prediction, we found five templates that met the fourth rule above (among top 50 hits using raw scores and among top 10 hits using reliability assessment scores). They are 1wod (13, 2), 1cyd-A (27, 3), 1enp (47, 5), 2cmd (39, 7), and 1qpz-A (28, 8), where the first number in a bracket indicates the rank of raw score and the second number shows the rank of reliability assessment score. The reliability assessment scores indicate that these templates have 40-50% probability of representing native-like folds. Interestingly, all five templates are structurally alignable with each other in FSSP. This observation convinced us that $T0087_1$ must share some common structural features with the five templates. We then generated structural models for all five templates and ran WHATIF to check the quality of the models. The model derived from 1qpz-A had the best quality. Hence, we chose this model, which turned out to be a good prediction as shown in Section 1.4.2.
7. **Consensus among homologous query sequences.** If several homologous query proteins all rank the same template among their top hits, it increases the chance of the template representing the native-like fold of the homologs in the query family. For example, we found six representative homologs in the same family of the CASP4 target T0100 (Pectin Methylesterase) in Pfam.[33] All of them (including T0100) rank multiple templates of the β-helix fold very high, as shown in Table 1.1. Indeed, T0100 belongs to the fold, as shown in Section 1.4.2.

The first row shows the query proteins T0100 and other SWISSPROT entries that belong to the same Pfam family of T0100 but have significant sequence variations from each other. The first column gives the templates belonging to the β-helix fold in the FSSP database. A number indicates the ranking by the raw scores, and an empty entry indicates that the ranking is not among top 20 hits.

Table 1.1. Raw score ranking of T0100 and its homologs.

proteins	T0100	pme_asptu	pme_aspac	pm21_lyce	pme_burso	pme_braba
1pcl	10			5	13	5
1air		11	7	9	1	19
1qcx-A	18					
1dbg-A				3		4
1bhe	20		17	11	12	3
1rgm	17	1	1	6	15	7
1qal-A	16			19		
1tyv	11	18		13		
1czf-A			18			

1.3.4. Manual Refinement

Although PROSPECT can predict protein structure with little human intervention, as described in Section 1.3.2, manual refinement may improve the quality of the predicted structure. Sometimes, the refinements need to be carried out iteratively. The user may try the following refinements.

1. **Using active sites to refine alignment.** When two proteins have a similar function, they may have the same type of active site. Aligning their active sites provides a good constraint, which can improve the accuracy of threading alignment. The target protein t0053 (CbiK protein with 264 residues) in CASP3 is a good example. The best template to represent the native-like fold of the query was 1akl (310 residues), with a sequence identity of 11.2%. Our threading program recognized 1akl as the best template for t0053, but it aligned only one out of its thirteen secondary structures correctly. It is known that 1akl has an active site at His-183. Using the BLOCK search,[36] we identified His-145 of t0053 as the corresponding active site. We then ran PROSPECT using the constraint that His-145 of t0053 is aligned with His-183 of 1akl. This improved the alignment accuracy. With this constraint, five secondary structures aligned correctly and seven more secondary structures aligned within a 4-residue shift from the structure-structure alignment by SARF.[26]

2. **Manual adjustment.** Currently we use MODELLER to generate a 3D structure after getting the threading alignment. MODELLER works reasonably well when a target and its template have significant sequence similarity, but it was not designed for a target without significant sequence similarity to its template. Hence, the quality of structural models generated from a direct run of MODELLER could be poor. One can improve the quality of a structural model by doing some refinements using loop modeling, multiple templates, etc. The user should also visually check structural models. If the user finds some obvious errors in a model, e.g., overlap of residues in space, he or she may fine-tune the alignment and rebuild a model.
3. **Docking domains.** When a query protein is partitioned into several domain sequences for separate threading jobs, different domains often have different native-like templates that cannot be structurally connected. The user may apply the protein-docking program GRAMM to build a structural model for the whole query protein by docking the domains together (see *the example of vitronectin* in Section 1.5). GRAMM[42] is particularly suitable for this type of application, since it allows inputs of low-resolution structures like our predicted domain structures. Docking is only necessary when the confidence level for the structure prediction of the domains is high. Otherwise, the docked structure would not be very meaningful.

1.3.5. Structure-Based Functional Inference

For many hypothetical proteins, fold recognition may provide useful information about their functions. Currently functional inference from predicted structures is not very reliable. Nevertheless, given the high cost of experimental approaches (e.g., mutation studies), structure-based functional inference is worthwhile before doing experiments. In many cases, the inference can suggest potential function or a type of function, (e.g., whether a protein binds DNA), and it may further suggest the function of each domain and important active sites. Still it is important to remember that any functional inference needs to be validated through experiments. Here we provide some directions for making inferences based on predicted structures.

1. **Suggesting possible functions in a fold.** Although proteins with the same fold may not have the same function, their functions are often related due to evolution. For example, in vitronectin, which will be shown in details in Section 1.5, we predicted a domain of β-propeller fold. The exact function of this domain is unknown. However, from other known structures in the fold, e.g., lectins and the WD-40 domains in G-protein signaling molecules, we know that the β-propeller fold often plays an important role in governing protein-protein interactions. Hence, we can suggest that the β-propeller domain of vitronectin may be involved in interactions with other proteins or oligomerization. One can suggest possible functions of the query from a predicted fold using the SCOP database,[43] which provides a hierarchical classification of proteins into families, superfamilies, and fold families. Generally, proteins of the same superfamily share common functions at the molecular level. By checking the functions of all the superfamilies in the predicted fold family, one may get an idea about possible functions of the query protein.
2. **Suggesting function type based on consensus of top hits.** Sometimes, the user may not be able to identify a specific fold for a query protein reliably, since the confidence level may be low or the query protein may represent a new fold. However, the top hits may mostly contain a specific type of proteins, e.g., DNA-binding proteins. It is possible that PROSPECT has captured some common features among this type of proteins. In this case, the consensus among the functions of the top hits may suggest the function of the query protein. For example, for the C-terminal domain of PTR3 protein in yeast (see details in Section 1.5), most top hits are signal transduction proteins. This is a good indication that the domain has a function in signal transduction. A PROSPECT user can easily check the functions of all the top-hit templates by using the Web-interface of the PROSPECT output, which lists the function of each template.
3. **Identifying function through motifs.** To further pin down the function of a query protein from an identified fold, one can apply functional motifs as additional constraints during the process of functional assignments. If the predicted structure contains a functional motif (conserved residues at a particular position in the 3D

structure) of a protein in a fold family, the query protein is likely to have the same function as the template containing the motif. Zhang *et al.*[44] have shown that structure-based motifs help to identify whether two proteins in the same fold family share the same function. The authors have constructed a database of functional motifs for known structures (e.g., EF-hand motif for calcium binding), called SITE. Currently, SITE contains identified functional motifs from about 50% of the SCOP superfamilies. A comparison of motif can be made between the proteins in the predicted fold family and the predicted model of the query protein to determine if the functional motif is conserved. One can also search the predicted protein structure against PROCAT,[45] which is a database of 3-D enzyme active site templates.

4. **Confirming function from electrostatics.** The user can calculate continuum electrostatics of a predicted structure using DELPHI[46] and visualize the electrostatics profile on the surface of the protein using GRASP.[47] Electrostatics profiles may confirm functional assignment or active site identification. For example, if a protein is suspected of being involved in DNA binding based on the results from PROSPECT, a strong electrostatic potential on a region of the protein surface may provide evidence of a DNA binding site.
5. **Studying relationship between genomes based on predicted folds.** The predicted structural folds can also be used to analyze the distinct features of a particular genome or a group of related genomes. In their recent analysis, Gerstein *et al.*[48] have shown that among all 339 identified structural folds from worm, *E. coli*, and yeast, 149 folds are shared by the three genomes. *E. coli* has 43 unique folds, not shared by the other two genomes, while worm and yeast have 35 and 8, respectively. Based on the number of shared folds between two genomes, the authors have constructed a fold tree of eight sequenced microbial genomes. The fold tree agrees well with the traditional phylogenetic tree, but it provides some unique perspective particularly when comparing genomes with low sequence similarities. Such a method may provide a unique approach for comparing genomes with low sequence similarity.

1.4. Performance of PROSPECT

In this section, we will provide the benchmarks of PROSPECT for threading accuracy so that the user can get an idea about what to expect from PROSPECT. One benchmark is tested on a set of protein pairs in PDB, with each pair known to share the same superfamily. The other benchmark is based on results from blind tests (CASP contests), which compares the performance of PROSPECT with the performance of other computer packages.

1.4.1. Testing of PROSPECT Using Known Structures in PDB

We have conducted a test[9] to evaluate the overall performance of PROSPECT, using 312 pairs of query-template proteins from the PDB database. Each pair belongs to the same super-family and have sequence identity of less than 30%. We put one protein from each pair into the query set and the other one into the template set. The goal of this study was to evaluate the ability of PROSPECT to correctly recognize the original pairing for each query protein, and how well the alignments compared to the structure-structure alignments of SARF.[26] We first trained the weights between different energy terms (see Section 1.2.2) of PROSPECT based on a subset of 175 pairs, and tested its performance on the remaining 137 (independent) pairs. PROSPECT recognized 72% of the templates correctly and aligned 74% of the structurally alignable residues correctly (defined as being aligned to within a 4-residue shift from aligned position by SARF) on the training set. PROSPECT recognized 69% of the templates correctly and aligned 73% of the structurally alignable residues correctly on the test set. In Table 1.2, the training and test sets are divided into bins of query-template pairs with different levels of sequence identity.

Table 1.2. Threading performance versus sequence identity.

seq. iden.	1-6%	7-9%	9-12%	13-15%	16-18%	19-21%	22-24%	25-27%	28-30%	Overall
fold recog.	0%	10%	15%	58%	78%	100%	90%	100%	100%	72%
alignment	68%	22%	58%	64%	73%	87%	89%	84%	89%	74%
fold recog.	0%	0%	37%	67%	53%	80%	100%	100%	100%	69%
alignment	39%	38%	49%	67%	67%	82%	89%	93%	98%	73%

The second and the third rows represent data on the training set, and the fourth and the fifth rows are for the test set. Each column represents a different range of sequence identity level. Each percentage represents the correct fold recognition rate or the rate of the correctly aligned residues.

1.4.2. Blind Test in CASP

We predicted and submitted structures for all the prediction targets during the blind tests CASP3 (43 targets in total) and CASP4 (also 43 targets in total), using PROSPECT. In each CASP contest, predictors are given a list of protein sequences whose structures have been solved experimentally (or are expected to be solved during the CASP prediction season) but are unpublished. The prediction teams submit five predictions for each target before a preset expiration date. Their prediction results are then evaluated against the experimental structures at the end of the prediction season. The performance in CASP represents the state of art for protein structure predictions. Shortly after the first version of PROSPECT was written, we tested PROSPECT in CASP3.[12] Twenty-three targets were in the category of fold recognition, and hence they were suitable for threading methods. Among them, we correctly identified the native-like folds for 11 prediction targets and predicted closely similar folds for another 5 targets among the 23 threading targets. For the 11 correctly identified folds, PROSPECT obtained good sequence-structure alignments for 9. After significant improvements were made on PROSPECT since CASP3, we participated in CASP4,[13] with very limited human involvement. The protocol described in Section 1.3 was used to make predictions. Most of the outputs from PROSPECT were used directly, and little time was spent on doing manual evaluation and refinement. Therefore, PROSPECT users can generally use our protocol, and they can expect a similar performance for other predictions. Overall, the prediction performance by PROSPECT in CASP4 was ranked the 6th among the 127 participating teams in the fold recognition category.[15]

Among the 43 targets (T0086-T0128) in CASP4, the structures of eight targets have not been made available to the predictors yet so that we cannot assess them. The CASP4 organizers/assessors classified the targets into six categories: (1) comparative modeling (CM), (2) comparative modeling/fold recognition (CM/FR), (3) homologous fold recognition (FR/H), (4) analogous fold recognition (FR/A), (5) fold recognition/new fold (FR/NF), and (6) new fold (NF). When the structure of a target could not be aligned to a single template, the target was divided into different domains, each as an individual prediction for the assessment. In total, 47 predictions for 35 targets were made available for assessment.

In the comparative modeling category (CM and CM/FR) of the CASP4 contest, PROSPECT used the correct templates (folds) in *model 1* (the most confident one among the five submitted models) for every target. In the CM category (eight targets in total), PROSPECT alignments were generally "perfect" (basically the same as the structure-structure alignment between the target and the template). The good alignments are confirmed by the small RMSDs for large numbers of alignable residues (the error is typically about 1 Å RMSD for roughly half of the protein, and less than 3 Å RMSD for almost all of the protein). On the seven CM/FR targets, the alignments were typically ambiguous using different sequence-sequence alignment methods. PROSPECT alignments are either "perfect" or close to the structure-structure alignment between the target and the template, with the exception of target T0103, where extensive alignment gaps existed. All the predictions in this category were also done with little human intervention. A structural model was produced using a PROSPECT command for each predicted threading alignment. In terms of the quality of atomic structures generated, our models were not as good as some of those generated using manual refinement and loop modeling. Nevertheless, the difference between our model and the best model in CASP4 was typically small.[15] In this way, PROSPECT provides a convenient tool which requires no comparative modeling experience to generate a draft model quickly.

The fold recognition category included FR/H and FR/A. CASP4 organizers evaluated the performance in this category using Sippl's evaluation scores,[15] where a score of 0 indicates an incorrect model and a score between 1 and 4 represents a different level of structural relationship between the model and the experimental structure: 1 for having structural similarity, 2 for finding correct fold, 3 for finding correct fold and having a medium alignment quality, and 4 for finding correct fold and having a good alignment quality. PROSPECT's performance in this category in CASP4 was substantially better than its performance in

CASP3. PROSPECT recognized correct folds (with an evaluation score of 2 points or more) as *model 1* for 8 out of 22 predictions. Of the 8 FR/H targets, "reasonable" alignments (scoring 3 points or more) were obtained for four targets. Of the 10 FR/A targets, where threading methods typically fail, PROSPECT recognized the correct fold for two targets, although the alignments are poor. Fig. 1.4 shows a comparison between our predicted model and the experimental structure of the target protein T0100, which had no detectable sequence similarity to any known structure. One can see that the overall fold of the protein was predicted correctly, although some regions of the protein in the prediction were incorrect. Such predictions can provide useful information for function analysis and for suggestions to further experiments.

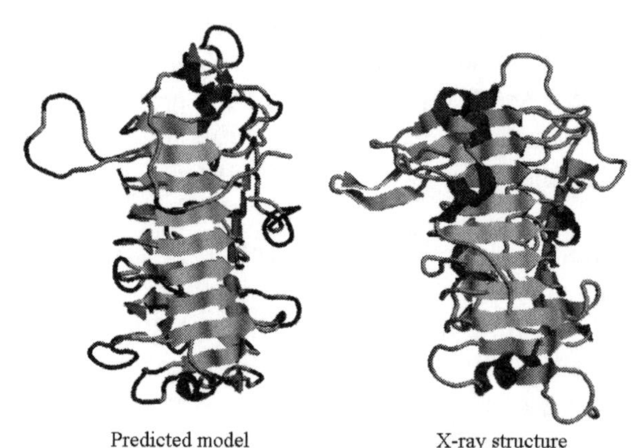

Figure 1.4. A comparison between the predicted structure by PROSPECT (left) and the experimental structure (right) of target t0100 in CASP4. The RMSD between the two structures is 5 Å for the core residues. The dark ribbons indicate α-helices, the gray strands are for β-sheets, and the lines represent loops.

The *new fold* category included FR/NF and NF. An FR/NF target has weak structural similarity to a known protein structure in PDB. The target and its closest structure in PDB have some similar secondary structures in the same arrangement, and they have similar topological connections. It is difficult to say whether the two proteins belong to the same fold due to many other differences. Therefore, Sippl's evaluation score for the fold recognition category was too strict to assess the prediction quality in the *new fold* category. A widely used assessment for this type of prediction is the RMS/Coverage graph,[49] as shown in Fig. 1.5c. A "good" model has a large number of residues that can be superimposed on the experimental structure for a relatively low RMSD, and thus the model can be represented by a line closer to the x-axis than almost all other predictions. Our models can be considered as "good" for four out of our 12 predictions in this category.

By visualizing our "good" predictions, we found that they all captured major features of the experimental structures, and appeared significantly better than almost all the models submitted by other groups. For example, in the prediction for $T0087_1$ (residues 2-194 in target T0087), we "correctly"predicted the portion 2-116 in *model 1*, as shown in Fig. 1.5a,b. We searched the experimental structure of T0087 against PDB using the DALI server[50] and located its closest structure, 1moq (glutamine amidotransferase). The structure of 1moq can be aligned to residues 2-133 in T0087, i.e., 61 residues in the compact domain 2-194 were not alignable. Hence, the PROSPECT prediction achieved nearly the best result that threading could offer in this case. The quality of the prediction for $T0087_1$ may not be sufficient for functional inferences. However, such predictions may provide good starting points for further computational refinements and experimental studies.

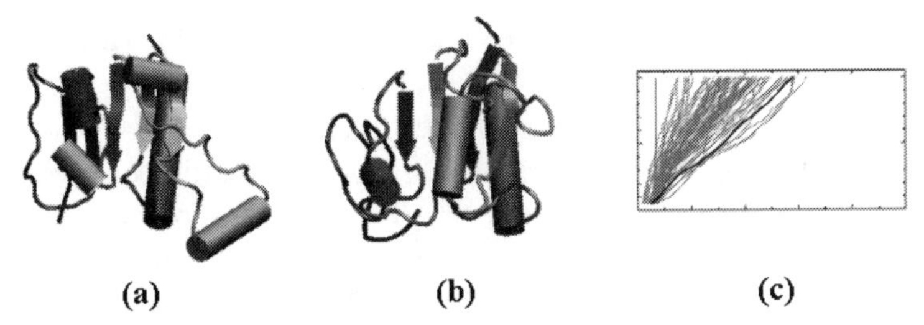

(a) (b) (c)

Figure 1.5. PROSPECT prediction for target $T0087_1$. The left two graphs show a comparison between the experimental structure (a) and our predicted one (b), where the cylinders indicate α-helices, the strands indicate β-sheets, and the lines represent loops. The plot in (c) shows the RMS/Coverage graph, where the x-axis (from 0 to 100%) is the maximum percentage of residues in a predicted structure that can be superimposed on the experimental structure for a given RMSD threshold (represented along y-axis, from 0 to 10 Å). The dark line in the plot represents our *model 1*, and the gray lines represent predictions of *model 1* submitted by other groups. The RMS/Coverage graph was taken from the CASP Web page available at http://predictioncenter.llnl.gov/.

1.5. Application of PROSPECT in Protein Studies

We have carried out a number of structure predictions through collaborations with experimental biologists. For each prediction, our collaborators gave us some experimental information as input for threading. Then we provided structural models and function annotation to our collaborators. Even when no direct functional information was found, fold recognition still provided useful guidance to experimentalists in developing further experiments for detailed functional investigation. This section gives three examples of our studies.

1.5.1. Human Vitronectin

We predicted the structure of human vitronectin protein through collaboration with Dr. Cynthia Peterson's laboratory at the University of Tennessee.[51] Vitronectin is a multifunctional glycoprotein found in blood and in the extracellular matrix. It interacts with a wide variety of ligands, such as glycosaminoglycans, heparin, collagen, plasminogen, plasminogen activator inhibitor-1 (PAI-1), serine protease inhibitor-protease complexes, and the urokinase-receptor.[52] No structure has been solved for vitronectin or any of its close homologs. Previous studies have suggested that the 459-residue protein contains three structural domains and a long unstructured linker (residues 54-130) between the N-terminal and central domains. Fold recognition and sequence-structure alignment were performed using PROSPECT for each of three structural domains, i.e., the N-terminal somatomedin B domain (residues 1-53), the central region which folds into a four-bladed β-propeller domain (residues 131-342), and the C-terminal heparin-binding domain (residues 347-459). The atomic structure of each domain was generated using MODELLER based on the alignment obtained from threading. Docking experiments between the central and C-terminal domains were conducted using the program GRAMM,[42] with limits on the degrees of freedom from a known inter-domain disulfide bridge. The docked structure has a large inter-domain contact surface and defines a putative heparin-binding groove at the inter-domain interface (see Plate 1.1a). We also docked heparin together with the combined structure of the central and C-terminal domains using GRAMM (see

Plate 1.1b and c). The predictions from the threading and docking experiments are consistent with experimental data on purified plasma vitronectin pertaining to protease sensitivity, ligand-binding sites, and buried cysteines. The computational study sheds some light on the mechanism of vitronectin function, and the detailed conformation of the heparin-binding site. It provides a road map for further experiments, such as mutation studies.

1.5.2. Human DNA-Activated Protein Kinase

DNA-activated protein kinase (DNA-PK) plays an important role in mammalian DNA repair.[53] Particularly genetic studies have shown that DNA-PK is required for repairing double-strand breaks, which are the primary cause of death in response to ionizing radiation and many anti-cancer agents. It is known that *in vitro* DNA-PK is activated by low concentrations of linear double-stranded DNA fragments or circular DNAs with single-to-double strand transitions, but not by closed, fully duplexed, circular DNA. In addition, it phosphorylates DNA-binding proteins including tumor antigen, single-stranded DNA-binding protein, the tumor suppressor p53, and many other transcription factors. DNA-PK is a very large protein, containing 4128 residues. The mechanism by which DNA activates DNA-PK is unclear. Through sequence-based search, it is found that its C-terminal domain (with about 400 residues) is similar to phosphatidylinositol 3-kinases, with known structure in PDB 1qmm-A. The other part of the protein has no significant sequence similarity to any known structure.

We carried out structure prediction for DNA-PK in collaboration with Dr. Carl Anderson's laboratory at Brookhaven National Laboratory. Fig. 1.6 shows our prediction results.

The C-terminal domain (residues 3551-4128) is shown in Fig. 1.6d. Experimental data indicated that there were two likely compact domains: one in residues 1156-1497, and the other in residues 2331-2636. PROSPECT results showed the best template for domain 1156-1497 was 1sky-B. The confidence for the prediction was about 50%. Many of top hits of this domain were DNA-binding proteins, including 1cr1-A (DNA primase/helicase). This suggested domain 1156-1497 may be related to DNA binding. For domain 2331-2636, the best template was 1fsz (cell division protein FTSZ homolog 1), which is related to signal transduction. The confidence for the prediction was about 60%. This domain could be related to signal transduction as well, but the indication is rather weak from threading. PROSPECT predicted that the rest of the protein, residues 1-1155, residues 1498-2330, and residues 2637-3550, mainly contain ARM repeats. Fig. 1.6c shows a structural model for the domain 2637-3550. The confidence for this model is about 60%. Interestingly, about half of the C-terminal domain is also formed by ARM repeats. ARM repeats are found in many kinase-related proteins. They often play a role in protein-protein interactions.

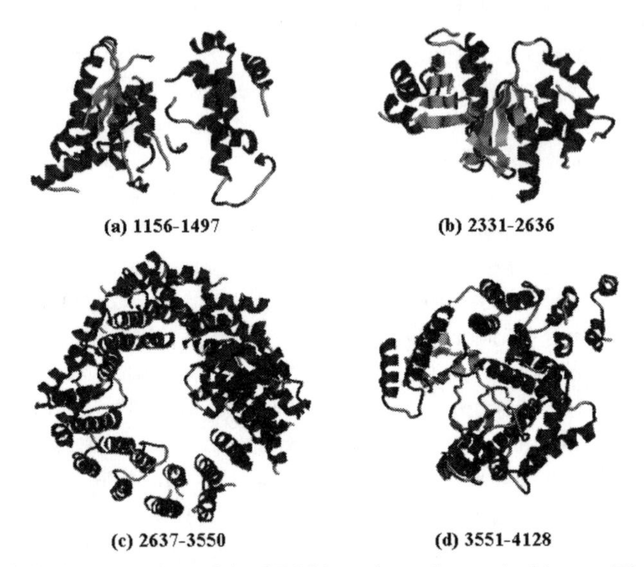

Figure 1.6. Structural model of DNA-activated protein kinase. The graphs in (a-d) show four domains of the protein, where the numbers indicate the residue ranges of the protein domains.

1.5.3. Yeast PTR3 Protein

In collaboration with Dr. Jeff Becker's laboratory at the University of Tennessee, we predicted a structure for the yeast PTR3 protein. This protein has 678 residues. It is an orphan gene, i.e., no protein has significant sequence similarity to it.

As annotated by SWISSPORT, PTR3 is a "regulator of the expression of the transporters BAP2, GAP1 and PTR2." However, the mechanism

of how PTR3 performs the function was unclear. PROSPECT results suggested this protein may have two major domains, one around 1-250, and the other around 260-678, since most of the top hits align with one of the domains, but not both. The top hits with good confidence for the C-terminal domain were mostly signal transduction proteins such as 1qqg (PH-PTB targeting region of IRS-1) and 1tbg (the heterotrimeric G-protein transducin). The C-terminal domain may be related to signal transduction as well. A structural model based on 1tbg is shown in Fig. 1.7b. For the N-terminal domain, PROSPECT confirmed a Zinc finger of C3HC4 type (RING finger) motif around residues 52-61 identified by PROSITE[34] (see Fig. 1.7a).
This motif often occurs at the DNA binding site of a transcription factor. Becker's group is now doing mutation studies on the zinc-finger sites based on our prediction.

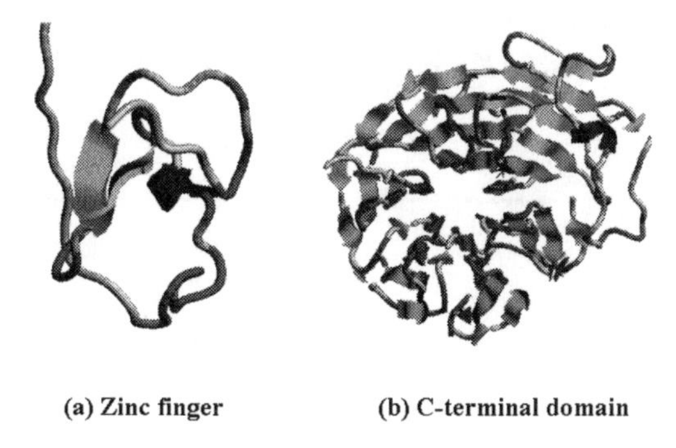

(a) Zinc finger (b) C-terminal domain

Figure 1.7. Structural model of yeast PTR3 protein.

1.6. Summary

Since PROSPECT 1.0 was released in December 2000, more than 60 labs worldwide have licensed it. Different labs use PROSPECT for different purposes. Some users are interested in the structural features of proteins with known functions so that they can get some new ideas to carry out further experiments. Others use PROSPECT to annotate novel genes for identifying drug targets. According to feedback from our users, PROSPECT is often very helpful in their research. In sharing some of our experiences of using PROSPECT in this chapter, we hope to help users improve their ability to better use the features offered by PROSPECT. Even without human evaluation or manual refinement, the results of PROSPECT can be very useful. However, by following the rules provided here, a greater amount of useful information may be obtained.

Acknowledgments

We thank Victor Olman, Philip LoCascio, Cynthia Peterson, Jeffrey Becker, and Carl Anderson for helpful discussions. We also thank Christal Secrest for a critical reading of this manuscript. This research was sponsored by the Office of Health and Environmental Research, U.S. Department of Energy, under Contract No. DE-AC05-000R22725 managed by UT-Battelle, LLC.

References

1. Altschul SF, Madden TL, Schaffer AA, Zhang J, Zhang Z, Miller W, Lipman DJ: **Gapped BLAST and PSI-BLAST: a new generation of protein database search programs.** *Nucleic Acids Res* 1997, **25**:3389–3402.
2. Li Z, Scheraga HA: **Monte Carlo-minimization approach to the multiple-minima problem in protein folding.** *Proc Natl Acad Sci USA*, 1987, **84**:6611–6615.
3. Skolnick J, Kolinski A: **Dynamic Monte Carlo simulations of a new lattice model of globular protein folding, structure and dynamics.** *J Mol Biol* 1991, **221**:499–531.
4. Bernstein FC, Koetzle TF, Williams GJB, Meyer EF, Brice MD, Rodgers JR, Kennard O, Shimanouchi T, Tasumi M: **The protein data bank: a computer based archival file for macromolecular structures.** *J Mol Biol* 1977, **112**:535–542.
5. Montelione GT, Anderson S: **Structural genomics: keynote for a human proteome project.** *Nature Struct Biol* 1999, **6**:11–612.
6. Karplus K, Barrett C, Hughey R: **Hidden Markov models for detecting remote protein homologies.** *Bioinformatics* 1998, **14**:846–856.
7. Bowie JU, Luthy R, Eisenberg D: **A method to identify protein sequences that fold into a known three-dimensional structure.** *Science* 1991, **253**:164–170.
8. Jones DT, Taylor WR, Thornton JM: **A new approach to protein fold recognition.** *Nature* 1992, **358**:86–89.

9. Xu Y, Xu D: **Protein threading using PROSPECT: Design and evaluation.** *Proteins* 2000, **40**:343–354.

10. Xu Y, Xu D, Uberbacher EC: **An efficient computational method for globally optimal threading.** *J Comput Biol* 1998, **5**:597–614.

11. Xu Y, Xu D, Crawford OH, Einstein JR: **A computational method for NMR-constrained protein threading.** *J Comput Biol* 2000, **7**:449–467.

12. Xu Y, Xu D, Crawford OH, Einstein JR, Larimer F, Uberbacher EC, Unseren MA, Zhang G: **Protein threading by PROSPECT: A prediction experiment in CASP3.** *Protein Eng* 1999, **12**:101–109.

13. Xu D, Crawford OH, LoCascio PF, and Xu Y: **Application of PROSPECT in CASP4: Characterizing Protein Structures with New Folds.** *Proteins **2001,*** **45 Suppl 5**:140–148.

14. Koehl P, Levitt M: **A brighter future for protein structure prediction.** *Nature Struct Biol* 1999, **6**:108–111.

15. **CASP4 Protein structure prediction issue.** *Proteins* 2001, **45 Suppl 5.**

16. Holm L, Sander C: **Dictionary of recurrent domains in protein structures.** *Proteins* 1998, **33**:88–96.

17. Holm L, Sander C: Mapping the protein universe. *Science* 1996, **273**:595–602.

18. Sali A, Blundell TL: **Comparative protein modelling by satisfaction of spatial restraints.** *J Mol Biol* 1993, **234**:779–815.

19. Kabsch W, Sander C: **Dictionary of protein secondary structure: Pattern recognition of hydrogen bonded and geometrical features.** *Biopolymers* 1983, **22**:2577–2637.

20. Bryant SH, Lawrence CE: **An empirical energy function for threading protein sequence through the folding motif.** *Proteins* 1993, **16**:92–112.

21. Gonnet GH, Cohen MA, Benner SA: **Exhaustive matching of the entire protein sequence database.** *Science* 1992, **256**:1443–1445.

22. Abagyan RA, Batalov S: **Do aligned sequences share the same fold?** *J Mol Biol* 1997, **273**:355–368.

23. Rost B, Sander C: **Prediction of protein secondary structure at better than 70% accuracy.** *J Mol Biol* 1993, **232**:584–599.

24. Hughey R, Krogh A: **Hidden Markov models for sequence analysis: Extension and analysis of the basic method.** *CABIOS* 1996, **12**:95–107.

25. Smith TF, Waterman MS: Comparison of biosequences. *Adv Appl Math* 1981, **2**:482–489.

26. Alexandrov NN: **SARFing the PDB.** *Protein Eng* 1996, **9**:727–732.

27. Xu Y, Xu D, Olman V: A **Practical Method for Interpretation of Threading Scores: An Application of Neural Network.** *Statistica Sinica (special issue in bioinformatics)* 2002, **12**(1), in press.

28. Bairoch A, Apweiler R: **The SWISS-PROT protein sequence data bank and its supplement TrEMBL in 1999.** *Nucleic Acids Res* 1999, **27**:49–54.

29. McGarvey PB, Huang H, Barker WC, Orcutt BC, Garavelli JS, Srinivasarao GY, Yeh LS, Xiao C, Wu CH: **PIR: A new resource for bioinformatics.** *Bioinformatics* 2000, **16**:290–291.

30. Hirokawa T, Boon-Chieng S, Mitaku S: **Classification and secondary structure prediction system for membrane proteins, bioinformatics.** *Bioinformatics* 1998, **14**:378–379.

31. Corpet F, Gouzy J, Kahn D: **Recent improvements of the ProDom database of protein domain families.** *Nucleic Acids Res* 1999, **27**:263–267.

32. Nielsen H, Engelbrecht J, Brunak S, von Heijne G: **Identification of prokaryotic and eukaryotic signal peptides and prediction of their cleavage sites.** *Protein Eng* 1997, **10**:1–6.

33. Bateman A, Birney E, Durbin R, Eddy SR, Finn FD, Sonnhammer ELL: **Pfam 3.1: 1313 multiple alignments match the majority of proteins.** *Nucleic Acids Res* 1999, **27**:260–262.

34. Hofmann K, Bucher P, Falquet L, Bairoch A: **The PROSITE database, its status in 1999.** *Nucleic Acids Res* 1999, **27**:215–219.

35. Attwood TK, Flower DR, Lewis AP, Mabey JE, Morgan SR, Scordis P, Selley J, Wright W: **PRINTS prepares for the new millennium.** *Nucleic Acids Res* 1999, **27**:220–225.

36. Henikoff JG, Henikoff S, Pietrokovski S: **New features of the blocks database servers.** *Nucleic Acids Res* 1999, **27**:226–228.

37. Nakai K, Horton P: **PSORT: A program for detecting the sorting signals of proteins and predicting their subcellular localization.** *Trends Biochem Sci* 1999, **24**:34–35.

38. Jones DT: **Protein secondary structure prediction based on position-specific scoring matrices.** *J Mol Biol* 1999, **292**:195–202.

39. Vriend G: **WHAT IF: A molecular modeling and drug design program.** *J Mol Graphics* 1990, **8**:52–56.

40. Laskowski RA, MacArthur MW, Moss DS, Thornton JM: **PROCHECK: A program to check the stereochemical quality of protein structures.** *J Appl Cryst* 1993, **26**:283–291.

41. *MDL Information Systems, CHIME.* MDL Information Systems, Inc. San Leandro, Carlifornia. 1999.

42. Vakser IA, Aflalo C: **Hydrophobic docking: A proposed enhancement to molecular recognition technique.** *Proteins* 1994, **20**:320–329.

43. Murzin AG, Brenner SE, Hubbard T, Chothia C: **Scop: A structural classification of proteins database for the investigation of sequences and structures.** *J Mol Biol* 1995, 247:536–540.

44. Zhang B, Rychlewski L, Pawlowski K, Fetrow JS, Skolnick J, Godzik A: **From fold predictions to function predictions: automation of functional site conservation analysis for functional genome predictions.** *Protein Sci* 1999, **8**:1104–1115.

45. Wallace AC, Laskowski RA, Thornton JM: **Derivation of 3D coordinate templates for searching structural databases: application to Ser-His-Asp catalytic triads in the serine proteinases and lipases.** *Protein Sci* 1996, **5**:1001–1013.

46. Honig B, Nicholls A: **Classical electrostatics in biology and chemistry.** *Science* 1995, **268**:1144–1149.

47. Nicholls A, Sharp KA, Honig B: **Protein folding and association: Insights from the interfacial and thermodynamic properties of hydrocarbons.** *Proteins* 1991, **11**:281–296.

48. Gerstein M, Lin J, Hegyi H: **Protein folds in the worm genome.** *Pac Symp Biocomput* 2000:30–41.

49. Hubbard TJ: RMS/coverage graphs: **A qualitative method for comparing three-dimensional protein structure predictions.** *Proteins* 1999, 37:15–21.

50. Holm L, Sander C: **Protein structure comparison by alignment of distance matrices.** *J Mol Biol* 1993, 233:123–138.

51. Xu D, Baburaj K, Peterson CB, Xu Y: **A model for the three-dimensional structure of vitronectin: Predictions for the multi-domain protein from threading and docking.** *Proteins* 2001, **44**:312–320.

52. Preissner KT. **Structure and biological role of vitronectin.** *Ann Rev Cell Biol* 1991, **7**:275–310.

53. Moll U, Lau R, Sypes MA, Gupta MM, Anderson CW: **DNA-PK, the DNA-activated protein kinase, is differentially expressed in normal and malignant human tissues.** *Oncogene* 1999, **18**:3114–3126.

2

Bayesian Approach to Protein Fold Recognition: Building Protein Structural Models from Bits and Pieces

Jadwiga Bienkowska, Hongxian He, Robert G. Rogers Jr., and Lihua Yu

Abstract

This chapter presents a protein fold-recognition method that selects the best fold model for a given protein sequence from a library of structural Hidden Markov Models (HMMs). The HMMs are built from protein structures following their modular decomposition into the secondary structure elements (SSEs) and representing those elements by a pre-designed set of SS submodels (HMMs of SSEs). An automated procedure for building the composite model from SSE submodels creates a distinct model for each functional family identified by SCOP. Since proteins often exist in different solvent exposure states (bound to interaction partners or unbound), we build a small number of models that represent such states of the protein structure. Additionally, when the family is characterized by a minimal pattern of strictly conserved residues, an HMM encoding the specific residue conservation is built from each structural HMM. The complete library of HMMs comprises all types of models built for each structural family. The primary measure of the compatibility between the query sequence and the model is the sum of probabilities of all sequence-to-model alignments, i.e. the total probability. The final compatibility of the sequence with the model is assessed across the library of models using a Bayesian formula. We show that fold predictions that use the total probability are more accurate than the fold predictions that use the more conventional primary measure, namely the probability of the optimal sequence-to-model alignment. We demonstrate that using the library enriched with pattern embedded HMMs improves the performance of fold recognition over the library that includes only structural HMMs.

2

Bayesian Approach to Protein Fold Recognition: Building Protein Structural Models from Bits and Pieces

Jadwiga Bienkowska, Hongxian He,
Robert G. Rogers Jr., and Lihua Yu

Boston University, Boston, Massachussetts

2.1. Introduction

Protein structure prediction is often viewed from the physics perspective. Physical chemistry methods, ranging from quantum mechanics to phenomenological energy potentials, well describe the structure of amino acids and their assembly into a protein sequence. The next-level theory, describing the dynamics of the long-range interactions among amino acids within a sequence, remains an as-yet-unattained goal of both physicists and chemists. Such new theory would predict how a protein sequence folds into its final three-dimensional (3D) shape. According to our current understanding of protein folding, some proteins can be kinetically trapped in a local minimum energy state. However, for most proteins the native structure is the 3D conformation with the lowest free energy. Hence, the problem of protein structure prediction may be formulated as finding the lowest free energy conformation of the sequence. In the threading approach to protein structure prediction, a library of previously known protein structures delimits the range of possible conformations of a protein, and one looks for the conformation with the lowest free energy within the library. In this chapter we present protein structure prediction as seen from the purely probabilistic, Bayesian, viewpoint.

Let us first review the usual assumptions of threading[1,2] and threading-inspired (non *ab initio*) methods for protein structure prediction. The threading approach reduces the protein structure prediction to identifying the best structural model for a protein sequence, among available alternatives. The foundation of threading methods is Boltzmann statistics,[3,4] which is used to derive the phenomenological free energy function E from the observed events' probabilities P, and to relate the structure with the lowest free energy to the most probable structure, P~exp(–E/kT), where events are observations of amino acids or their constituent atoms in particular 3D environments inside proteins. Those 3D environments are defined by other amino acids in the protein and the surrounding solvent. Detailed definitions of the free energy function vary, but the task of other threading algorithms is to find the optimal free energy configuration of the sequence aligned to a structural model. This is usually stated as finding the optimal sequence-to-structure alignment. The free energy function is equivalent to the alignment scoring function, the same as is used in sequence-to-sequence alignment methods.[5,6] Adopting this analogy, we consider the free energy to be equivalent to the sequence-to-structure alignment score. However, it should be noted that when long-range interactions are included in the scoring scheme, finding the optimal sequence-to-structure alignment is a much harder problem than sequence-to-sequence alignment, and it is known to be NP-complete.[7]

Boltzmann statistics dictates that the optimal sequence-to-structure alignment should give an estimate of the lowest free energy of the conformation of the query sequence restrained by the structural model. The comparison among optimal free energies for each model structure is used to select the best model for the query sequence. However, threading restricts the sampling of the conformational space, and it is well known that the distribution of threading scores is not Boltzmannian.[8] Ultimately, the selection of the best structure model compatible with the sequence depends on the statistical weight assigned to the optimal sequence-to-structure-alignment score. Different threading approaches use different heuristic methods to calculate the statistical weights assigned to the optimal scores[8-19] but there is one common feature: selection of the most probable structure model depends on the optimal sequence-to-structure alignment scores. In contrast, we propose the structure prediction method that is stated as two independent problems: first, recognizing the correct fold model, and second, finding the optimal alignment to the selected model. The Bayesian approach attempts to resolve the first problem—the recognition of the correct fold. In this approach, the probability of observing a given structural model for a sequence is not interpreted as free energy, but according to Bayes' formula, fold recognition is measured by a posterior probability of observing a model given the query sequence[20-22] and all possible alignments.

A statistical approach to protein structure prediction has to start with a clear distinction between a protein structure and a structure model. A protein structure is a list of 3D coordinates of atoms in the protein, while a structure model is a mathematical representation of the structure by abstracting 3D coordinates into mathematical objects. Those objects represent the protein structural features, such as secondary-structure elements (α-helix, β-strand) and other structural properties (Plate 2.1) at different levels of detail. The abstraction generally permits the structural model to represent many similar structures.

The Bayesian approach is motivated by three observations. First, we recognize that protein structures can be organized into distinct classes or folds.[23-25] Therefore, it should be possible to build a specific model for each structural fold. As a result, protein structure prediction can be formulated as a task of recognizing a fold model with the highest posterior probability given a protein sequence. Second, a structural model can be seen as an ensemble of similar structures (Plate 2.2). By definition, the optimal sequence-to-model alignment has the highest probability, but sub-optimal alignments may, and often do, have comparable probabilities.[15,21,22] The globally optimal alignment to a model is equivalent to the optimal alignment to one structure; and the sub-optimal alignments can be interpreted as optimal to the other structures from the ensemble. Alternatively, sub-optimal alignments can be interpreted as optimal under minor fluctuations in the alignment-scoring function. Third, when compared to the optimal alignment, the sub-optimal alignments represent shifts in the placement of the secondary structure elements onto the query sequence. Such shifts in the alignment usually coincide with the periodic pattern of solvent exposure of the secondary structure elements. Often, threading the query sequence onto a "correct" structural model has many such sub-optimal alignments. This observation suggests that including these sub-optimal alignments in a fold recognition algorithm should help in recognizing the correct fold. The measure of the sequence-model compatibility that takes into account the above observations is the sum over probabilities of all sequence-to-model alignments. The total alignment probability yields a rigorous conditional probability of observing a sequence given a structure model.[21,22]

Like any general fold recognition scheme, the Bayesian approach requires a fold/structure model library, a measure of compatibility of a sequence with a model, and an algorithm to calculate such measure for each model. We use the total alignment probability as the measure of compatibility of a sequence with a model. Given the prerequisites above, the model most compatible with a sequence is selected following Bayes' formula (see Section 2.6).

The choice of the mathematical representation of protein structures influences the choice and the design of algorithms. Conversely, the available algorithms influence the choice of mathematical representation. An algorithm to calculate the total alignment probability has been developed for Hidden Markov Models (HMMs), and the Bayesian approach uses an HMM as its mathematical representation of protein structure.[26,27] HMMs of protein structures encode single amino acid structural preferences and can be seen as equivalent to a set of structural profiles.[1,20] Consequently, HMMs do not model interactions between atoms or structural elements explicitly.

Our structure models are designed following a set of *a priori* chosen rules. This distinguishes them from HMMs that are trained on a set of examples following the machine learning approach.[13] A distinct name—Discrete State-space Models (DSMs)—was chosen to reflect this difference. Previous work on stochastic modeling of protein structures using DSMs described a set of manually-designed models for protein structural classes.[20] Due to the exponential growth of the Protein Structure Databank (PDB),[28] the manual design of DSMs for each fold has become impractical. Moreover, the manually-designed models are idealized structural fold representations. Here we present a method for the automated design of a DSM from any PDB structure with most of the backbone atoms present. Our models directly represent the observed features of protein structures.

To build a library of unique fold models, we use the SCOP structural classification.[24] The model library described here is restricted to single protein domains and excludes membrane proteins and irregular structures. Membrane proteins require special treatment since they have solvent exposure preferences that are very different from soluble proteins. Irregular structures have low content of regular secondary structure and are difficult to model by an automated design that is based on local structural preferences. Models for membrane proteins and irregular structures are constructed manually and will not be described in this chapter.[29]

The rest of this chapter is organized as follows. First, we introduce basic concepts of the DSM modeling of protein structure. Second, we describe the automated analysis of the protein structure from the PDB and its representation as a structural template. Third, we describe the translation of the structural template into the DSM and the design of the DSM submodels. Fourth, the automated embedding of the homologous sequences' profile in a structural DSM is described. Fifth, the Bayesian approach to protein fold recognition is formulated and alternative approaches to calculating prior model probabilities are discussed. The last section presents results of fold recognition using the Bayesian approach and its comparison with other methods.

2.2. Fundamentals of DSMs and HMMs

The DSM approach[20] to protein structure modeling has many advantages: flexibility, generality, high degree of abstraction, and the possibility of building models for hypothetical structures. A few experts designed the original models by hand according to their understanding of protein structure families. However, with the number of deposited protein structures in the PDB surpassing 15,000 (April 2001) and the number of distinct protein folds approaching 600 (SCOP 1.53), it has become difficult to update DSM libraries manually. Therefore, we have recently developed a method to construct DSMs automatically from determined protein structures. This method inherits many of the merits of the original DSMs, such as dividing structural states into several classes and using expert prior structural knowledge for model design to reduce the number of free parameters.[22]

2.2.1. Representation of Protein Structure by a DSM

Residue positions

In the DSM approach the hidden states of the model represent the structural positions of residues in the protein structure. Therefore, they are also called structural states, denoted as s. Each structural state is characterized by its secondary structure and solvent exposure or distinct loop/turn assignment. The amino acids, labeled a, are the observed states emitted from the structural states of the model. Each structural state s is represented by a vector of probabilities of emitting (or being aligned to) any amino acid a. A DSM, just like any HMM, is composed of N structural states s that are connected to each other and form a Markov chain. For example, a simple representation of the protein structure by a "trivial" Markov chain is a one-to-one mapping of structural positions to hidden states (see Plate 2.3). A non-trivial HMM represents many possible trivial chains of hidden states. An alignment of amino acids to residue positions from a structure corresponds to an alignment of amino acids to the hidden states of the model.

Secondary-structure elements

The automated design of a DSM from a protein structure relies on the modularity of protein structure. In the simplest approximation, each protein structure is composed of three basic structural elements: α-helices, β-strands, and loops (Plate 2.1). The most conserved region in a protein structure is usually composed of the secondary-structure elements (SSEs)—α-helices and β-strands—and is called the structural core. Loops connecting secondary structure elements are usually the most variable parts of the structure among homologous proteins. The three-dimensional packing of the SSEs determines the 3D structure of the protein, which is reflected, in part, by the solvent exposure of structural positions in each SSE. Solvent exposure indicates which positions are buried inside the protein and interact with other buried positions, however, their mutual pairing is not represented.

First, we analyze the structure (a PDB file) and assign the secondary-structure elements (SSEs) and their types. Each SSE is a list of structural positions with their associated features, such as solvent exposure. A structural submodel represents each SSE and particular features observed in a given element such as its length and solvent exposure patterns. The *a priori* defined rules are applied to build such submodels from SSEs. Those rules represent our prior knowledge about the variations observed in structural elements among homologous proteins. For example, in agreement with the observed variations among homologs, the submodels representing loop elements have the most variability encoded, while the secondary structure submodels encode a limited variability.

The final DSM is constructed by assembling the structural submodels in the order they are observed along the protein sequence (see Plate 2.4). The protein structure is represented as a ribbon diagram using RASMOL. In the lower linear representation, red circles represent helix submodels, green triangles represent strand submodels and gray ovals represent loop submodels. The thin black arrows represent transition probabilities from one submodel to the next. Thick blue arrows represent the mapping of structural elements onto submodels. The anticipated variations in the protein structure are encoded in each secondary-structure submodel. N-terminal and C-terminal loop submodels allow an additional amphipathic helix to be added at either end of a structural domain.[22]

The DSM is a mapping from structural elements to structural submodels. The design of basic submodels can be easily changed without interfering with the design of the final DSM. One can also envision designing more specific submodels that are built from the super-secondary-structure elements such as β-hairpin or helix-turn-helix motifs.

2.2.2. Mathematical Representation of a DSM

A DSM is defined by two matrices: a state-to-state transition matrix Φ, and an emission probability matrix H. Φ contains conditional probabilities $\phi(s|s')$ of passing from a structural state s' to any other state s. H contains conditional probabilities $h(a|s)$ for each amino acid a to be emitted by a given structural state s. The state-to-state transition probabilities are assigned directly from the protein structure and also model the structural variations. We distinguish a small number (much smaller than the number of structural states in a typical HMM) of structural state classes CL that are represented by conditional probability distributions

$h_0(a|CL)$ and are stored in a matrix $\mathbf{H_0}$. $\mathbf{H_0}$ is precalculated from a selected set of representative structures as described in section 2.4.1. The H-matrix for any given model is then constructed over all modeled states by $h(a|s) = h_0(a|CL_s)$, where CL_s is the class of state s.[20,22] The H matrix representing the trivial DSM from Plate 2.3 is equivalent to a structural profile.[1] The negative logarithm of H-matrix elements is proportional to the free energy function (or scoring function) E ~ –log(H).

In order to construct a DSM, it is necessary to (a) define a set of states s based on the residue positions in the PDB structure, (b) define state classes CL for each of those positions, (c) define the emission probabilities for each state class by populating the $\mathbf{H_0}$ matrix, and (d) define the allowed state transitions and their associated probabilities by populating the $\mathbf{\Phi}$ matrix. Subtasks (a) and (d) require making higher-level model-representation choices, and are described in section 2.4. Given the decision to use a fixed set of state classes, (c) can be computed in advance, as is discussed in section 2.4.1. The subtask (b) is executed during the submodel design described in section 2.4.2.

2.2.3 Measures of Compatibility of a Protein Sequence with a DSM

In order to assess the compatibility of the protein sequence with a model defined by H and $\mathbf{\Phi}$ matrices, we should know the answer to these three questions. (1) What is the probability of generating a sequence by a given model, P(**seq**|Model)? (2) What is the most probable alignment of the sequence to that model? (3) What is the probability of that alignment?

The answer to the first question is given by the total alignment probability. Algorithms to calculate the total alignment probability are known as the "Filtering" algorithm[26] or "Forward-Backward" algorithm.[27,30] The answer to the second and third questions is provided by the calculation of the optimal sequence-to-structure alignment. The algorithm to calculate the optimal alignment probability P(**seq**|Model, optimal-alignment) was proposed byViterbi[31] and it is equivalent to the dynamic programming algorithm.[5,6] The P(**seq**|Model, optimal-alignment) is the primary measure of compatibility of the sequence with a model that is used in sequence-to-sequence alignments as well as in traditional, or Boltzmann, threading methods. According to Boltzmann statistics, the minimal free energy is expressed as E ~ –log(P(**seq**|Model, optimal-alignment)). In contrast, the Bayesian approach to fold recognition uses the P(**seq**|Model) as the primary measure of compatibility of the sequence with a model. In section 2.7.1 we show that, for DSMs, the total alignment probability gives more accurate fold predictions than the optimal alignment probability. Some combination of the two measures may prove to be more powerful yet.

2.3. Automated Generation of Protein Structural Templates

PDB format files contain a wealth of detail about the placement of atoms within a protein structure, along with varying amounts of higher-level information. DSMs, on the other hand, describe protein structures in very general terms. Most of the structural states in a DSM correspond to the residue positions and are described by only a few bits of broad structural information. In addition to these residue-representing hidden states, there are a small number of pre-assigned hidden states representing the anticipated structural variations among homologous proteins. Therefore, in order to create a DSM from a PDB entry, one must first extract the appropriate descriptive information from the PDB entry, creating a "structural template," and then translate the resulting structural template into a DSM. This section describes the first part of the process; section 2.4 explains how to generate a DSM from the structural template.

Of course, much of the PDB detail must simply be thrown away, not just in order to use restricted per-residue information, but because including too much information is likely to result in a model that is too specific to a single sequence or structure (see article[32] for a structure modeling approach that includes very detailed information.)

In this section, we first consider the criteria for selecting information from PDB coordinates in light of the constraints imposed by constructing a DSM as an HMM as described above. We then survey some of the useful quantities that can be extracted from PDB coordinates in light of the above criteria, finishing with a discussion of how we chose to compute these quantities to produce our DSM library.

2.3.1. Criteria for Selecting Structural Information

Criteria for selecting quantities to be computed from PDB coordinates fall into two broad classes. These can be thought of as two sets of hurdles all candidate criteria must pass before being considered useful. The first set of hurdles selects criteria that are useful for structural model-making in general, while the second set selects those that are suitable for DSMs in particular.

General criteria

The general criteria are as follows:

1. Selected quantities must be independent of the coordinate system that the PDB entry happens to use. Angles are therefore ideal from this standpoint. Distances between atoms are also suitable.
2. Perhaps most important, no such quantity should depend on the identity of the amino acids at any sequence position. Doing so would bias the resulting model away from distant homologs, and might even disable recognition of close homologs (but see Section 2.5, which reintroduces sequence information in a controlled way). This criterion can always be met easily for most measures by converting the entire sequence to poly-alanine, i.e., substituting a single β-carbon for the actual side-chain, before extracting the desired quantity. Subsection 2.3.3 below is an example of this point.

DSM-related criteria

In order to be useful for making a DSM, the information culled from a PDB structure must meet the following additional criteria:

1. The information must be per-residue, i.e., must be a function of a single residue and its immediate neighborhood. (Other characteristic quantities can be defined that depend on two or more residues, such as the distance between β-carbons. However, these will not be discussed here, since such quantities cannot be readily used in HMMs.)
2. If not already discrete, the information must be amenable to discretization. For scalar quantities, one or more thresholds must be chosen.

Most difficult of all, the resulting structural states must have distinct residue statistics or, in other words, be reasonably informative. A rigorous method for selecting informative structural states for a scalar quantity is described in article.[33]

2.3.2. Candidate Structural Quantities

The following list itemizes some quantities that have been used to varying degrees for the purpose of structural model construction. This list is not intended to be exhaustive but enumerates properties that we have considered.

- Dihedral angles. Unfortunately, the backbone ϕ and ψ angles[34] are not as predictive of secondary structure as H-bonding information, which is more global.
- Backbone H-bonding patterns.[35]
- Local chain torsion (i.e. chirality of a 4-residue window).[35]
- Counts of neighboring atoms (backbone and β-carbons).[32]
- Solvent-accessible area.[36]
- Visible volume.[32,33,37]

These can be thought of as "primary measurements" that are taken directly from the structure. Although H-bonding is not strictly a local quantity, it is essential for accurate characterization of β-sheets.

A higher-level view of secondary structure is provided by the DSSP program,[35] which has become the standard tool for automatic extraction of secondary structure information from the atomic coordinate data. DSSP uses a set of simple local rules to abstract discrete secondary structure states (i.e., describing strand/helix membership). The most basic rule is that for a hydrogen bond, which is defined geometrically in terms of the distance between the backbone carbonyl oxygen (O) and the amino nitrogen (N). Higher-level concepts such as "n-turns" and "bridges" are then defined as patterns of hydrogen bonding. An n-turn is simply an H-bond between residues i and i+n, where n is 2, 3, or 4. A bridge involves a pair of H-bonds, and can be either parallel or antiparallel. Finally, helices are defined as patterns of adjacent n-turns, and strands and sheets are defined as patterns of adjacent bridges.

For example, for a residue at sequence position i to be classified as "H" (β-helix), one of the following must be true:

1. The residue must be H-bonded to its neighbor at i+4, and its neighbor at i–1 must be H-bonded to i+3; or

2. The residue must be H-bonded to its neighbor at $i-4$, and its neighbor at $i+1$ must be H-bonded to $i-3$; or
3. The residue must lie in between two such pairs of H-bonded residues (i.e. it must be one of residues $i+1$ or $i+2$ in the first case, or $i-1$ or $i-2$ in the second).

As a consequence of this, the minimum length of a helix as reported by DSSP is four, consisting of residues numbered from i through $i+3$. (Although the residues at $i-1$ and $i+4$ form part of the helix H-bonding pattern, they are not considered H residues themselves.) Longer helices are identified by repetitions of the minimal H-bonding pattern; so longer helices need not exhibit a perfect i:$i+4$ H-bonding pattern. Similar rules are defined for sheets. Notice that DSSP defines a helix entirely in terms of patterns of backbone H-bonding, which in turn is defined purely in terms of O-to-N distances. DSSP's great strength is that these rules have withstood almost two decades of scrutiny by the molecular biology community, having replaced manual assignment of secondary structures by the PDB depositors.

Table 2.1 lists the DSSP state codes and rules that define them. These rules are interpreted in the order given; if more than one applies, the first is chosen. (Based on article.[35]) While DSSP uses a blank (ASCII SPC character) to indicate that a residue does not qualify under any of its state assignment rules, we prefer to use an L for 'loop' in order to avoid possible ambiguity.

Table 2.1. DSSP state codes and their definitions.

Letter	Name	Definition
H	(3.6_{13}) α-helix	Must be within two or more consecutive 4-turns, e.g. residues i through $i+3$ are α-helical if $i-1$ is H-bonded to $i+3$, and i is H-bonded to $i+4$.
B	Isolated β-bridge residue	Must not have a bridged neighbor that qualifies it for β-strand status. Bridge partner is identified in BP1 or BP2 column.
E	Strand ("extended")	Has a bridge partner and at least one neighbor bridged in parallel or antiparallel.
G	3_{10} α-helix	Must be within two or more consecutive 3-turns.
I	(4.4_{16}) π-helix	Must be within two or more consecutive 5-turns.
T	Turn	Within a 3-turn, 4-turn, or 5-turn that has no n-turn neighbor that would qualify them for H, G, or I status.
S	Bend	Local curvature greater than 70 degrees, measured as the angle between α-carbons at $i-2$, i, and $i+2$.
blank	None	Meets none of the criteria above.

Note that the above definition of "turn" is quite broad; it includes any residue i that happens to be within an isolated (i.e., not paired and therefore helical) n-turn, where n is 2, 3, or 4. DSSP "T" residues therefore include residues in a wide range of structural situations. DSSP also omits labeling the flanking H-bonded pair as part of the turn, which is consistent with its treatment of helices, but is inconvenient for our purposes. We discuss how to remedy this in section 2.3.3, below.

2.3.3. Classification of Structural States

The process of selecting a number of structural states is a tradeoff between meeting the selection criteria (section 2.3.1) and finding enough unbiased data in the PDB to populate the selected states. Based on these tradeoffs, DSMs were originally designed using two secondary structure categories (helix and strand), plus a binary exposure distinction (buried versus exposed), giving rise to four secondary structure state classes. Four DSSP-based turn states were employed, labeled $turn_1$, $turn_2$, $turn_3$, and $turn_4$, and were used in situations where the (human) model-builders believed that a characteristic turn might appear in the structure. Adding an exposure-independent loop state to cover all other non-helix, non-strand residues gave nine state classes altogether.[20] The number of structures deposited in the PDB made it possible to compile adequate statistics for all residues in all structure/exposure state classes. Valid statistics require independent structures, the number of which grows more slowly than the number of total structures in the PDB. Also, increasing the number of state classes beyond a certain number does not provide an increase in the information content of the resulting probability distribution.[33] This allows us to increase the number of state classes used, but only modestly. In helices and in strands, we use distinct solvent exposure threshold values to define solvent exposure states. After analyzing the information content of the amino acid probability distributions,[33] we decided to use three solvent exposure states in both helices and strands. In addition to the loop state class, we have introduced two state classes to model tight-turns and four state classes to model β-turns.[34] The two-residue tight-turn has strongly characteristic probability distributions with glycine being the most probable amino acid at both positions (see Table 2.2).

Similarly, positions two and three in a four-residue β-turn have very specific probability distributions. However, distributions characterizing positions one and four in a β-turn are identical to a coil-loop state distribution (see Table 2.2). Thus, the total number of state classes used in our current DSMs is 13.

Table 2.2. Amino acid (AA) emission probabilities for the structural state classes used in the DSM $H_0(a|CL)$. The state classes are: H_b—helix buried, H_{pb}—helix partially buried, H_e—helix exposed; S_b—strand buried; S_{pb}—strand partially buried; S_e—strand exposed; tight-turn states T_{t1}, T_{t2}; β-turn states T_{b1}, T_{b2}, T_{b3}, T_{b4}; and coil state C.

State / AA	HH_b	HH_{pb}	HH_e	SS_b	SS_{pb}	SS_e	TT_{t1}	TT_{t2}	TT_{E1}	TT_{E2}	TT_{E3}	TT_{E4}	CC
A	.147	.083	.100	.081	.044	.040	.030	.023	.074	.066	.046	.074	.074
C	.017	.004	.001	.029	.018	.006	.003	.003	.021	.011	.013	.021	.021
D	.022	.050	.082	.017	.043	.057	.165	.144	.069	.069	.098	.069	.069
E	.033	.099	.148	.013	.068	.096	.079	.032	.047	.065	.046	.047	.047
F	.071	.035	.009	.081	.049	.027	.008	.006	.034	.019	.024	.034	.034
G	.042	.021	.031	.058	.035	.043	.300	.422	.109	.090	.280	.109	.109
H	.014	.029	.017	.018	.030	.020	.017	.006	.024	.012	.022	.024	.024
I	.098	.049	.023	.127	.070	.051	.008	.006	.036	.022	.006	.036	.036
K	.018	.108	.125	.017	.084	.095	.048	.030	.056	.066	.045	.056	.056
L	.176	.095	.043	.122	.067	.050	.009	.009	.061	.035	.021	.061	.061
M	.034	.023	.013	.024	.019	.011	.008	.003	.015	.008	.015	.015	0.15
N	.018	.040	.060	.017	.040	.049	.147	.089	.062	.050	.115	.062	.062
P	.010	.015	.029	.013	.014	.031	.036	.012	.076	.199	.014	.076	.076
Q	.026	.075	.086	.018	.039	.054	.024	.026	.035	.022	.039	.035	.035
R	.029	.088	.072	.022	.066	.052	.026	.024	.038	.034	.040	.038	.038
S	.036	.039	.066	.046	.058	.078	.064	.070	.077	.102	.093	.077	.077
T	.041	.051	.045	.043	.089	.116	.006	.071	.071	.052	.037	.071	.071
V	.098	.051	.034	.167	.104	.081	.009	.005	.052	.038	.008	.052	.052
W	.019	.015	.005	.020	.016	.012	.002	.006	.012	.012	.013	.012	.012
Y	.048	.033	.011	.067	.047	.029	.014	.014	.032	.028	.024	.032	.032

"Smoothing" DSSP to remove non-characteristic features

The DSSP output is adjusted only slightly, mostly to remove states that are not modeled by the DSMs. In outline, the algorithm is as follows:

1. All S, B, and I states are changed to L. Besides the difficulty of getting good statistics for these less frequent states, they are much less likely than H and E to be characteristic of related structures.
2. Up to three consecutive G states (3-10 helix) immediately after a helix are changed to H. These might easily appear as an extra turn of the helix in a related structure, and so are better modeled as H rather than L.
3. The residues flanking each pair of 3-turn T residues are relabeled as the first and fourth residues of the turn, if they are not already labeled as strand or helix.
4. Strands of less than three consecutive residues and helices of less than five consecutive residues are all changed to L (the DSSP minimum is two for strands and four for helices).

5. All remaining non-helix, non-strand states (including leftover G's and T's) are changed to L.

The resulting states are used for counting helix, strand, loop, and turn statistics for the sake of creating the H_0 matrix described in section 2.4.1. The probability distributions for these states are shown in Table 2.2. Note that, since explicit turn states are not used in the generated DSMs, all T's are changed to L at the outset, and step 3 (T relabeling) can be skipped in actual DSM generation.

Classification of solvent exposure states

Solvent exposure for a structure is computed by first replacing all actual residues by artificial residues with one β-carbon. Effectively, all side-chains are pruned back to the β-carbon, and glycines are "promoted" to alanine by adding a β-carbon. All β-carbons are then assigned a radius of 2.1 Å, as opposed to the standard radius of 1.9 Å for the backbone carbons, hence the nickname "fat alanine" for these artificial residues. Finally, the exposed surface of each residue is measured using the method developed by Eisenberg and MacLachlan.[36] The result is a value between zero and 140, nominally in $Å^2$, though it is difficult to interpret as such. Since these artificial residues are identical, the above solvent exposure value can be used as a measure of the degree to which each position is exposed in the protein, and is comparable across all positions in all structures.

The quantitative exposure value is subsequently discretized to one of three categories defined as 'buried', 'partially buried', or 'exposed', using empirically chosen cut-off values that yield the best discrimination among amino acid types. We currently use the binning threshold values of 15 and 40 for helix residues, and 20 and 40 for strand residues. The optimal thresholds can be rigorously selected using the method described in article.[33]

Combining structure, exposure, and loop geometry

The structural template produced by the foregoing analysis can thus be described as a series of symbols <x>, where each <x> can now be defined as a contiguous run of H or E residues. Each H or E residue comes with an associated solvent exposure value, which will be used to define the DSM state for that structural position, as described in section 2.4.2 below. The only information required for the loop module is the distance between the end of one SSE and the start of the next (see Section 2.4.2), calculated from the coordinates of the carbonyl carbon of the last residue in an element and those of the amino nitrogen of the first residue of the following element.

The following list summarizes the steps taken during the generation of a structural template:

1. Run the DSSP program on the PDB entry.
2. Filter the DSSP result and postprocess it in order to produce a list of elements.
3. Compute Eisenberg "fat alanine" exposure for the PDB entry.
4. Combine the element information, the exposure information and the original PDB atom coordinates of the start and end backbone atoms of each element.

The representative structure generates more than one structural template when the PDB records that protein as a multimeric or heteromeric protein complex or a multidomain protein. In that case, multiple templates are built to represent alternative patterns of solvent exposure for that protein domain. For multimeric/multi-domain proteins, the exposure pattern is calculated for a monomer alone and with various numbers of multimers/domains present. For heteromeric proteins, the exposure pattern is calculated with and without heteroatoms. The domain identification is taken from the SCOP structure classification database.[24]

2.4. Automated Design of a Structural DSM from a Structural Template

In this section we describe how we translated information from the structural template and information about the structural variations in close homologs into DSM submodels. The submodels represent basic structural elements. The last step of the DSM construction combines the submodels into a final DSM.

2.4.1. Design Principles

We applied two basic principles to the automated DSM construction:

1. No training set. For every known structure, a DSM can be built regardless of how many similar structures are known. This allows

us to build models for protein structure families even if only a single structure is available.

2. Modular design. This design principle simplifies model construction and allows for the generality of models, which permits the recognition of sequences belonging to the same structural family.

Our ability to build DSMs using only one structure is achieved by using a limited number of structural state classes with precalculated emission probabilities (H_0 and H matrices) and by assigning transition probabilities (Φ matrix) according to prior knowledge. Modular design is achieved by using physically motivated structure submodels. Each submodel is built in a way that allows anticipated variations among homologs.

Assigning structural-state emission probabilities

Each non-loop position in a protein structure is converted into a structural state of a DSM by automatic generation of a structural template from a structure (see Section 2.3.3). The probability of an amino acid being observed in a particular structural state class, which is recorded in the H_0-matrix (see Table 2.2), is precalculated from a set of 319 nonhomologous globular, water-soluble, ingle-domain proteins that are classified as different α, β, α/β or $\alpha+\beta$ folds in SCOP. Each structure in this training set is then converted to a string of structural states. By counting the frequency of each amino acid appearing in each structural state class, and normalizing the emission probabilities over the 20 amino acids, we generated a 20-by-13 emission probability matrix H_0.

Assigning transition probabilities

In addition to classifying structural positions of a protein fold into a limited number of structural state classes, we also need to assign transition probabilities to allow possible structural variations and to satisfy the no-training-set requirement. We generally use the 'maximal uncertainty' principle to assign transition probabilities. We allow equally probable length variation in the beginning and end of each secondary structure submodel, as well as equally probable transitions from the end of one structural submodel to the following alternative structural submodels. We will give detailed illustration of these general principles in the following sections on secondary-structural submodels and construction of DSMs from structural templates.

2.4.2. Secondary Structure Submodels

We represent each protein structure by a set of three basic substructures that are consecutive along the one-dimensional sequence: α-helix, loop (including both loop and turn), and β-strand. We have constructed the DSM structural submodels to represent the above three secondary structure elements, which are then used as building blocks to generate the final DSMs.

α-helix submodel

In the helix submodel (Fig. 2.1), the structural state (buried, partially-buried, or exposed helix) of each structural position is determined as described in section 2.3.2.

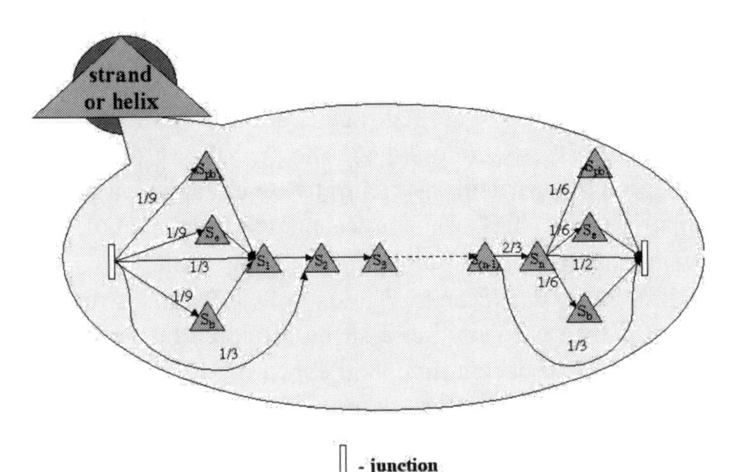

Figure 2.1. Design of the helix or strand submodel with n residue positions. Small triangles represent the hidden states of the DSM strand submodel. S_i denotes the solvent exposure state of the *i*-th position. S_{pb}, S_e, S_b are partially buried, exposed, and buried states that model the strand extension by one residue. Arrows connecting the states represent the nonzero state-to-state transition probabilities. The numbers associated with arrows give values of transition probabilities that are different from one. Arrows without numbers associated represent transition probabilities equal to one.

A linear DSM is constructed by connecting individual states together with a transition probability of 1 from the last-encountered state in a chain to the next state. The observed variations among homologous structures are encoded by allowing an extension/deletion by one position at both ends of the secondary structure. Following the "maximal uncertainty" principle, it is equally probable (with probability 1/3) to start/end the SSE at the exact beginning/end position and to extend or to shorten the SSE by one position. Since we have no prior knowledge of what the extension state will be, other than the fact that it is a helix state, we allow that state to be exposed, partially buried, or buried with equal probability of 1/9, which sums to 1/3.

β-strand submodel

The β-strand submodel is constructed in the same way as the α-helix submodel as shown in Plate 2.5 (see description in Section 2.4.4).

Loop module

We use "loop" as a general term to represent anything other than the two well-defined secondary structure elements: α-helices and β-strands. It can be a two-residue tight-turn, a four-residue β-turn or a long loop. Since loop regions are the most flexible regions in protein structures and are the most variable parts among homologous proteins, we allow maximal variability in the loop submodel. Instead of recording the structural states faithfully, as we do for α-helix and β-strand submodels, we classify loops into 3 categories according to the geometric distance between the ends of the two consecutive secondary structure elements that bound the loop. The structural template file records the 3D coordinates of the beginning (x_b^i, y_b^i, z_b^i of the nitrogen atom of the first residue) and ending (x_e^i, y_e^i, z_e^i of the carboxyl atom of the last residue) of each secondary structure element SS_i. The distance between two secondary structure segments can then be defined as

$$d(SS_i, SS_{i+1}) = \sqrt{(x_e^i - x_b^{i+1})^2 + (y_e^i - y_b^{i+1})^2 + (z_e^i - z_b^{i+1})^2} .$$

We have analyzed a set of protein structures and determined the maximal distance between consecutive secondary structure elements that can be connected by two special loop types: a tight-turn or β-turn.

If $d < 4.3$ Å, the extremely short distance indicates that the connector can be a tight turn or short loop, which is modeled by loop submodel type a (Fig. 2.2). We allow the very short two-residue tight-turn, the 4-residue β-turn and 5-to-10-residue loop to occur with equal probability of 1/3.

The 5-to-10-residue loop has a uniform probability for loops with lengths between 5 and 10 and has exponentially decreasing probability for loops longer than 10.

If 4.3 Å < d < 10.5 Å, a two-residue tight-turn will not be able to span the physical distance even though the distance is still short enough to be spanned by a β-turn. Loop submodel type b (Fig. 2.2) models this type of loop. In type b loop model, a 4-residue β-turn and a 5-to-10-residue loop are equally probable with a probability of 1/2 each.

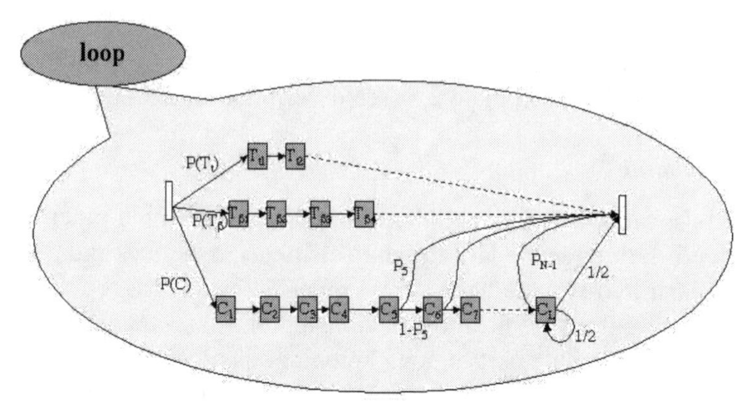

Figure 2.2. Design of a loop module. Rectangles represent the hidden states in the loop module. Labels T_{t1}, T_{t2}, $T_{\beta 1}$, $T_{\beta 2}$, $T_{\beta 3}$, $T_{\beta 4}$, and C_i indicate classes of states. All C_i states are characterized by the same amino acid emission probability distribution C given in Table 2.2. Loop type a has $P(T_t) = P(T_\beta) = P(C) = 1/3$. Loop type b has $P(T_t) = 0$ and $P(T_\beta) = P(C) = 1/2$. Loop type c has $P(T_t) = P(T_\beta) = 0$ and $P(C) = 1$. The probabilities P_n are calculated to represent a uniform length distribution between L_{min} and L_{max}. In this figure $L_{min} = 5$ and $L_{max} = L$ where L is the length of the loop.

If d > 10.5 Å, the long distance is a good indicator that the two neighboring secondary structure elements can only be connected by a long loop. Loop submodel type c (Fig. 2.2) models this long loop. The minimum length of loop c is five. This loop is modeled to have a uniform length distribution between $L_{min}(= 5)$ and L_{max}, and exponentially decreasing probability for loops longer than L_{max}. L_{max} is equal to 10 for loops shorter than 10 residues and is equal to the observed loop length for loops longer than 10. We use only one loop state with a transition back to itself to model the exponentially decreasing probability of loops longer than L_{max}.

2.4.3. Construction of DSM from the Structural Template

As the reader may have noticed in Figs. 2.1 and 2.2, we use a junction (represented as a white rectangle in figures) at the beginning and end of each submodel. The junction is not a structural state and has no emission probability vector associated with it. It is an artificial symbol created to allow the modular construction of DSMs.[20] Junctions help to assemble submodels into a complicated DSM following simple rules. From the structural template, we can build all the secondary structure submodels as described in this section. Then we connect all the structure submodels and eliminate the junctions to produce a final DSM model. This model is then used as any ordinary HMM model.

Let matrix $\phi(s|s')$ represent one submodel's internal state transitions. The complete submodel is constructed by adding the junction j_1 that marks the beginning and junction j_2 that marks the end of the submodel. Two additional matrices $\phi(s|j_1)$ and $\phi(j_2|s)$ hold the transition probabilities of passing from the beginning junction to any state in the submodel, and probabilities of passing from any state in the submodel to the end junction. The transitions of passing from junction to junction within the same submodels are set to zero $\phi(j_2|j_1) = \phi(j_1|j_1) = \phi(j_2|j_2) = \phi(j_1|j_2) = 0$. Only the transition probabilities between the end junctions of one submodel to the beginning junction of the next one are set to one (see Plate 2.4.), other junction-to-junction transition probabilities are zero.

The assignment of state-to-state, state-to-junction, junction-to-state and junction-to-junction transition probabilities for all submodels and between submodels generates four matrices: Φ_{ss}, Φ_{sj}, Φ_{js}, and Φ_{jj}. The complete model is represented by the Φ calculated according to the equation

$$\Phi = \Phi_{ss} + \Phi_{sj}\,(\mathrm{I} - \Phi_{jj})^{-1}\,\Phi_{js}$$

where I is the identity matrix. Junctions and matrices describing the transitions to, from and between junctions are used only during the model-building process.

2.4.4. Using Structural Alignments and Multiple Structural Templates in Building a DSM

The alignment of analogous structures can be utilized for the design of an umbrella DSM to capture the structural variations observed within a family of analogous structures. In addition to the length variation of structural elements, the variations among analogous structures often include insertions, deletions and substitutions of some structural elements as well as variation in the hydrophobicity profiles of SSEs that are conserved. A starting point for the automatic model building is a structural alignment that represents all structural variations observed within a family. Alternative transitions between basic structural elements would represent alternative realizations of the same structural fold (Plate 2.5) and our knowledge about the probability of different realizations. In Plate 2.5 the "plus" symbols indicate the combination of information about solvent exposure from both structures with the information from the structural alignment. The PDB codes for protein and chain identification represent cystatin (1CEWI) and monellin (1MONA). The above alignment was taken from the Dali/FSSP database.[38] The bottom of the picture is a schematic representation of the model.

2.5. Automatic Pattern Embedding in a DSM

The design of structural models described in the previous sections does not encode any information about the native sequence of the protein structure in order to avoid bias toward that sequence or its close homologs. However, recently there has been ample evidence that including some sequence information considerably improves fold recognition in other threading approaches.[14,16,18,39] Thus, the combination of information about structural preferences with information about amino acid conservation at key functional positions proves to be more powerful in identifying distant functional homologs than either class of information by itself. It was demonstrated previously that a combination of a minimal functional pattern with a structural DSM provides a very sensitive method for the identification of functional homologs with a similar structure and limited sequence similarity.[22,40] Those functional patterns have previously been selected and embedded in the DSMs manually. We describe below a set of rules for automated minimal pattern selection and embedding.

The selection of a minimal functionally diagnostic pattern to embed relies on a multiple sequence alignment among representative homologous proteins. Such a set of aligned homologous sequences can be represented as a positional sequence profile,[41,42] where each profile is described by a set of vectors giving the probability of observing each of the 20 amino acids at each alignment position. Profiles can be aligned with the structural model (see Plate 2.6) provided that at least one of the profile-defining proteins aligns with the model. This is straightforward if the model itself was constructed from the structure of one of the profile-defining proteins. In the method presented here, the DSM design is based directly on the determined structure of one of the proteins used in profile generation. Thus, each position in the profile can be directly associated with a structural-model position. We assume that a residue completely conserved across a wide range of homologs corresponds to a key functional position. Thus the simplest pattern embedding in a structural model should be applied to such conserved positions. This approach avoids imprinting sequence similarities onto positions that have structural preferences but show limited or inconsistent sequence similarity across a wide range of homologs.

The automation of pattern embedding in a DSM requires three steps. The first step is to construct a DSM from the structure of the protein (or domain) of interest as described in section 2.4. The second step is to generate the sequence profile for the functional family of that protein/domain. The third step is to combine the sequence profile with the DSM. Below we present simple rules to select the positions from the profile to be embedded in a DSM—only those positions with strictly conserved amino acid—as a proof-of-concept of the approach. Alternative rules to select the positions for embedding can be easily envisioned by the reader, such as the selection of conserved hydrophobic residues, etc.

2.5.1. Automated Pattern Generation and Selection

For the profile generation from a set of homologous sequences we use the PIMAII algorithm described in the article.[42] This algorithm iteratively aligns diverse homologous sequences while constructing the profile. The probabilities of observing different amino acids in a particular position are calculated as mixtures of amino acid prior probabilities with those in the profile alignment.

We select the sequences used for generating the profile from a set of homologs of the functional domain. The homologs are identified by a PSI-BLAST[41] search against the nonredundant protein database at NCBI. We run searches up to 5 durations in order to minimize the inclusion of false positives. Hits with an E-value less than 10^{-10} are recorded and a short list (less than 6 sequences) is selected as follows:

1. Select all hits with the alignment covering at least 85% of length of the seed sequence and containing gaps in more than 10% of the aligned positions.
2. From the hits with E-values less than 10^{-50} select sequentially those that differ by 10^{-5} in E-value. From the hits with E-values greater than 10^{-50} and less than 10^{-10} select those that differ by 10^{-3} in the E-value.
3. Select up to 6 sequences (minimum 2). If the aligned region of the hit sequence does not cover the entire query sequence, extend it at both ends by 10% of the length of the seed sequence.

This selection speeds up the profile generation step by excluding sequences that are very similar to each other and sequences with very long insertions.

We characterize the profile using the information content density of the profile. The density of the information content of a profile $D(IC)$ is defined as:

$$D(IC) = \frac{1}{N}\left(-\sum_{k}\log(L_k + 1) + \sum_{i=1}^{N}\sum_{a} P_i(a)\log(P_i(a)/B(a))\right) \qquad (2.1)$$

where N is the number of positions in the profile, $P_i(a)$ is the amino acid probability distribution at the profile position i, $B(a)$ is the amino acid background probability distribution and L_k is the length of the k-th gap in the profile. The profile is considered useful as a functionally diagnostic pattern for embedding in a structural model if the density of the information content is less than 0.2 amino acid identity equivalents. This assures that profiles that represent domains with highly conserved sequence are not considered. Those domains can be unambiguously recognized by sequence comparison methods like BLAST, hence do not require the use of the more powerful DSMs.

2.5.2. Automated Pattern Embedding in a DSM

For each profile position i, the information content is calculated according to the formula:

$$IC_i = \sum_a P_i(a)\log(P_i(a)) \tag{2.2}$$

The information content at each position gives a numerical description of the level of amino acid conservation. Selecting for this value provides a filter for the profile positions that will be embedded in a DSM. In the current scheme, information content greater than –2.008 corresponds to a strictly conserved amino acid position and only those positions are chosen for embedding. For those positions, the profile probabilities replace structural amino acid emission probabilities in the corresponding DSM hidden state. The threshold value of the information content depends on the precalculated priors.[42] The information content has a negative value since it is the negative of entropy $-(-P\times\log(P))$. Often it is compared to reference state information content, which we have omitted here.

The alignment of the seed domain to the sequence profile implies the alignment of the structural template positions (and corresponding DSM hidden states) to the profile positions (see Plate 2.6). For a position labeled i in the structural template, the amino acid probability distribution is given by P_i with the information content IC_i. $CL_s(i)$ is the structural state class of the position i. The pattern-embedding step is realized by the following reconstruction of the H-matrix elements $h(a|s_i)$ corresponding to the hidden state s_i of the position i.

$$h(a|s_i) = \begin{cases} P_i(a) & if \quad IC_i \geq -2.008 \\ H_0(a|CL_s(i)) & if \quad IC_i < -2.008 \end{cases} \tag{2.3}$$

The pattern-embedded DSM (PATDSM) is similar to the DSM generated just from a structure. However, there are a few differences. The length distribution of loop submodels reflects the observed length of loops in the profile. The amino acid emission matrix represents both the amino acid emission probabilities of structural states in nonconserved positions and the profile probabilities for conserved positions. The states that model the length variation of structural elements remain unchanged.

2.6. A Bayesian Approach to Fold Recognition

We state the fold recognition as a problem of finding the model from the library with the highest posterior probability, given the query sequence.[21,22] According to Bayes, the posterior model probability is given by the product of the conditional probability of the sequence given the model and the prior probability of the model.

$$\mathrm{P}(\mathrm{Model}_i \mid \mathbf{seq}) = \frac{\mathrm{P}(\mathbf{seq} \mid \mathrm{Model}_i) \cdot \mathrm{P}(\mathrm{Model}_i)}{\sum_{j=1}^{K} \mathrm{P}(\mathbf{seq} \mid \mathrm{Model}_j) \cdot \mathrm{P}(\mathrm{Model}_j)}, \tag{2.4}$$

where $P(\mathrm{Model}_i)$ is the prior probability assigned to a Model_i over K models. We discuss in detail different approaches to assign *prior* model probabilities in section 2.6.2.

2.6.1. The Filtering Algorithm

The probability of observing a sequence given a model $P(\mathbf{seq}|\mathrm{Model})$ is calculated by the Filtering algorithm.[26] This algorithm is expressed iteratively and requires the initial hidden state probability distribution $\mathbf{X}(s)$ where s are the hidden states of the HMM. The adaptation of the algorithm to the protein fold recognition problem[20] selects the $\mathbf{X}(s)$ such that initial probability of the first state in the model is set to 1 and initial probabilities of other states are set to 0: $\mathbf{X} = (1, 0 \ldots 0)$. Let $\mathbf{seq} = (a_1, \ldots, a_t, \ldots, a_L)$ be a sequence of amino acids and let $\mathbf{S} = (S_1, \ldots, S_t, \ldots S_L)$ be one fixed sequence of states, where L is the length of the sequence. The capital symbols S_t indicate the distinction between the fixed sequence state from the general indexing of the HMM's states s_i, $i = 1 \ldots \mathrm{N}$. N is the total number of hidden states in the model. The probability of observing a fixed sequence of states $\mathbf{S}$ given an HMM is equal to:

$$\mathrm{P}(\mathbf{S} \mid \mathrm{Model}) = x(S_1) \cdot \phi(S_2 \mid S_1) \cdot \phi(S_3 \mid S_2) \cdots \phi(S_L \mid S_{L-1}) \tag{2.5}$$

The probability of observing a sequence $\mathbf{seq}$ given a fixed sequence of states $\mathbf{S}$ from an HMM is given by:

$$\mathrm{P}(\mathbf{seq} \mid \mathbf{S}, \mathrm{Model}) = h(a_1 \mid S_1) \cdot h(a_2 \mid S_2) \cdots h(a_L \mid S_L) \tag{2.6}$$

The joint probability of observing an amino acid sequence and the fixed sequence of hidden states is:

$$\mathrm{P}(\mathbf{seq},\mathbf{S}\,|\,\mathrm{Model}) = \mathrm{P}(\mathbf{seq}\,|\,\mathbf{S},\mathrm{Model}) \cdot \mathrm{P}(\mathbf{S}\,|\,\mathrm{Model})\,. \tag{2.7}$$

The total probability of observing a sequence given the model is the sum over all possible state sequences. This is calculated iteratively by introducing the probability of the partial observation of the sequence up to position t and state s_i in that position:

$$\mathrm{q}_t(i)=P(a_1,a_2...,a_t\,,\, S_t=s_i|\mathrm{Model})\,. \tag{2.8}$$

The total probability is calculated as the product of the conditional probabilities of observing the amino acids a_{t+1} in position $t+1$ given that a_t was observed in position t $P(a_{t+1}|a_t)$, given a model.

$$\mathrm{P}(\mathbf{seq}\,|\,\mathrm{Model})=\prod \mathrm{P}(a_1)\mathrm{P}(a_2\,|\,a_1)\cdot \mathrm{P}(a_3\,|\,a_2)\cdots \mathrm{P}(a_{t+1}\,|\,a_t)\cdots \mathrm{P}(a_L\,|\,a_{L-1})\,. \tag{2.9}$$

Potential computational float underflow problems are avoided by calculating the logarithm of the total probability. This is done by the following algorithm proposed and implemented by White:[26]

1) Initialization

$$q_1(i) = x(s_i)\cdot h(a_1\,|\,s_i), \quad 1\le i\le N, \tag{2.10a}$$

$$P(a_1) = \sum_{i=1}^{N} q_1(i) \quad . \tag{2.10b}$$

2) Iteration, for $l = 1$ to L

$$q_{t+1}(j) = h(a_{t+1}\,|\,s_j)\cdot(\textstyle\sum_{i=1}^{N} q_t(i)\cdot\phi(s_j\,|\,s_i)), \quad 1\le i\le N\,, \tag{2.11a}$$

$$P(a_{t+1}\,|\,a_t) = \sum_{i=1}^{N} q_{t+1}(i)\,. \tag{2.11b}$$

Normalization step:

$$q_{t+1}(i) \to q_{t+1}(i)/P(a_{t+1}\,|\,a_t), \quad 1\le i\le N\,. \tag{2.12}$$

3) Termination

$$\log(\mathrm{P}(\mathbf{seq}\,|\,\mathrm{Model})) = \begin{array}{c}\log(P(a_1)) + \log(P(a_2\,|\,a_1)) + \log(P(a_3\,|\,a_2) + \\ + \cdots + \log(P(a_L\,|\,a_{L-1}))\,.\end{array} \tag{2.13}$$

2.6.2. Prior Model Probabilities

In the Bayesian formula (2.4), the model likelihood $P(\text{Model}_i|\mathbf{seq})$ measures the degree to which the model Model_i is compatible with the observed sequence. The prior probability $P(\text{Model}_i)$ expresses how plausible we thought the alternative models were prior to the observation of any particular sequence. The assignment of priors is usually a subjective matter, where many different priors can be tried and each particular set corresponds to a different hypothesis about how alternative models are related. For example, one can assign equal priors to all models to reflect the maximum uncertainty about each model. This approach is usually applied when there is no preference to any particular model; therefore *the models are compared and ranked in the light of the sequence data alone.* However, one can certainly employ some methods to work out the best priors for a specific problem based on expert knowledge.

Hierarchical method of assigning priors

A hierarchical scheme of assigning priors was used previously in DSM analysis of structural classification.[20] A pre-defined set of tertiary structural models was organized hierarchically, following the classification of structural domains by Jane Richardson.[23] Four super-classes were defined at the top level according to the occurrence of α-helices and β-strands: α, β, α–β, and irregular. Within each super-class, more highly constrained domain types were modeled as macro-classes. Within each macro-class, models with size variations were further defined. Accordingly, the prior probability was uniformly distributed over the four super-classes, uniformly distributed over macro-classes within each super-class, and uniformly distributed over size variants within each macro-class. This assignment of priors reflects the maximum prior uncertainty among and within the classes, and prevents bias from over-represented structural classes.

For the automatically generated DSMs, we have investigated several different ways to assign model priors. In our DSM library, the structural models are organized according to the SCOP classification, which uses the hierarchy of *class*, *fold*, *superfamily*, and *family*. Moreover, under each family, the bottom level of the SCOP hierarchy, there are one or more representative models, each with a distinct solvent-exposure profile. The first option of prior assignment is to apply the previous approach: The priors are assigned in a hierarchical manner, with uniform priors over the folds within each class, uniform over the superfamilies within each fold, uniform over the families within each superfamily, and uniform over models with different solvent exposure profiles within each family.[22] For example, the prior probability of a DSM classified by its class, fold, superfamily, and family is given as:

$$\mathrm{P(Model_j)} = \frac{1}{\#\,\mathrm{classes}} \times \frac{1}{\#\,\mathrm{folds/class}} \times \frac{1}{\#\,\mathrm{superfamilies/fold}} \times \\ \times \frac{1}{\#\,\mathrm{families/superfamily}} \times \frac{1}{\#\,\mathrm{models/family}}. \tag{2.14}$$

The posterior probability of observing a unique structural fold given a sequence is then calculated by:

$$\mathrm{P(fold} \mid \mathbf{seq}) = \sum_{\mathrm{Model}_j \in \mathrm{family} \in \mathrm{superfamily} \in \mathrm{fold}} \mathrm{P(Model}_j \mid \mathbf{seq}). \tag{2.15}$$

Here the underlying assumption is that alternative models are mutually exclusive and all-inclusive at each level being considered, i.e., SCOP class, fold, superfamily, and family. In other words, we consider these models as completely independent structural hypotheses.

Maximum method of assigning priors

The above assumption is incorrect at the SCOP superfamily and family levels and when pattern-embedded models are included in the library. Since all models representing the same SCOP fold are effectively overlapping, they cannot be regarded as alternative hypothesis in the Bayesian formula. Therefore, we devised a second way to assign priors. Here, the hypothesis space is composed of distinct SCOP folds. Since several DSMs represent a fold, we select only the DMS with the highest likelihood $P(\mathbf{seq}|\mathrm{Model}_i)$ as the fold representative for a given sequence:

$$\mathrm{P}(\mathbf{seq} \mid \mathrm{fold}) = \mathrm{Max}\left\{ \mathrm{P}(\mathbf{seq} \mid \mathrm{M}_i);\ \mathrm{M}_i \in \mathrm{fold} \right\}. \tag{2.16}$$

The fold $P(\mathrm{fold}_k)$ priors are assigned uniformly over all folds. Posterior fold probabilities are calculated as:

$$P(\text{fold}_k|\,\mathbf{seq}) = \frac{P(\mathbf{seq}\,|\,\text{fold}_k)\cdot P(\text{fold}_k)}{\sum_{j=1}^{N_f} P(\mathbf{seq}\,|\,\text{fold}_j)\cdot P(\text{fold}_j)} \quad . \tag{2.17}$$

where N_f is the total number of folds. We call this approach to assigning priors the "maximum" method.

We found empirically that the second approach gives better fold recognition results (see Table 2.6) with the DSM library representing all SCOP families. For the DSM library consisting of only the superfamily level models, the two methods perform comparably. The second approach is also more appropriate when the library of automatically generated structural DSMs is combined with a library of pattern-embedded DSMs.

Triage method of assigning priors

One potential problem arises with the direct use of library of DSMs for each SCOP family in combination with the Bayesian approach; the presence of a large number of models increases the likelihood that the correct model will exhibit a small (<0.5) posterior probability relative to competing models. In order to account for this we have devised a third method to assign priors, a "triage" method.[29] The triage method classifies protein sequences into major structural classes prior to predicting the most probable fold within that class (He, McAllister, and Smith, private communication). This method relies on a set of high-level, manually designed generic DSMs, which model major structural classes M_{cl}, such as all-α, all-β, α/β, irregular or membrane proteins. Each generic DSM is an idealization of a major structural class and is designed to represent general structural features of a particular class while allowing a large range of anticipated secondary structure variations.

Since the structural classes defined by any generic DSMs do not universally correspond to the SCOP classification, we use the following method to classify the automatically generated DSMs into different structural classes.

We classify each automatically generated DSM by assigning a structural class to its template sequence. The class of the sequence is given by the class of generic DSM, which has a posterior probability greater than 0.3 (empirical value) for that sequence. Although such class prediction is not perfect it will likely be similar for related sequences.

For a query sequence, we first predict its structural class using the set of generic DSMs. If the most probable generic DSM has a posterior probability greater than 0.3 (empirical value) then its corresponding structural class is selected. Otherwise, the top two structural classes are chosen. Once the sequence is assigned to a structural class, the subset of automatically generated DSMs belonging to that structural class(es) is chosen for the fold prediction. It means that the prior probabilities for models belonging to other structural classes are set to zero. We then select the top-scoring model from each fold and the prior probabilities are assigned uniformly over the folds. Finally, we calculate the posterior probability for each selected fold and identify the most probable one. This assignment is similar to the maximum method described above, the only difference being that the structural hypothesis space is modeled by a set of fold-level DSMs limited to a particular structural class rather than by all fold models in the entire DSM library.

2.7. Results

2.7.1. Comparing the Bayesian Approach and Total Alignment Probability with Other Methods

First, we compare the performance of two fold recognition methods, the Viterbi method and the Filtering method. In order to minimize the dependence of the results on assignment of priors and domain dissection, the comparison was done using the library of DSMs representing different superfamily and single-domain protein models. The model prior probabilities are assigned using the hierarchical method described in section 2.6.2. The Viterbi method[31] uses the probability of the optimal sequence-to-model alignment $P(\mathbf{seq}|\text{Model, optimal-alignment})$ as the sequence-model probability. The Filtering method[26,27] uses the sum of probabilities over all possible sequence-to-model alignments $P(\mathbf{seq}|\text{Model})$ as the sequence-model probability. We equate the Viterbi method with the Boltzmann approach. However it should be noted that other approaches relying on the optimal alignment use various methods to evaluate the significance of the optimal sequence-to-structure alignment.

Table 2.3. Summary of results of self-recognition experiments. The ratio is the number of correctly first ranked fold models (true positives—TP) versus the total number of models at first rank (true and false positives—TP+FP) with the posterior probability greater than 0 or 0.5. The sensitivity of the fold recognition TP/(TP+FP) is shown in parentheses. In this case the false positives and false negatives are equal.

Method	P(first rank model)>0	**P(first rank model)>0.5**
Viterbi-optimal alignment	110/188 (58%)	**107/168 (64%)**
Filtering-total alignment	152/188 (80%)	**145/161 (90%)**

In self-recognition experiments we threaded 188 single-domain protein sequences through the DSM library (see Table 2.3). In the structural homolog prediction experiments, we threaded 71 sequences, representing ten SCOP superfamilies that had been used to evaluate the Recursive Dynamic Programming Threading (RDPT) algorithm recently developed by Thiele, Zimmer, and Lengauer[15] (see Table 2.4). On average, the Filtering method (or Bayesian approach) is about 40% more sensitive at fold recognition than the Viterbi method. The RDPT method reported a success rate of 56% for the structural analog recognition.

Table 2.4. Summary of the structural homolog threading experiments. The ratio is the number of the correctly first ranked fold models versus the total number of models at first rank with the probability greater than 0 or 0.5. The sensitivity of the method is shown in parenthesis.

Method	P(first rank model)>0	**P(first rank model)>0.5**
Viterbi-optimal alignment	31/71 (44%)	**26/66 (39%)**
Filtering-total alignment	33/71 (46%)	**32/53 (60%)**

Additional comparison of the performance of the Viterbi and the Filtering methods can be found in article.[22] The performance of the DSM method using the purely structural model library was also assessed during the recent CASP4[43] prediction contest (group name BMERC). At the time of the CASP4 meeting, the average score over 18 submitted predictions placed the DSM method among the top 20 groups. The sensitivity and the specificity of fold recognition by DSMs are improved by averaging threading results over a set of homologous sequences. This approach was applied to CASP4 blind predictions of protein structure, but for the sake of clear comparisons it was not used in results presented here.

2.7.2. Results of Automatic Pattern Embedding

The current version of our DSM library was constructed by selecting one representative structure from each SCOP[24] fold, version 1.48. This library represents 305 single domain folds, 117 of which come from multidomain proteins. According to the selection rules presented in section 2.5.1, satisfactory profiles were generated for 300 out of 305 domains in that library. On average, 22% of the positions in these profiles are conserved. A DSM, in addition to modeling residue positions from the structure, contains hidden states that do not correspond to particular positions in the structure. Therefore, fewer than 22% of hidden states from a pattern embedded DSM represent profile positions.

To test the performance of DSMs with automatically embedded patterns we selected three sets of protein domain pairs from the SCOP database. For each of 305 structural domains represented in our DSM library, we selected a structural analog as defined by the SCOP family, superfamily and fold classifications. We additionally checked the CATH[25] database to see whether those pairs were identified to have the same CATH topology using the structural database comparison done by Hadley and Jones.[44] Finally, we selected a test set of protein pairs that are classified as structurally similar by two independent approaches. This set consists of 493 same-family pairs, 61 same-superfamily pairs and 42 same-fold pairs.

Using structural analogs as query sequences and 305 DSM-library-defining sequences as the search database, we checked how many pairs are identified by a simple BLAST search. A generous E-value of 10^{-5} was used as a cutoff for the BLAST homolog recognition. We then removed those homologs from the initial set thus reducing the test set to the difficult-to-recognize pairs. The final test set includes 131 same-family pairs, 59 same-superfamily pairs and 42 same-fold pairs.

Using the above test set, we compared three fold recognition approaches: PSI-BLAST search, automatically generated structural

DSMs, and automatically generated structural DSMs with sequence pattern embedded. The PSI-BLAST searches were done in two steps. First, for each pair, we used the structural analog sequence as the query to search against the "non-redundant" protein database from NCBI. The search was run for 5 iterations or until convergence, with the E-value for inclusion set to 10^{-3}, and the resulting position-specific scoring matrix (PSSM) was saved. Second, we used the calculated PSSM as an input to search against the PDB sequence database. This search was run for 10 iterations, effectively till convergence in all cases. The PSI-BLAST search was considered successful if the correct DSM-defining protein was recognized in the second step with an E-value better than the cutoff. The comparison of the structural-analog recognition performance is summarized in Table 2.5.

Table 2.5. Comparison of the fold-recognition performance for difficult-to-recognize structural-analog pairs. The cutoff value indicates the highest E-value for PSI-BLAST search that was considered as analog recognition and the lowest P-value of the first ranking fold in the DSM fold recognition. The columns labeled TP indicate recognition of true positives and FP indicates false positives, i.e., the recognition as structural analogs of the proteins that are classified by both SCOP and CATH as having a different fold. Courtesy of IEEE[47].

	Structural-analog recognition method											
	PSI-BLAST				**Automatically designed structural DSMs**				**Automatically designed structural DSMs with embedded pattern**			
Cutoff score	E = e-03		**E = e-10**		P = 0		**P = 0.5**		P = 0		**P = 0.5**	
SCOP	TP	FP	**TP**	**FP**	TP	FP	**TP**	**FP**	TP	FP	**TP**	**FP**
131 family pairs	65	29	**65**	**29**	34	97	**25**	**43**	82	49	**70**	**12**
59 superfamily pairs	11	28	**9**	**26**	6	53	**4**	**34**	11	48	**6**	**14**
42 fold pairs	0	11	**0**	**10**	5	37	**3**	**22**	5	37	**4**	**19**
Total	77	68	74	65	45	187	**32**	**99**	98	134	80	45

The pattern-embedded DSM fold-recognition method competes favorably with PSI-BLAST, giving a greater number of true positives and lower number of false positives. In the PSI-BLAST search, we considered as false positives only those proteins that are classified as a different fold by both SCOP and CATH. Only one false positive per query sequence was counted regardless of how many were identified, in order to compare with the DSM method that makes a single prediction for each sequence. The proteins with E-values better than the cutoff but not classified in SCOP were not considered as false positives, so the true rate of false positives for PSI-BLAST could be even higher. The automated pattern embedding considerably improved the recognition of structural-analogs when compared with the automatically generated structural DSMs. The overlap of true positive predictions between the PSI-BLAST (with cutoff 10^{-10}) and pattern embedded DSMs (with probability higher than 0.5) is: 46 family pairs and 2 superfamily pairs. This confirms that the pattern-embedded DSM method is partially driven by the recognition of structural similarities since some of the structural analogs detected differ from those detected by PSI-BLAST and the total number of false positives is smaller.

2.7.3. Comparison of Different Assignments of Prior Probabilities

Table 2.6 shows the results of fold prediction using four different schemes of prior assignments. The result is expressed as the number of sequences with correct fold prediction versus the total number of sequences for which the prediction was made. We used a simple binary decision rule; a prediction was made only if the most probable fold has a posterior probability greater than 0.5. In the "Hierarchy" method, the prior probabilities were assigned hierarchically according to Eq. (2.14) and the posterior probability for each fold was computed according to Eq. (2.15). In the "Maximum" method, for each fold, only the model with the highest model likelihood P(**seq**|Model$_i$) was selected. The prior probabilities were uniformly assigned to the selected models (or folds), then the fold posterior probabilities were calculated over the selected set (Eq. (2.16) and Eq. (2.17)). In the "Triage" method, the query sequence was first assigned to one or two structural classes via a set of high-level generic DSMs as described in Section 2.6.2. Then uniform priors were assigned to models with the highest model likelihood from each fold within the assigned classes and the fold prediction was made. The "Uniform"method assigned the model priors uniformly over the entire DSM library. For this comparison we used a library of 1282 models that represent 882 SCOP families and 358 folds. No pattern-embedded models were included in the library. As we expected, the hierarchy method did not perform well for such a library. The Uniform method had the highest sensitivity, however, this method had considerably fewer true positives than the other methods. Overall, the Triage method performed the best, giving the most true positives with a sensitivity slightly worse than the Maximum method.

Table 2.6. Comparison of different approaches to assigning prior probabilities. Results of fold self-prediction for 882 sequences whose structures have been used to construct the DSM library. The library consists of 1282 automatically generated DSMs without pattern embedding. The total number of positives (TP+FP) indicates the predicted folds/models that had the posterior probability greater than 0.5.

	Maximum	**Hierarchy**	**Triage**	**Uniform**
TP/(TP+FP)	412/482	371/443	457/538	356/391
Sensitivity	85.5%	83.7%	84.9%	91.0%

2.8. Conclusion

The automation of model construction and controlled pattern embedding in the purely structural DSMs provides the possibility of constructing a large library of models for protein functional domains. One advantage of the DSM design approach is that only one representative structure is needed to construct a model for the homologous family. This feature is particularly appealing since Structural Genomics projects[45] aim to generate one structure for most functional families not yet represented in the PDB. The targeted sequence similarity among the members of the family is around 30% (and above) for those families. This makes the Structural Genomics targets ideal for the construction of the pattern-embedded DSM library. For such targets only few key residues conserved in the same structural context define the functional family, just as the pattern embedded DSMs do.

The DSM method is very sensitive at identifying distant homologs, but there are still a few aspects of protein structure modeling that need to be addressed. First, the specificity of the library of DSMs can be improved by including additional competing models, such as a set of generic DSMs.[20,29] Those models would represent structures of proteins that do not have a structural analog in the current PDB-based library, such as irregular structures and membrane proteins. The second and the most pressing aspect, is to design a method to construct umbrella models covering the large variations within each structural family. Such models should be more sensitive in recognizing proteins with similar structure but without functional similarity. Third, we expect that minimal profiles generated from PROSITE[46] motifs will perform better in distant homolog recognition than the automatically generated profiles presented here. The PROSITE motif preferences would replace a smaller number of structural preferences in DSMs than profiles described in this paper. Such minimal sequence profiles should produce models that are better suited for detection of structural analogs with a conserved (minimal) functional pattern.

Acknowledgments

We would like to thank Temple Smith for providing a stimulating environment for independent research at BMERC and his many contributions to the DSM approach. We wish to acknowledge the pioneering role of Jim White in introducing the DSM and HMM modeling in protein structure prediction. We also thank Scott Mohr, Gregg McAllister and Prashanth Vishwanath (Vishy) for careful reading of the manuscript and many helpful discussions and suggestions.

References

1. Bowie JU, Luthy R, Eisenberg D: **A method to identify protein sequences that fold into a known three-dimensional structure.** *Science* 1991, **253**:164–170.

2. Jones D, Taylor W, Thornton J: **A new approach to protein fold recognition.** *Nature* 1992, **358**:86–89.

3. Sippl MJ: **Boltzmann's principle, knowledge-based mean fields and protein folding.** *J Comput-Aided Mol Des* 1993, **7**:473–501.

4. Sippl MJ: **Knowledge-based potentials for proteins.** *Curr Opin Struct Biol* 1995, **5**:229–235.

5. Needleman SB, Wunsch CD: **A general method applicable to the search for similarities in the amino acid sequence of two proteins.** *J Mol Biol* 1970, **48**:443–453.

6. Smith TF, Waterman MS: **Identification of common molecular subsequences.** *J Mol Biol* 1981, **147**:195–197.

7. Lathrop RH: **The protein threading problem with sequence amino acid interaction preferences is NP-complete.** *Protein Eng* 1994, **7**:1059–1068.

8. Bryant SH, Altshul SF: **Statistics of sequence-structure threading.** *Curr Opin Struct Biol* 1995, **5**:236–244.

9. Maiorov VN, Crippen GM: **Contact potential that recognizes the correct folding of globular proteins.** *J Mol Biol* 1992, **227**:876–888.

10. Miyazawa S, Jernigan RL: **Residue-residue potentials with a favorable contact pair term and unfavorable high packing density term, for simulation and threading.** *J Mol Biol* 1996, **256**:623–644.

11. Lathrop RH, Smith TF: **Global optimum protein threading with gapped alignment and empirical pair score functions.** *J Mol Biol* 1996, **255**:641–665.

12. Rost B, Schneider R, Sander C: **Protein fold recognition by prediction-based threading.** *J Mol Biol* 1997, 270:471–480.

13. Karplus K, Sjolander K, Barrett C, Cline M, Haussler D, Hughey R, Holm L, Sander C: **Predicting protein structure using hidden Markov models.** *Proteins* 1997, Suppl, **1**:134–139.

14. Jones DT: GenTHREADER: **An efficient and reliable protein fold recognition method for genomic sequences.** *J Mol Biol* 1999, **287**:797–815.

15. Thiele R, Zimmer R, Lengauer T: **Protein threading by recursive dynamic programming.** *J Mol Biol* 1999, **290**:757–779.

16. Rychlewski L, Jaroszewski L, Li W, Godzik A: **Comparison of sequence profiles. Strategies for structural predictions using sequence information.** *Protein Sci* 2000, **9**:232–241.

17. Fischer D: **Hybrid fold recognition: Combining sequence derived properties with evolutionary information.** *Pac Symp Biocomput* 2000:143–154.

18. Panchenko AR, Marchler-Bauer A, Bryant SH: **Combination of threading potentials and sequence profiles improves fold recognition.** *J Mol Biol* 2000, **296**:1319–1331.

19. Kolinski A, Betancourt MR, Kihara D, Rotkiewicz P, Skolnick J: **Generalized comparative modeling (GENECOMP): A combination of sequence comparison, threading and lattice modeling for protein structure prediction and refinement.** *Proteins* 2001, **44**:133–149.

20. Stultz CM, White JV, Smith TF: **Structural analysis based on state-space modeling.** *Protein Sci* 1993, **2**:305–314.

21. Lathrop RH, Rogers Jr. RG, Bienkowska JR, Bryant BKM, Buturovic LJ, Gaitatzes C,

Nambudripad R, White JV, Smith TF: **Analysis and algorithms for protein sequence-structure alignment.** In: *Computational Methods in Molecular Biology*, Salzberg S, Searls D, Kasif S, eds. Amsterdam, Netherlands: Elsevier Press, 1998:227–283.

22. Bienkowska JR, Yu L, Rogers Jr. RG, Zarakhovich S, Smith TF: **Protein fold recognition by total alignment probability.** *Proteins* 2000, **40**:451–462.

23. Richardson J: **The anatomy and taxonomy of protein structures.** *Adv Prot Chem* 1981, **34**:167–339.

24. Murzin A, Brenner SE, Hubbard T, Chothia C: **SCOP: A structural classification of proteins database for the investigation of the sequences and structures.** *J Mol Biol* 1995, **247**:536–540.

25. Orengo CA, Michie AD, Jones S, Jones DT, Swindells MB, Thornton JM: **CATH—A hierarchic classification of protein domain structures.** *Structure* 1997, **5**:1093–1108.

26. White JV: **Modeling and filtering for discretely valued time series.** In: *Bayesian Analysis of Time Series and Dynamic Models.* Spall JC, ed. New York, NY USA: Marcel Dekker, 1988:255–283.

27. Rabiner LR: **A tutorial on hidden Markov models and selected applications in speech recognition.** *Proc IEEE* 1989, **77**:257–286

28. Berman HM, Westbrook J, Feng Z, Gilliland G, Bhat TN, Weissig H, Shindyalov IN, Bourne PE: **The protein data bank.** *Nucleic Acid Res* 2000, **28**:235–242.

29. He H, McAllister G, Smith TF: **Triage Protein Fold Recognition.** *Personal Communication.*

30. Baum LE, Egon JA: **An inequality with applications to statistical estimation for probabilistic functions of a Markov process and to a model for ecology.** *Bull Amer Meteorol Soc* 1967, **73**:360–363.

31. Viterbi AJ: **Error bounds for convolutional codes and an asymptotically optimal decoding algorithm.** *IEEE Trans Information Theory* 1967, **IT-13**, April, 260–269.

32. Bienkowska JR, Rogers Jr. RG, Smith TF: **Filtered neighbors threading.** *Proteins* 1999, **37**:346–359.

33. Bienkowska JR, Rogers Jr. RG, Smith TF: **Performance of threading scoring functions designed using new optimization method.** *J Comput Biol* 1999, **6**:299–311.

34. Creighton TE. *Proteins: Structures and Molecular Properties.* New York: W.H. Freeman and Company, 1993.

35. Kabsch W, Sander C: **Dictionary of protein secondary structure: Pattern recognition of hydrogen-bonded and geometrical features.** *Biopolymers* 1983, **22**:2577–637.

36. Eisenberg D, MacLachlan AD: **Solvation energy in protein folding and binding.** *Nature* 1986, **319**:199–203.

37. Lo Conte L, Smith TF: **Visible volume: A robust measure for protein structure characterization.** *J Mol Biol* 1997, **273**:338–348.

38. Holm L, Sander C: **Touring protein fold space with Dali/FSSP.** *Nucleic Acid Res* 1998, **26**:316–319.

39. Kelley LA, MacCallum RM, Sternberg MJE: **Enhanced genome annotation using structural profiles in the program 3D-PSSM.** *J Mol Biol* 2000, **299**:501–522.

40. Yu L, White JV, Smith TF: **A homology identification method that combines sequence and structure.** *Protein Sci* 1998, **7**:2499–2510.

41. Altschul SF, Madden T, Schaffer A, Zhang J, Zhang Z, Miller W, Lipman DJ: **Gapped BLAST and PSI-BLAST: A new generation of protein database search programs.** *Nucleic Acid Res* 1997, **25**:3389–3402.

42. Das S, Smith TF: **Identifying nature's protein lego set in analysis of amino acid sequences.** *Adv Protein Chem* 2000, **54**:159–183.

43. *CASP4*. http://PredictionCenter.llnl.gov/casp4/CASP4.html.

44. Hadley C, Jones DT: **A systematic comparison of protein structure classifications: SCOP, CATH, and FSSP**. *Structure* 1999, **9**:1099–1112.

45. Gershon D: **Structural genomics—From cottage industry to industrial revolution.** *Nature* 2000, **408**:273–274.

46. Hofmann K, Bucher P, Falquet L, Bairoch A: **The PROSITE database, its status in 1999.** *Nucleic Acid Res* 1999, **27**:215–219.

47. Bienkowska J, He H, Smith TF: **Automated pattern embedding in protein structure models.** *IEEE Trans Intelligent Systems* 2000, **16**:21–25.

3

Three-Dimensional Structure Prediction Using Simplified Structure Models and Bayesian Block Fragments

Jun Zhu and Roland Lüthy

Abstract

The goal of the work described in this chapter is to fold proteins into simplified three-dimensional structures constructed from small fragments cut out of a representative set of known three-dimensional structures. Using the sequence of the protein to be folded, which we will call the 'target protein,' a Bayesian sequence profile (Zhu, Lüthy et al. 1999) is constructed. This sequence profile is then used to build a database of fragments with high likelihood to be similar to the target protein by searching the sequences of a collection of representative three-dimensional protein structures. The three-dimensional protein structures and fragments are represented in a simplified form (Levitt 1976) as a sequence of angle pairs, one angle pair per residue. An initial model with random angle pairs is constructed and then refined by an iterative procedure. In each iteration a fragment from the database is inserted into the three-dimensional structure model and the new structure is evaluated using a set of knowledge based scoring functions. A Monte Carlo criterion is used to decide whether to accept the change before the next iteration.

3

Three-Dimensional Structure Prediction Using Simplified Structure Models and Bayesian Block Fragments

Jun Zhu and Roland Lüthy

Amgen Inc., Thousand Oaks, California

3.1. Introduction

The following chapter describes a method to fold proteins into simplified three-dimensional structures constructed from small fragments cut out of a representative set of known three-dimensional structures. Using the sequence of the protein to be folded, which we will call the 'target protein', a Bayesian sequence profile[1] is constructed. This sequence profile is then used to build a database of fragments with a high likelihood of being similar to the target protein by searching the sequences of a collection of representative three-dimensional protein structures. The three-dimensional protein structures and fragments are represented in a simplified form[2] as a sequence of angle pairs, one angle pair per residue. An initial model with random angle pairs is constructed and then refined by an iterative procedure. In each iteration a fragment from the database is inserted into the three-dimensional structure model and the new structure is evaluated using a set of knowledge based scoring functions. A Monte Carlo criterion is used to decide whether to accept the change before the next iteration.
Fig 3.1 below is an overview of the procedure:

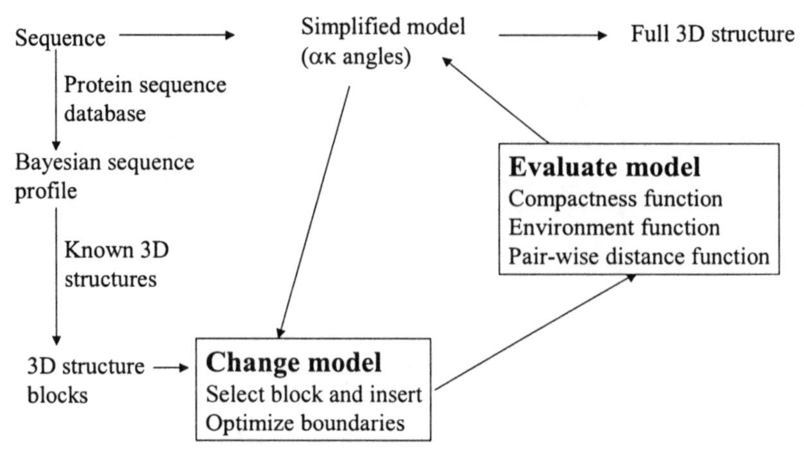

Figure 3.1. Overview of the procedure.

There are two types of interactions to be considered for protein folding, local and non-local interactions. The local interactions are between amino acids that are close in the sequence, typically less than 6 residues apart and they are thought to be important for local substructures such as alpha helices and turns. Non-local interactions are between residues further apart, and are thought to stabilize local structures in an arrangement that produces a minimum energy state. We used short structure blocks as our building blocks, which were selected based on their sequence information, with the assumption that the blocks would represent the correct local structures. We then assembled these blocks into the complete structure attempting to achieve an energy minimum. To calculate the energy of a structure we use knowledge based energy terms.

The sequence to local structure prediction can be accomplished by searching the target sequence against the sequence database derived from known structures. In general, multiple sequence alignment methods are more sensitive than pairwise methods in remote sequence similarity search. Here we choose Bayesian sequence profiles as our sequence similarity search method based on the following reasoning: (1) Bayesian sequence profiles perform better than popular multiple sequence alignment methods in searching remote sequence similarities among structural related sequences.[1] (2) Bayesian profile is a natural block-based method, which only aligns regions of high confidence. (3) Each alignment block reported by Bayesian profile has a confidence value that can be explored during block assembling process.

3.2. Methods

3.2.1. Simplified Backbone Angle Representation of 3D Structures

Protein structures were represented using a sequence of angle pairs. One pair consists of κ, the angle between three consecutive C_α atoms, and α the torsion angle between four consecutive C_α atoms, shown in Fig. 3.2:

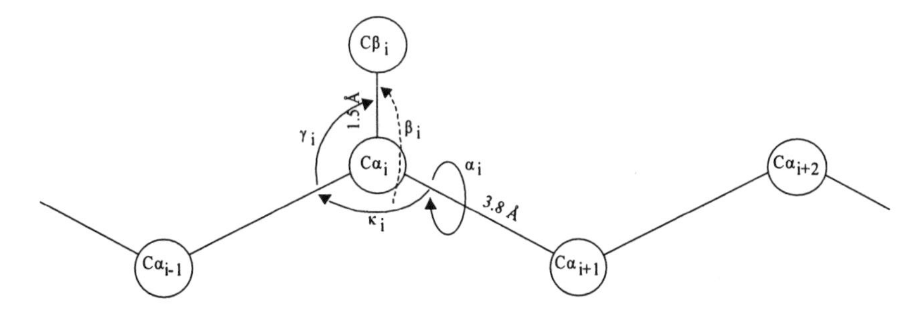

Figure 3.2. The three-dimensional structure of a protein can be represented as a sequence of ακ pairs.[2]

From the sequence of ακ pairs coordinates for the C_α and C_β can be reconstructed using the following procedure:

Place $C_{\alpha 1}$ atom at origin.
Place $C_{\alpha 2}$ atom at a distance of 3.8 Å along the *x*-axis.
Place $C_{\alpha 3}$ atom at a distance of 3.8 Å and κ_2 in *xy* plane.
For atoms $C_{\alpha i}$ where $i = 4$ to n place atom *i* at distance 3.8 from atom $i-1$ using torsion angle α_{i-2} and angle κ_{i-1}.
Place C_β atoms for residues 2 to n–1 using constants for γ, the angle between $C_{\alpha i-1}$–$C_{\alpha i}$–$C_{\beta i}$, β, the angle between $C_{\beta i}$ and the $C_{\alpha i-1}$–$C_{\alpha i}$–$C_{\alpha i+1}$ plane, and the C_α–C_β distance. No C_β atoms were generated for the first and last residue.
The constants were derived from a representative set of known protein structures which pair-wise sequence similarity less than 35% (PDB35):

C_α–C_α distance	3.80 ± 0.04 Å
C_α–C_β distance	1.53 ± 0.04 Å
γ	120 ± 13°
β	48 ± 19°

New structures were generated by copying the α and κ angles from blocks of known structures and inserting them at the alignment position of the block with respect to the target sequence.

3.2.2. Block Selection

3.2.2.1. Bayesian Profiles

Bayesian profiles[1] are similar to the sequence profile method.[3] The Bayesian profile $M(i,a)$ is the probability of amino acid a at position i and can be expressed as follows:

$$M(i,a) = \frac{pseudo * \theta_a + \sum_{n,R_j^{(n)}=a} P(A_{i,j} = 1 \mid Q, R^{(n)})}{\sum_a pseudo * \theta_a + \sum_n P(A_{i,j} = 1 \mid Q, R^{(n)})} \tag{3.1}$$

where $P(A_{i,j} = 1 \mid Q,R^{(n)})$ is the marginal posterior alignment distribution of a query sequence Q and a sequence $R^{(n)}$; $A_{i,j}$ is an alignment matrix; *pseudo* is a pseudo count; θ_a is the prior probability of observing amino acid a. $P(A_{i,j} = 1 \mid Q,R^{(n)})$ can be obtained using Bayes sequence aligner.[4] Bayesian profiles differ from general profile methods in the weighting scheme: (1) each individual residue in each sequence is weighted differently depending on $P(A_{i,j} = 1, R_j^{(n)} \mid Q,R^{(n)})$, such that high confidence alignments will contribute more; (2) all residues in a column of the model are weighted differently from other columns depending on $\Sigma_n P(A_{i,j} = 1 \mid Q,R^{(n)})$, such that the prior information (the pseudo counts) will contribute less to columns with high confidence alignments.

After obtaining the Bayesian profile, it can be compared with sequences similar to the pairwise Bayes sequence aligner by substituting the general relation matrices (PAM[5] or BLOSUM series[6]) with the Bayesian profile. Then the likelihood of the profile and a sequence conditioned on the alignment is:

$$\log P(M,R \mid \theta, A) = \sum_j \theta_{R_j} + \sum_{i,j} A_{i,j} \log M(i,R_j) \tag{3.2}$$

Whether or not a sequence R and a Bayesian profile $M(i,a)$ are related is measured by the Bayesian evidence:

$$BayesianEvidence = 1 - \sup_{P(k)}\{P(k>0 \mid M,R)\} \tag{3.3}$$

where $P(k|M,R)$ is the posterior probability of a k-block alignment:

$$P(k \mid M,R) = \frac{\sum_{A} P(M,R \mid A,k)P(A \mid k)P(k)}{\sum_{k}\sum_{A} P(M,R \mid A,k)P(A \mid k)P(k)} \tag{3.4}$$

The Bayesian evidence has similar meaning as a p-value. The smaller the value of the Bayesian evidence, the higher is the confidence that the profile $M(i,a)$ and the sequence R are related.

If a sequence and a Bayesian profile are related, you may want to see the alignment of the sequence and the profile. The alignment can be described using the marginal posterior alignment distribution as follows:

$$P(A_{i,j}=1 \mid M,R) = \frac{\sum_{k}\sum_{A_{i,j}=1} P(M,R \mid A,k)P(A \mid k)P(k)}{\sum_{k}\sum_{A} P(M,R \mid A,k)P(A \mid k)P(k)} \tag{3.5}$$

The marginal posterior probability can be shown as a function of the sequence position as in Fig. 3.3.

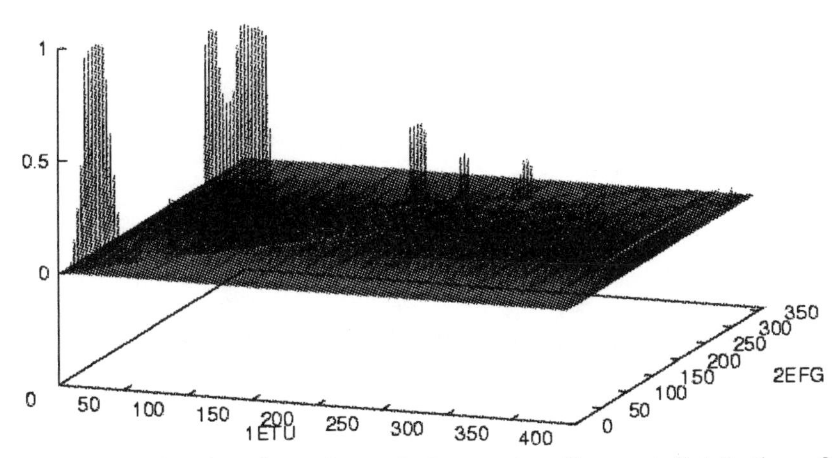

Figure 3.3. The plot view of marginal posterior alignment distribution of Eq.(3.5) for sequences of structure 1ETU versus 2EFG. The high peaks correspond to conserved blocks.

There are several high confidence peaks corresponding to conserved blocks. The conserved blocks can be displayed as shown in Fig. 3.4:

```
Rank=2, peak= 0.997 at (21,21)
   7 ERTKPHVNVGTIGHVDHGKTTLTAA   31
            ..:::::::::::
            ::::::::::::::
           .::::::::::::::::.
        ...:::::::::::::::::::
      ..::::::::::::::::::::::
   7 YDLKRLRNIGIAAHIDAGKTTTTER   31

Rank=4, peak= 0.389 at (33,70)
   32 ITTVLAKTY   40
      :::..
      :::::::..
   69 VTTCFWKDH   77

Rank=1, peak= 0.999 at (102,104)
   76 HYAHVDCPGHADYVKNMITGAAQMDGAILVVAATDGPMPQTREHILLGRQVGVPY  130
           .::::::.          .:::::::::::::::.
          ::::::::::........:::::::::::::::::::
          :::::::::::::::::::::::::::::::::::::.
         .::::::::::::::::::::::::::::::::::::::.
      ..:::::::::::::::::::::::::::::::::::::::::::::::::::::
   78 RINIIDTPGHVDFTIEVERSMRVLDGAIVVFDSSQGVEPQSETVWRQAEKYKVPR  132

Rank=3, peak= 0.645 at (136,137)
132 IVFLNKCDMVDDEELLELVEMEVR  155

     .::::::
    :::::::::.
    :::::::::::::...........
133 IAFANKMDKTGADLWLVIRTMQER  156

Rank=5, peak= 0.366 at (174,262)
164 PGDDTPIVRGSALKALEGD  182
        .::::::::.
      ..::::::::::..
252 DLKITPVFLGSALKNKGVQ  270

Rank=6, peak= 0.218 at (201,281)
191 ELAGFLDSYIPEPERA  206

          ..:::::.
271 LLLDAVVDYLPSPLDI  286

Rank=7, peak= 0.182 at (243,299)
234 RGIIKVGEEVEI  245
    ......::::::
290 KGTTPEGEVVEI  301
```

Figure 3.4. The Bayesian alignment blocks for sequences of structures 1ETU versus 2EFG. Each position in an alignment block has a confidence value that corresponds to marginal posterior alignment distribution of Eq. (3.5).

Each conserved block has a peak confidence value (highest marginal posterior alignment probability within the block) and a block length.

What we are looking for are long conserved blocks with high peak confidence values.

3.2.2.2. Building Bayesian Profiles

To build a Bayesian profile for a target sequence, you first need to collect a set of sequences that are similar to the target sequences. At this stage, we want to include as many remote similar sequences as possible even if many unrelated sequences are also included. In recruiting similar sequences one can use transitive BLAST[7] or simply use BLAST[8] with a high cutoff e-value. Here we used BLAST with a cutoff E-value of 500: The target sequence was searched against a non-redundant protein sequence database using BLAST. All sequences with an E-value less than 500 were selected, and are collectively referred as E500 sequences.

Then, the target sequence was searched against the collected E500 sequences using Bayes aligner with similarity matrices BLOSUM35-100. The sequences with Bayesian evidence less than 1e-5 were selected and used to build the initial profile according to Eq. (3.1). The profile was then iteratively refined by comparing it to the E500 sequences using the Bayes aligner according to Eq. (3.3). After the first iteration the sequences with Bayesian evidence less than 1e-4 were selected and used to refine the Bayesian profile. This step was then repeated with Bayesian evidence cutoff set to 1e-3 and 1e-2.

3.2.2.3. Bayesian Blocks Selection

The Bayesian profile for the target sequence was searched against a selection of sequences from PDB[9] with less than 45% sequence identity between any pair (PDB45). We use the Bayes aligner and alignments were generated for each sequence in PDB45. The generated Bayesian alignments are block-based alignments where only highly conserved regions are aligned. The confidence that the Bayesian profile for the target sequence and each sequence in the database are related is measured by the Bayesian evidence. Each residue in an alignment has a confidence value that is described by the marginal posterior alignment distribution.

An alignment block was included in the blocks selection for folding if it met the following criteria:

- Bayesian evidence for the sequence is less than 0.1;
- The highest confidence level of residues in the alignment block is higher than 0.5;
- Length of consecutive high confidence (>0.4) region is larger than 7.

All the blocks that met these criteria were collected in a database and later used as building blocks in the structure generation process.

3.2.3 Energy Functions

After the structure blocks are selected, we assemble them together. To evaluate different arrangements of structure blocks, we need to derive an objective energy function. There are two main approaches: systematic potential functions based on chemical properties[10] or knowledge-based potential functions extracted from known structures. We used the second approach, which is also used by Baker's group.[11,12] We used similar terms as Simon *et al.*[12] to divide functions into three categories: sequence-independent functions, first and second order sequence-dependent functions. The sequence independent functions are the compactness function, the helix-helix interaction function and the β-strand-β-strand interaction function:

Compactness function (radius of gyration):

The hydrophobic force is expected to drive proteins to adopt a compact structure rather than extended forms. One function that measure compactness is the radius of gyration R:

$$R_{gyration} = 0.5 * \sqrt{\frac{\sum_i (\hat{\mathbf{r}}_i - \hat{\bar{\mathbf{r}}})^2 * mass_i}{\sum_i mass_i}}$$

where $\hat{\mathbf{r}}_i$ is the position of amino acid i, $\hat{\bar{\mathbf{r}}}$ is the structure center, and is the *mass* of amino acid i. We used a similar function as Simon *et al.*[11] to score the compactness of a structure:

$$E_1 = -\log P(structure) = R^2_{gyration} + clashweight * numberOfClashes$$

where a clash is recorded when a distance between any two C_α atoms is less than 3.2 Å. Simulation experiments in literature[11] and our own show that native structures are more compact than random coils without residue clashes.

Helix-helix interaction function:

$$E_2 = -\log P(structure) = -\sum_{i,j} \log p(r_{i,j}, angle_{i,j} \mid h_i, h_j, sep)$$

where $r_{i,j}$ and $angle_{i,j}$ are the distance and orientation between the two helices h_i and h_j. The inter C_α-C_α distance criterion[13] is used for helix definition: An ideal helix of six residues with the angles $\alpha = 54°$ and $\kappa = 92.2°$ is compared to all six-residue windows in the structure. A helix consists of a stretch of at least 6 residues with each six-residue window having a RMSD of less than 0.5 Å compared to the ideal helix. The axis of a helix is defined by the centers of the first three C_α atoms and the last three C_α atoms.

β-strand-β-strand interaction function:

$$E_3 = -\log P(structure) = -\sum_{i,j} \log p(r_{i,j}, angle_{i,j} \mid \beta_i, \beta_j, sep)$$

A β-strand consists of at least 4 residues. It must meet one of the two following criteria:[13]

(1) torsion angle criterion: $\alpha < -100°$ or $\alpha > 120°$ and $\kappa > 95°$;

(2) inter C_α-C_α distance criterion: $|\mathbf{r}_{j+k} - \mathbf{r}_{j-k}| < 6.5$ Å for $k = 0,1,2..n$ and $|i - j| > n + 1$ for parallel beta-sheet; $|\mathbf{r}_{j+k} - \mathbf{r}_{j-k}| < 6.5$ Å for $k = 0,1,2,...n$ and $|i - j| > 3$ for antiparallel beta-sheet.

The axis of the β-strand is defined by the center of the first and last two C_α atoms.

The helix-helix and beta-strand-beta-strand interaction functions are the same as Simons *et al.*[12]

The next two terms, the environment function and the pair-wise distance functions are the first order sequence dependent interaction functions.

Environment function:

Assuming that all atoms are independent of each other, we can express *P(sequence | structure)* as follows:

$$E_4 = -\log P(sequence \mid structure) = -\sum \log p(a_i \mid \varepsilon_i)$$

This function is similar to the environment score used in 3D profiles[14] if ε_i is the solvent accessibility. Here we describe ε_i in terms of the number of C_β atoms within a sphere of 10 Å around the C_β atom of residue a_i.

We put the number of C_β into 50 bins (1,2,3…50+). $n_i(k)$ is the number of observations of amino acid a_i in environment of k C_β atoms. Then the marginal probability of amino acid a_i is $p(a_i) = \frac{\sum_k n_i(k)}{\sum_i \sum_k n_i(k)}$, and the probability of a_i given an environment is $p'(a_i \mid \varepsilon_k) = \frac{n_i(k)}{\sum_i n_i(k)}$. To deal with sparse data, a pseudo count is set to $\sigma = 0.005$. Then the modified probability of a_i given environment ε_k is

$$p(a_i \mid \varepsilon_k) = \frac{1}{(1+\sigma * \sum_k n_i(k))}(p(a_i) + \sigma * (\sum_k n_i(k)) * p'(a_i \mid \varepsilon_k))$$

Pair-wise distance function:

If we assume independence of each pair of C_α, we can express *P(sequence | structure)* as

$$E_5 = -\log P(sequence \mid structure) = -\sum_i \log p(a_i, a_j \mid R)$$

where R is the distance between the C_β atoms of the two amino acids a_i and a_j. This function is similar to mean force used by Sippl and others.[15] The distances were put into 100 bins of width = 1 Å. $n_{i,j}(k)$ is the number of observations of two amino acids a_i and a_j in k^{th} bin. The prior probability of a pair is $p(a_i, a_j) = \frac{\sum_k n_{i,j}(k)}{\sum_i \sum_j \sum_k n_{i,j}(k)}$, and the probability to see a pair with a distance R_k is $p'(a_i, a_j \mid R_k) = \frac{n_{i,j}(k)}{\sum_i \sum_j n_{i,j}(k)}$. With sparse data the pseudo count is set to $\sigma = 0.005$ and the modified probability is

$$p(a_i, a_j \mid R_k) = \frac{1}{(1+\sigma * \sum_k n_{i,j}(k))}(p(a_i, a_j) + \sigma * (\sum_k n_{i,j}(k)) * p'(a_i, a_j \mid R_k)) .$$

Any assumption of independencies of E_4 or E_5 is over-simplified. To capture the second order interaction term, joint probability functions for pair-wise interactions in environments and sequence-dependent pair-wise interactions can also be used.

Environment specific pair-wise distance function:

$$E_6 = -\log P(sequence \mid structure)$$
$$= -\sum_{i,j} (\log p(a_i, a_j \mid \varepsilon_i, \varepsilon_j, R_{i,j}) - \log p(a_i \mid \varepsilon_i, R_{i,j}) - \log p(a_j \mid \varepsilon_j, R_{i,j}))$$

Sequence specific pair-wise distance function:

$$E_7 = -\log P(sequence \mid structure)$$
$$= -\sum_{i,j} (\log p(a_i, a_j \mid R_{i,j}, sep) - \log p(a_i \mid R_{i,j}, sep) - \log p(a_j \mid R_{i,j}, sep))$$

For more details on the two second order interaction terms see Simons *et al.*[12] In our experience these two terms did not contribute significantly in distinguishing correct structures from incorrect structures. This could be due to the lack of 3D structure data needed to populate the large number of bins for these terms.

Extracting potential functions from database of known structures

Energy terms (2)-(7) were derived from PDB45-selected. The detail binning information is listed in Table 3.1. To deal with small number of observations, we used pseudo count factor 0.005 proposed by Bauer and Beyer.[16]

Table 3.1. The detail binning information for energy terms.

	No. of neighbors	$R_{i,j}$	$r_{i,j}$	Sep
$P(r_{i,j}, angle_{i,j} \mid h_i, h_j, sep)$			0-8Å, 8-16 ...72-80	0-1, 2-10, 10+
$P(r_{i,j}, angle_{i,j} \mid \beta_i, \beta_j, sep)$			0-8Å, 8-16 ...72-80	0-1, 2-10, 10+
$P(a_i \mid \varepsilon_i)$	1,2.....49, 50+			
$P(a_i, a_j \mid R_{i,j})$		0,1,2...100		
$P(a_i, a_j \mid \varepsilon_i, \varepsilon_j, R_{i,j})$	0-16, 16+	0-7Å, 7-10, 10-12, 12+		
$P(a_i, a_j \mid R_{i,j}, sep_{i,j})$		0,1,2...100		0-21, 21-36, 36+

Estimating weights for different functions

The energy function is proportional to negative log of *P(sequence,structure)*. As energy functions $E_1 - E_7$ listed above are not independent, we cannot simply add them together. The overall energy function can be expressed as a weighted sum of terms $E_1 - E_7$:

$$\begin{aligned} E(structure) &= -\log P(sequence, structure) \\ &= w_1E_1 + w_2E_2 + w_3E_3 + w_4E_4 + w_5E_5 + w_6E_6 + w_7E_7 \end{aligned} \tag{3.6}$$

To estimate the weights, we generated more than 18,000 random structures for all-α structures, then we did linear regression fitting of the energy functions versus RMSD to the native structure. All random structures had a radius of gyration within 15% compared to the native structures and they did not have clashes among residues. The energy term versus RMSD distributions for calbindin (PDB code 5ICB) are shown in Fig. 3.5. From Fig. 3.5 we can see that the second order interaction function terms E_6 and E_7 do not show RMSD dependencies. Only the environment function E_4 shows a clear correlation with RMSD. While the non-specific pair-wise distance function E_5 shows a slight correlation with RMSD. Thus, we set w_1, w_2, w_3, w_5, w_6, and w_7 to zero, and set

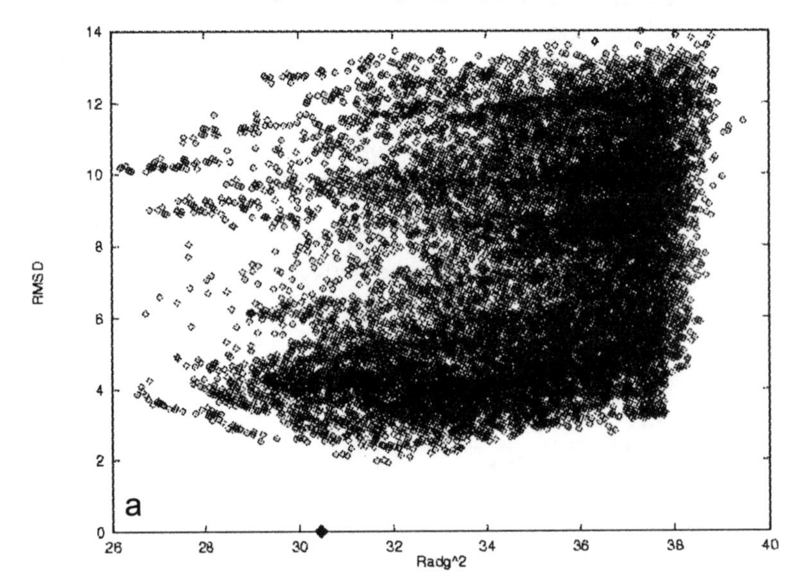

Figure 3.5a. RMSD *versus* energy function terms for randomly generated structures of 5ICB. E_1: the square of the radius of gyration. The diamond is the score for the native structure.

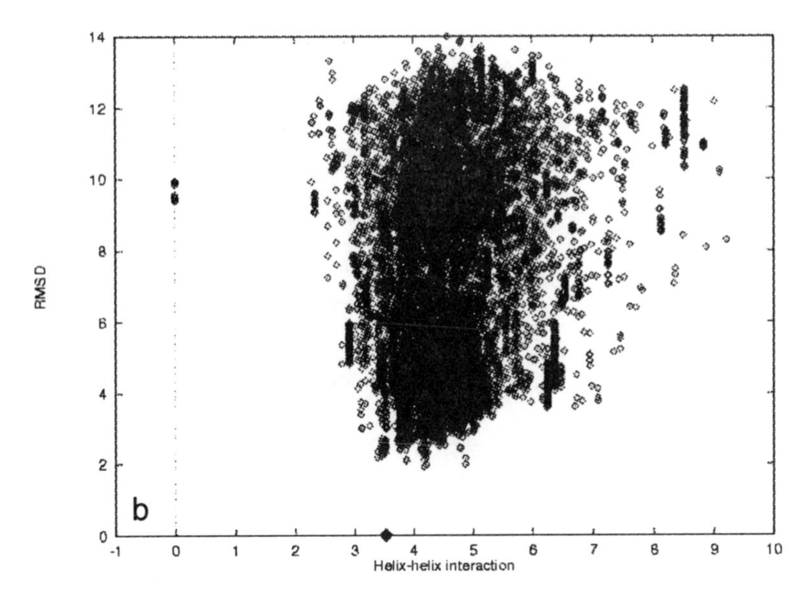

Figure 3.5b. RMSD *versus* energy function terms for randomly generated structures of 5ICB. E_2: the helix-helix interaction. The diamond is the score for the native structure.

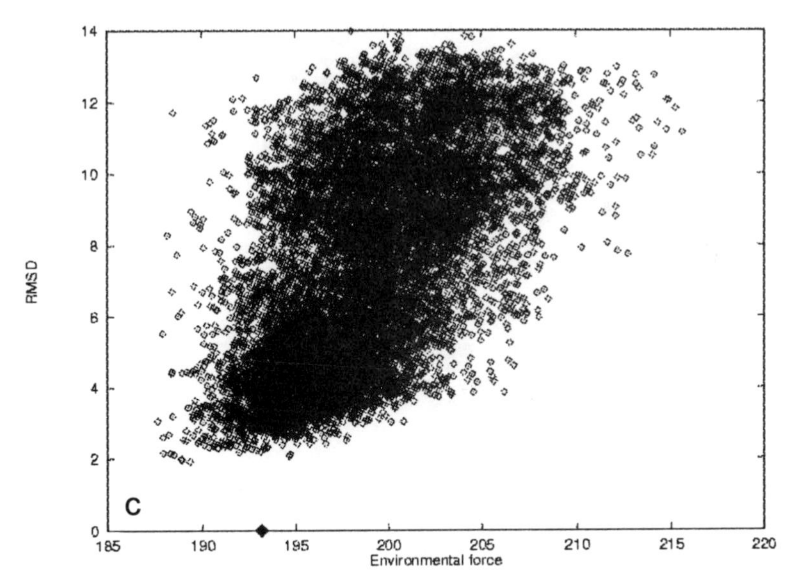

Figure 3.5c. RMSD *versus* energy function terms for randomly generated structures of 5ICB. E_4: environment function. The diamond is the score for the native structure.

w_4 to 1.

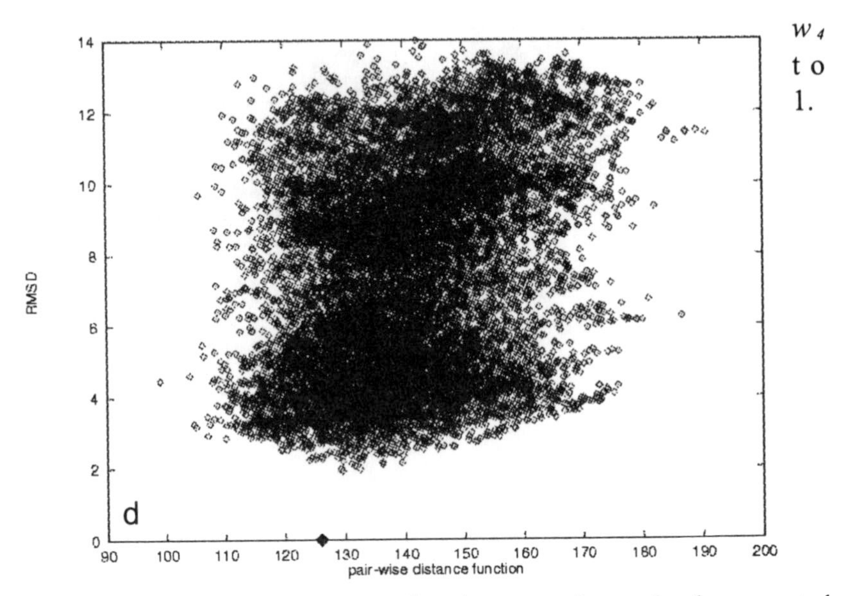

Figure 3.5d. RMSD *versus* energy function terms for randomly generated structures of 5ICB. E_5: pair-wise distance. The diamond is the score for the native structure.

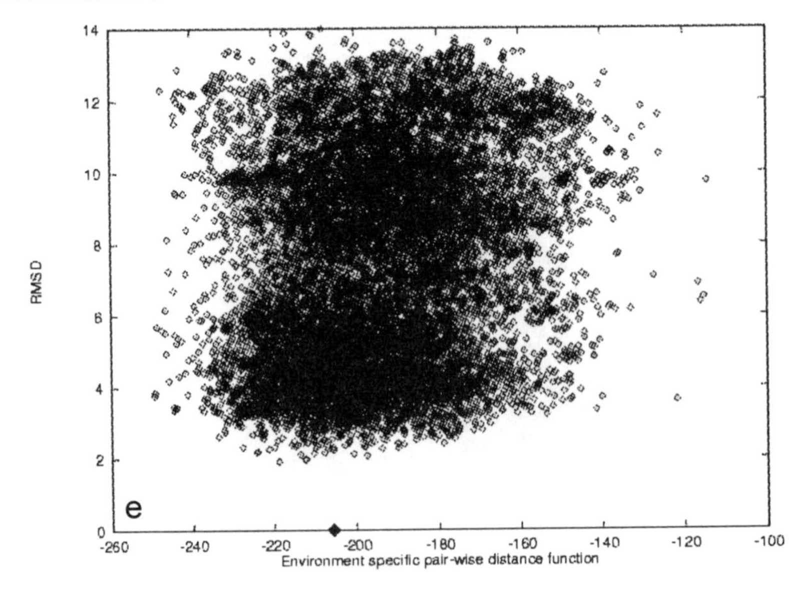

Figure 3.5e. RMSD *versus* energy function terms for randomly generated structures of 5ICB. E_6: environment specific pair-wise distance function. The diamond is the score for the native structure.

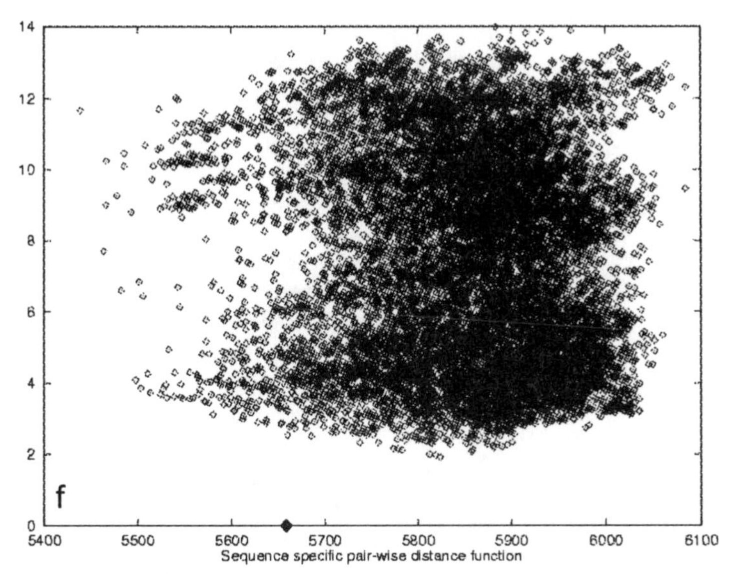

Figure 3.5f. RMSD *versus* energy function terms for randomly generated structures of 5ICB. E_7: sequence specific pair-wise distance. The diamond is the score for the native structure.

The above scoring function is for evaluating predicted structures. During folding process, we still need E_1 to drive the predicted structure to a compact state, so w_1 is set to 1.

3.2.4. Energy Minimization

A Monte Carlo simulated annealing process[17] that minimized the scores using the energy functions was used to generate structures. 5000 iterations are performed to search for a minimum state. The initial clash weight is 100, and it increases linearly to 500 in 400 iterations, and stays constant for the rest of the iterations. The starting temperature is 10, it linearly decreases to 0.01. At each iteration, one block is drawn from the block database and replaces the same region in the model. The $\alpha\kappa$ angles at the two hinge positions are sequentially optimized toward a local minimum state. Whether the update is accepted or not is determined by the amount of energy change and the current temperature.

Loop regions which consist of transition positions that are not helix or beta-strand are hard to model using sequence information alone.

There may not be any proper model in our block database for a given loop region. To avoid being trapped in a local minimum due to inadequate local models, we choose one of following steps in every 50 iterations:

(1) Randomly choose a transition position and mutate $\alpha\kappa$ angles associated with it, then sequentially optimize the angles to minimize the energy function;

(2) Randomly choose a core region that is surrounded by two transition positions. Keep the core region unchanged and add back flanking region at both sides, one piece at a time (pieces are regions that are delineated by transition positions).

3.2.5. Using Information from Bayesian Blocks

One of the advantages of using Bayesian blocks is that each Bayesian block has confidence information about the block: each alignment of a Bayesian profile and a PDB structure has an overall Bayesian evidence value for the relatedness of the target profile and the sequence from the PDB structure. Each block in the alignment additionally has a peak value indicating the confidence of the block alignment.

To use this information we sorted all Bayesian blocks according to a *blockscore:*

$$blockScore = -10\log(BayesianEvidence) + PeakConfidence * blockLength\,.$$

If a Bayesian block overlaps with blocks ahead of it, it will be moved to the end of the block list. Then we examine the Bayesian blocks starting at the top of the list. If we have a Bayesian block with Bayesian evidence less than 0.005, peak confidence value larger than 0.80 and length longer than 10 amino acids, we will keep this block in the predicted structure fixed. Blocks that are not fixed in the target structure and are longer than 15 residues are broken into partially overlapping small pieces of random length between 10 and 15 residues. As a result, the Bayesian block library consists of fixed blocks and small blocks from the Bayesian blocks.

To make fixed blocks in the predicted structure stable, we apply the following rules: (1) we do not select any residues in the fixed blocks for refinement; (2) if a replacement block partial overlaps with a fixed block, then the overlapping region is overwritten by the fixed block and the end point of the replacement block is shifted to the end of the fixed block.

Secondly we assign each position in a Bayesian block a normalized confidence value

$$normalized\_confidence = -10\log(BayesianEvidence) + Confidence * blockLength\ .$$

If this block is used in the predicted structure, then the normalized confidence value is propagated to the corresponding position in the structure. In each iteration, a position in the target was selected for refinement. The position was selected according to the confidence value of the position. The positions with less confidence have a higher probability to be selected for refinement (the probability is proportional to 1/*normalized confidence value*). After the position is chosen, a Bayesian block containing the position is selected from the Bayesian block library based on the block score. The higher the block score, the higher the probability that the block is selected (the probability is proportional to the block score).

3.2.6. Enforcing Secondary Structures

The secondary structure prediction for the target sequence can be obtained using experimental methods or one of many secondary structure prediction methods (see article[18] for review). We also can use secondary structure information derived from the Bayesian blocks. Each position in the target sequence is assigned to one of the three choices: a—α-helix; b—β-strand; u—unknown. At the beginning of the folding process the predicted structure is initialized according to secondary structures. Then, the blocks with high block confidence are added. In each iteration, if the end point of a selected block falls in the middle of a secondary structure element, according to the initial secondary structure information, the end point is pushed to the end of the secondary structure element, and the secondary structure is preserved.

3.3. Examples

Bacteriocin AS-48 from *E. faecalis*, target T0102 from the CASP4 experiment,[19] is a good example for testing folding programs. It is a cyclic 70 amino acids long polypeptide with 5 α-helices and no β-strand. The secondary structure of the protein was determined by NMR[20] and we took the results and enforced the secondary structure during the folding process. 100 structures were generated. The radius of gyration distribution of the structures is shown in Fig. 3.6.

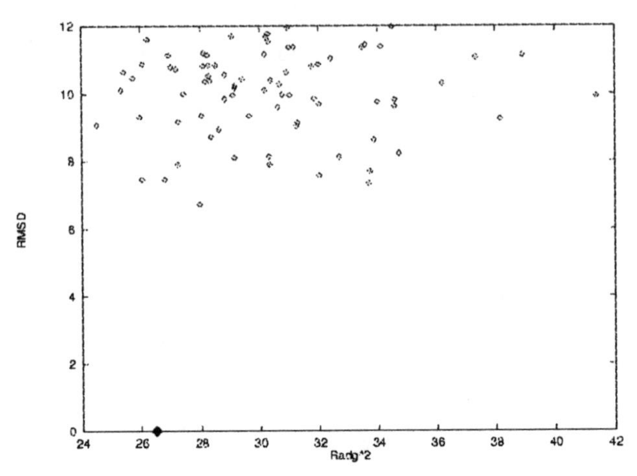

Figure 3.6. RMSD versus square of radius of gyration for predicted structures of target T0102. The diamond is the score for the native structure.

The environment function score distribution is shown in Fig. 3.7.

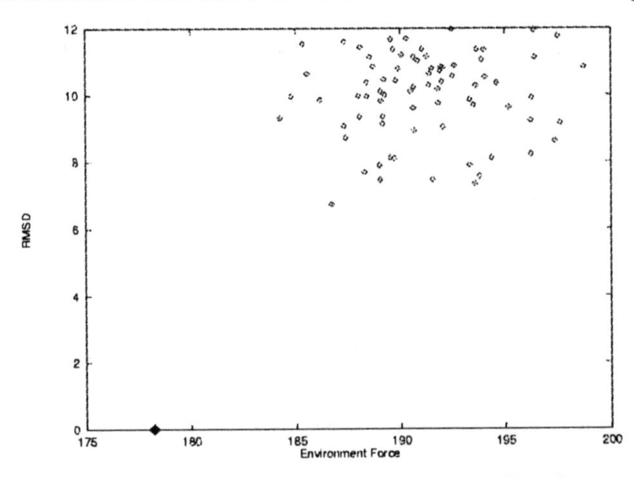

Figure 3.7. RMSD versus energy functions for predicted structures of target T0102. The structure with the lowest energy score has RMSD 9.3 Å. The predicted structure with the lowest RMSD (6.7 Å) is of the sixth lowest energy score. The diamond is the score for the native structure.

The structure with the lowest energy score has a RMSD of 9.3 Å compared to the experimental structure.[21]

The superposition of the model and the experimental structure is shown in Plate 3.1. The model with the best RMSD (6.7 Å) has the sixth lowest environment function score. The superposition of this model and the experimental structure is shown in Plate 3.2. In this structure, all 5 helices have the same orientation as in the experimental structure.

Another example is Spo0A, the trans-activating domain of the sporulation response regulator of *B. Stearothermophilus*. This was target T0098 in CASP4. It is an all α-helix protein and had no known similar three-dimensional structure at the time. The environment function score distribution is shown in Fig. 3.8.

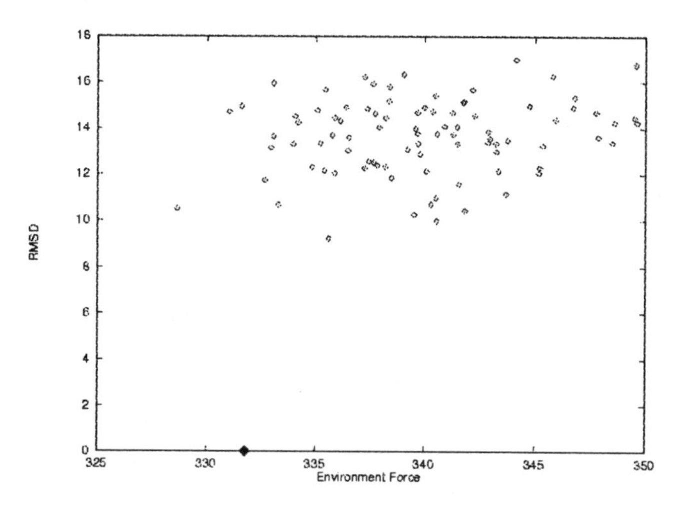

Figure 3.8. RMSD versus energy score for predicted structures of target T0098. The structure with the lowest energy score has RMSD 10.5 Å. The lowest RMSD among the predicted structures is 9.2 Å. The diamond is the score for the native structure.

The structure with the best score has RMSD of 10.5 Å. This structure aligned with the experimental structure[22] is shown in Plate 3.3. In the Bayesian block library for this target there were two high confidence blocks that were fixed in the predicted structure. The two blocks are both from structure 1WDC with Bayesian evidence 0.003, shown in Fig. 3.9. From Plate 3.3, we can see two well aligned segments (red and green) that exactly correspond to the two fixed high confidence blocks.

```
Rank=1, peak= 0.984 at (24,16)
  12 TSIIHEIGVPAHIKGYLYLREAIAMVYHDI    41
                 .:::::.
               .::::::::::.
               ::::::::::::::...
              :::::::::::::::::::..
          ....::::::::::::::::::::::::
   4 RLSKIISMFQAHIRGYLIRKAYKKLQDQRI    33

Rank=2, peak= 0.840 at (68,40)
  62 TASRVERAIRHAIEVAW    78

          ..::::::
         :::::::::
        .::::::::::
        :::::::::::::....
  34 GLSVIQRNIRKWLVLRN    50
```

Figure 3.9. The two high confidence blocks which are fixed in predictions for T0098.

References

1. Zhu J, Lüthy R, Lawrence CE: **Database search based on Bayesian alignment.** *ISMB* 1999, 297–305.

2. Levitt M: **A simplified representation of protein conformations for rapid simulation of protein folding.** *J Mol Biol* 1976, **104**(1):59–107.

3. Gribskov M, McLachlan AD, Eisenberg D: **Profile analysis: detection of distantly related proteins.** *Proc Natl Acad Sci USA* 1987, **83**(13):4355–4358.

4. Zhu J, Liu JS, Lawrence CE: **Bayesian adaptive sequence alignment algorithms.** *Bioinformatics* 1998, **14**(1):25–39.

5. Dayhoff MO, Schwartz M: **Matrices for detecting distance relationship.** In: *Atlas of Protein Sequence and Structure,* vol.5, suppl. 3, M. O. Dayhoff, ed. Washington, DC: National Biomedical Research Foundation: 1978:353–358.

6. Henikoff S, Henikoff JG: **Amino acid substiturion matrices from protein blocks.** *Proc Natl Acad Sci USA* 1992, **89**:10915–10919.

7. Neuwald AF, Liu JS, Lipman DJ, Lawrence CE: **Extracting protein alignment models from the sequence database.** *Nucleic Acids Res* 1997, **25**(9):1665–1677.

8. Altschul SF, Madden TL, Schaffer AA, Zhang J, Zhang Z, Miller W, Lipman DJ: **Gapped BLAST and PSI-BLAST: A new generation of protein database search programs.** *Nucleic Acids Res* 1997, **25**(17):3389–3402.

9. Berman HM, Westbrook J, Feng Z, Gilliland G, Bhat TN, Weissig H, Shindyalov IN, Bourne PE: **The Protein Data Bank.** *Nucleic Acids Res* 2000, **28**:235–242.

10. Pillardy J, Czaplewski C, Liwo A, Lee J, Ripoll DR, Kazmierkiewicz R, Oldziej S, Wedemeyer WJ, Gibson KD, Arnautova YA, Saunders J, Ye YJ, Scheraga HA: **Recent improvements in prediction of protein structure by global optimization of a potential energy function.** *Proc Natl Acad Sci USA* 2001, **98**(5):2329–2333.

11. Simons KT, Kooperberg C, Huang E, Baker D: **Assembly of protein tertiary structures from fragments with similar local sequences using simulated annealing and Bayesian scoring functions.** *J Mol Biol* 1997, **268**(1):209–225.

12. Simons KT, Ruczinski I, Kooperberg C, Fox BA, Bystroff C, Baker D: **Improved recognition of native-like protein structures using a combination of sequence-dependent and sequence-independent features of proteins.** *Proteins* 1999, **34**:82–95.

13. Levitt M, Greer J: **Automatic identification of secondary structure in globular proteins.** J *Mol Biol* 1977, **114**:181–293.

14. Bowie JU, Lüthy R, Eisenberg D: **A method to identify protein sequences that fold into a known three-dimensional structure.** *Science* 1991, **253**:164–170.

15. Hendlich M, Lackner P, Weitckus S, Floeckner H, Froschauer R, Gottsbacher K, Casari G, Sippl M: **Identification of native protein folds amongst a large number of incorrect models. The calculation of low energy conformations from potentials of mean force.** *J Mol Biol* 1990, **216**(1):167–180.

16. Bauer A, Beyer A: **An improved pair potential to recognize native protein folds.** *Proteins* 1994, **18**(3):254–261.

17. Press WH, Teukolski SA, Vetterling WT, Flannery BP: *Numerical Recipes.* 1992, Cambridge University Press.

18. Rost B, Sander C: **Structure prediction of proteins—Where are we now?** *Curr Opin Biotechnol* 1994, **5**(4):372–380.

19. *CASP4. (Fourth Community Wide Experiment on the Critical Assessment of Techniques for Protein Structure Prediction, Asilomar Conference Center, December 3–7, 2000.)* Conf. Proc., Asilomar, California, 2000.

20. Langdon GM, Bruix M, Galvez A, Valdivia E, Maqueda M, Rico M: **Sequence-specific 1H assignment and secondary structure of the bacteriocin AS-48 cyclic peptide.** *J Biomol NMR* 1998, **12**(1):173–175.

21. Gonzalez C, Langdon GM, Bruix M, Galvez A, Valdivia E, Maqueda M, Rico M: **Bacteriocin AS-48, a microbial cyclic polypeptide structurally and functionally related to mammalian NK-lysin.** *Proc Natl Acad Sci USA* 2000, **97**(21):11221–11226.

22. Lewis RJ, Krzywda S, Brannigan JA, Turkenburg JP, Muchova K, Dodson EJ, Barak I, Wilkinson AJ: **The trans-activation domain of the sporulation response regulator Spo0A revealed by X-ray crystallography.** *Mol Microbiol* 2000, **38**(2):198–212.

4

Protein Structure Prediction Using Hidden Markov Model Structural Libraries

Igor Tsigelny, Yuriy Sharikov, and Lynn F. Ten Eyck

Abstract

Use of hidden Markov models (HMM) constructed on the basis of structural alignments is discussed. An example system HMM-SPECTR is given with the description of different types of HMMs based on structural alignments. Principles of decision making in the system based on HMM libraries are discussed.

4

Protein Structure Prediction Using Hidden Markov Model Structural Libraries

Igor Tsigelny, Yuriy Sharikov, and Lynn F. Ten Eyck

University of California, San Diego, La Jolla, California

4.1. Introduction

More than 80% of new protein structures with relatively small sequence similarity to solved structures nevertheless adopt an already known protein fold.[1,2] This makes it possible to use a set of comprehensive HMM libraries in predicting a majority of possible protein folds, on the basis of sequential information only. Hidden Markov models (HMM) are widely used for biological sequence recognition.[3,4] One of the largest libraries of sequential HMMs for different protein structural domains is included in the Protein Families database (Pfam).[5] The profile HMMs built from the Pfam alignments are used for automatically decision making if a submitted protein sequence belongs to an existing protein family. Most current methods of using HMMs to search for distant homologs of proteins, are based on sets of pairwise alignments of protein sequences to a query sequence. For example, the SAM program[6] uses a BLAST search with the initial sequence producing the sets of potential homologs, which are then used to construct corresponding HMMs. This approach gives very promising results and is further refined in newer versions of the program. The HMM search method[7] based on the SAM concept, uses a library of sequence based HMMs (covering all proteins of known structure) covering 820 SCOP superfamilies, represented by specified groups of HMMs. The programs Pfam and TIGRFAM[8] assign specific proteins of known structure (extracted from the PDB database) as the structural representative of each of the sequence sub-libraries.

In many cases, pairs of proteins can have significant structural alignments and less than 15% sequence identity.[9] Here begins the challenge of creating HMM libraries based on structural alignments rather than sequence alignments. One of the approaches in this direction is the combination of sequential and structural information in the same HMMs. Bystroff *et al.*,[10] use libraries of HMMs consisting of Markov states which emit four categories of symbols. The first corresponds to amino acid residues (sequence), the second to three-state secondary structure, the third to backbone angles (discrete values in ϕ-ψ space), and the fourth to 'local tertiary structure elements' like hairpins, etc. This development leads to the discretization of the data. Such an approach actually uses the fuzzy logic to receive the discrete values and then uses the libraries of discrete objects for computation. It is more efficient from the computational point of view and makes it possible to use HMMs.

4.2. Structural Hidden Markov Model Libraries

The underlying assumption behind the use of sequence similarity as a tool for predicting three dimensional structure of proteins is that similar sequences will have similar folds, provided one can find the right definition of "similar." The difficulty is that the definition of "similar" appropriate for measuring evolutionary distance, for example, may not be appropriate for measuring structural conservation. The libraries of hidden Markov models described here are devoted to tertiary structure prediction on the basis of *structural* multiple alignments instead of *sequence* multiple alignments. These alignments are constructed automatically, using only geometric information about the proteins. The Combinatorial Extension of the common path (CE) Algorithm and the corresponding program,[11] both developed by Shindyalov and Bourne, became the foundation for implementation of this idea. If one submits an initial protein structure to this program, he/she receives a set of pairwise structural alignments of proteins which are extracted from the PDB database and have predefined structural similarity to the initial 'title' protein. We also used, for preparation of some structural alignments, the program described in Chapter 18 of this book. As the 'title proteins', we used typical representatives of each of SCOP superfamily.[12] The set of pairwise structural alignments contained only those proteins for which the Z-scores of the structural comparison with the 'title protein' was above a predefined threshold. Multiple pairwise alignments constructed using CE contained from 40 to 800 proteins, depending on the 'inhabitance' of the structural set. The number of proteins in each alignment depends on the threshold chosen for the Z-score to limit similarity of these proteins to the title protein. We required that these alignments include all members of the selected SCOP superfamily. This requirement sometimes forced the use of relatively low Z-score thresholds, which in turn gave multiple alignments that included additional proteins with structures sufficiently close to the title protein, but not included in this SCOP superfamily or family. The resulting sets of pairwise alignments are thus not strictly superfamily alignments. An alternative name for such alignments, and subsequent HMMs built on their basis, could be 'structural attractor alignments' and 'structural attractor HMMs'. Our goal thus becomes to select the representatives of a specific fold, or set of similar folds (attractor), instead of finding representatives of a specific family of proteins. We created the HMMs corresponding to each set of structural attractor alignments having as cores each superfamily for the main classes of folds: all alpha proteins (α), all beta proteins (β), alpha and beta proteins (α/β), alpha and beta proteins ($\alpha+\beta$), multi-domain proteins (α and β), coiled coil proteins, and 'small proteins'. We also created alignments for a number of specific families.

We employed three libraries of HMMs. All of the libraries were built using the HAMMER[13] package. The first library was built using the alignments from CE for each "title" protein. The basic assumptions for this library is that the CE algorithm has produced the optimal alignments and that the structural relationships in SCOP are correctly assigned. Seven HMMs are generated for each alignment, corresponding to different settings of the HAMMER "*g*-filter" parameter.

The second is derived by relaxing the assumption that the CE algorithm has optimally aligned the structures. The HMMs in the second library are further trained by adjusting the alignments to increase the specificity of the HMMs for the title proteins. This cyclic training process is done by using the initial HMM to create the next set of pair-wise alignments, which in turn is used to prepare the HMM for the next step.[14] This process is repeated three times and all of the steps are retained in the library. The HMMs for each title protein are an ordered set of increasing sensitivity. The search procedure for the second library selects those HMMs for which the score of a probe sequence improves with the training stage, reflecting an underlying similarity to the behavior of the title protein on which the HMMs were trained.

The third library is obtained by relaxing the assumption that the SCOP classification is the best possible structural classification. This is done by examining the CE scores for the alignments and noting that the distribution often contains discontinuities, and in many cases the discontinuities represent clear structural sub-families. An automated procedure is used to partition the structures according to the discontinuities in the CE Z-score. Each subset is used to generate a "partial" HMM. This gives a "spectrum" of HMMs for each of the initial title proteins, and is the origin of the name HMM-SPECTR for the package.

4.3. Decision Tree

The HMM-SPECTR program contains the following blocks (see Fig. 4.1): the libraries of HMMs and alignments; the HMM-probe comparison program; the secondary structure prediction and comparison program; and the dominant protein selection program.

4.3.1 Search for the Best HMM

A probe sequence is compared with each HMM from each library. The statistically soundest choices normally arise if the first library gives an unambiguous choice, in which case the fold of the probe sequence can be assumed to be the same as a correctly aligned member of a correctly assigned family of folds. When the scores from the initial library are ambiguous we look for consensus between the three libraries. To select

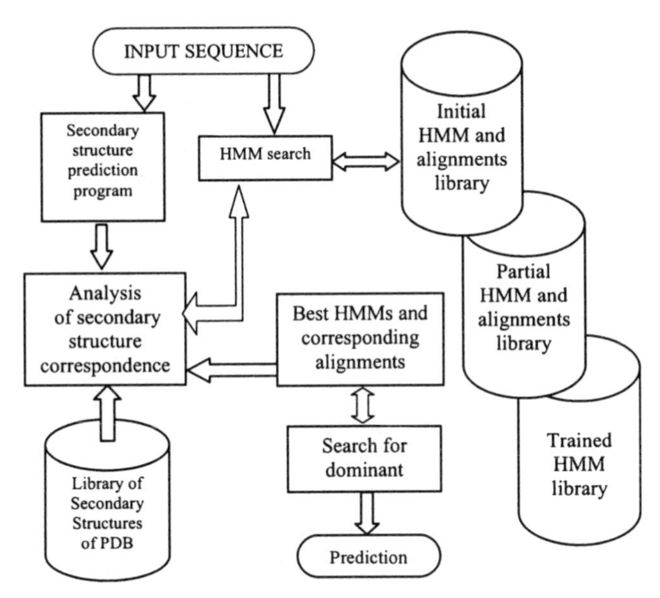

Figure 4.1. HMM-SPECTR architecture.

the best final solution, we also compare the predicted secondary structure of the probe protein with the secondary structures of the best 10 proposed predictions using the DSSP library.[15] We use a method based on pattern recognition techniques (manuscript in preparation) for secondary structure prediction.

Table 4.1 illustrates the effectiveness of our HMM training procedures on CASP 4 protein targets T0109, T0100, and T0087. The training procedures significantly increase the scores and, what is even more important, the length of predicted protein structures. The training procedures do not improve the scores and lengths in all cases. This often means that the initial HMM is prepared properly and does not need the further correction of the alignment.

Table 4.2 shows the search results obtained using each of the three different HMM libraries for CASP4 target T0120. Higher scores are obtained on trained and partial libraries rather than by using the initial library. This is not always the case. In some cases the initial library gives the same results as the rest of the libraries. Table 4.2 and Fig. 4.2 show how the scores are distributed on different layers of the HMM. The position of the maximum score in layer 4 corresponds to the proteins clustered in this layer of the HMM. The use of partial library is

Table 4.1. Scores for finding of proteins targets of CASP 4 using their sequences on the entrance to HMM-SPECTR 1.02.[18] Courtesy of Oxford University Press.

Target	Step of training	Score	Length of predicted protein sequence
T0109	0	-7.3	1
T0109	1	-2.5	40
T0109	2	1.8	43
T0109	3	3.0	44
T0100	0	-3.7	4
T0100	1	2.2	65
T0100	2	2.9	81
T087	0	-0.7	11
T087	1	3.7	67
T087	2	11.3	90

more significant in the case of weak families of proteins, when the number of close representatives in that family or superfamily is limited, and the initial alignment may contain a number of distantly related proteins.

Table 4.2. Scores for finding of proteins targets of CASP 4 using their sequences on the entrance to HMM-SPECTR 1.02.

Library of trained HMMs		Initial HMM library with different *g*-filters		Partial HMM library	
Step	**Score**	***g*-filter value**	**Score**	**Layer**	**Score**
0 (No training)	4.0	0.1	5.7	1	2.0
1	**10.5**	0.3	4.7	2	-1.7
2	10.5	0.4	7.7	3	1.5
3	6.6	0.5	8.4	4	**12.0**
		0.6	**9.1**	5	7.8
		0.7	-2.8	6	0.8
		0.9	-13.2	7	1.4
				8	-6.8
				Entire alignment	4.3

Table 4.3 shows the first ten results of the search using the initial library with a fixed *g*-filter. One can see that the alignments 1bj5100, 1av1100, and 1ifn371, having the first places in their libraries (Table 4.2), have quite humble 4th, 5th, and 10th places in the case of the initial library and fixed *g*-parameter for the target T0120. Often the global best solutions are not among the first 50-100 best solutions found using the single initial library with the fixed filter parameter. This clearly demonstrates that

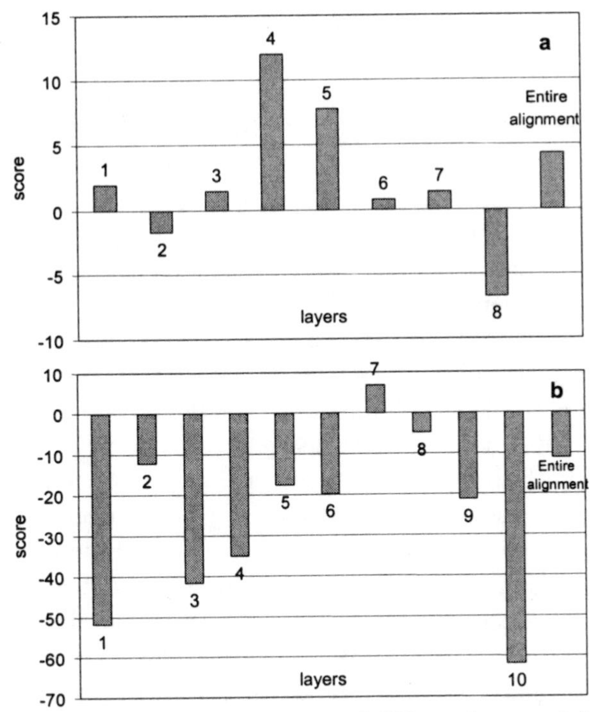

Figure 4.2. Scores of correspondence of different layers of the partial HMM to target protein putative structure: **a)** Target T0120, **b)** Target T0097.

one has to employ the libraries of trained HMMs and partial libraries.

Table 4.3. First ten results of the initial HMM library search for the target T120 with the fixed *g*-filter value.

HMM name	Number of proteins in HMM	Score
1fum161	161	5.8
1vdf100	100	5.2
1qbz100	100	4.6
1ifn371	371	4.3
1bj5100	100	4.0
1aq5100	100	4.0
1czq100	100	3.8
2spc99	99	3.5
1b3q156	156	3.2
1av1100	100	3.1

4.3.2. Searching within the Structural Alignment

In most general terms the structural alignments are described by the following parameters: the title protein which was used to construct the alignment; the number and boundaries of proteins included; and their sequences. Despite all precautions taken to choose as a 'title' protein the most 'respected' representative of a family or a superfamily, the title protein of the highest scoring HMM will frequently not be the desired putative structural homolog. In our decision tree we examine as the putative predictions three proteins—the title protein, the dominant protein, and the protein with sequence closest to that of the probe protein. The title protein is the protein upon which the alignments in the HMM are based. The dominant protein is the protein closest in sequence to the consensus sequence of the HMM. Table 4.4.1 shows dominant proteins for the different layers of partial alignment 1ifn371 found in the search for probe protein T0120, and compares their structures to the published structure of target protein 1FU1:A using the CE server.[11] The entire HMM, containing all layers, predicted the wrong boundaries of the putative target, rendering the following steps of the decision making useless. It is also noteworthy that the title protein and the protein with sequence most similar to the sequence of the probe protein also gave the wrong results for the unpartitioned HMM. This example shows that use of a title protein only, or the HMM without partitions, can be misleading. Tables 4.4.1 and 4.4.2 show the scoring of different layers of the partial HMMs for the targets T0120 and T0097. Fig. 4.2 shows the sharp maxima of the scores for the specific layers of the partial HMMs.

Table 4.4.1. Dominant proteins in the partial HMM found for the probe protein T0120.

Layer	Score of the layer	Dominant protein in the layer	Boundaries of target protein proposed by HMMSPECTR	Structural Comparison by CE with the real target protein 1FU1:A		
				RMSD (A)	Z-score	Length
4	**12.0**	1EI3:F	131-173	**1.0**	**4.7**	**42**
5	7.8	2NCD:A	131-173	1.2	4.7	42
1	2.0	2BU0:C	132-174	1.3	4.7	42
3	1.5	1AN2:A	173-214	No correspondence		
7	1.4	1DEQ:B	175-214	No correspondence		
6	0.8	1EXI:A	173-214	No correspondence		
2	-1.7	1QBZ:B	172-214	No correspondence		
8	-6.8	2SBL:B	11-55	6.8	0	24
Entire HMM		1QUU:A	179-214	No correspondence		
Title protein		1IFN	179-214	No correspondence		
Best sequence.		1AA0	179-214	No correspondence		

Tables 4.5.1 and 4.5.2 show the scoring obtained using initial HMMs with the different *g*-filters. All these results were assessed by structure comparison to the real target proteins 1FU1:A and 1G7D:A using the

Table 4.4.2. Dominant proteins in the partial HMM found for the probe protein T0097.

Layer	Score of the layer	Dominant protein in the layer	Boundaries of target protein proposed by HMMSPECTR	Structural Comparison by CE with the real target protein 1G7D:A		
				RMSD (A)	Z-score	Length
7	**6.8**	1RCC	18-92	**1.7**	**4.1**	**38**
8	-5.0	1EER:A	51-103	3.5	2.8	32
2	-12.3	1FOH:C	34-104	4.1	2.6	32
5	-17.7	1QLE:B	48-104	4.2	2.8	32
6	-19.8	1ALM:A	6-95	4.9	1.6	40
9	-21.3	1HGF:B	13-92	3.6	2.6	40
4	-35.1	1BBH:B	4-105	6.2	2.6	65
3	-41.6	1HMD:D	16-105	5.4	3.5	57
1	-51.8	1LE2	2-93	4.7	3.3	80
10	-61.9	1JSW:A	2-80	4.8	3.3	72
Entire HMM	-11.2	1NBB:B	2-73	3.6	3.1	56

CE algorithm.[11] In the case of T0120, the parameters of predictions (RMSD, Z-score, and length) obtained using initial HMM and partial HMM are mostly the same. In the case of T0097, partial HMM gives better results. These results confirm the necessity of running parallel searches in all libraries and of comparing all available results. They also show the reliability of the scoring system for the further decision making.

Table 4.5.1. Search within the best initial HMM for the target T0120.

g-filter	HMM-score	Boundaries predicted by HMMSPECTR	Dominant protein	Comparison with 1FU1:A by CE		
				RMSD (A)	Z-score	Length
0.1	5.7	171-214	No correspondence (wrong boundaries)			
0.3	4.7	138-171	1CZQ:A	0.8	4.4	33
0.4	7.7	138-178	1QBZ:C	0.7	4.4	34
0.5	8.4	133-180	1AV1:B	1.0	4.7	45
0.6	**9.1**	132-182	1AV1:B	**1.0**	**4.7**	**46**
0.7	-2.8	186-300	1AV1:B			
0.9	-13.3	149-304	1AV1:B	1.0	3.7	27
No filter	3.1	129-180	1EI3:D	3.6	2.6	24

Table 4.5.2. Search within the best initial HMM for the target T097.

g-filter	HMM-score	Boundaries predicted by HMMSPECTR	Dominant protein	Comparison with 1FU1:A by CE		
				RMSD (A)	Z-score	Length
0.1	-1.1	69-88	1FRV:B	2.5	2.0	16
0.3	7.6	63-99	1FRV:B	2.8	2.3	24
0.4	7.4	63-100	1FRV:B	4.4	2.6	32
0.5	4.6	53-100	1FRV:B	3.7	2.3	32
0.6	**5.8**	49-101	1SKN:P	**3.3**	**3.1**	**32**
0.7	2.4	40-101	1SKN:P	4.3	2.0	40
0.9	4.4	25-101	1SKN:P	6.2	2.3	48
No filter	4.4	72-100	1FRV:B	2.8	2.3	24

4.4. Program Testing

We tested our program against all of the protein targets proposed in CASP4 competition.[16] The data warehouse included libraries of structural alignment and HMMs constructed on the basis of proteins publicly available in the Protein Data Bank before the CASP4 meeting. The newest fully automated versions of HMM-SPECTR 1.02 and 1.02ss (with the secondary structure prediction) produced better results than the best result reported at CASP4, either by RMSD or by length (or both) in 64% (HMM-SPECTR 1.02) and 79% (HMM-SPECTR 1.02ss) of the cases.

Table 4.6 shows the effect of the complexity of targets on the final results of HMM-SPECTR. The 1 to 5 scale of complexity of the targets16 is used for these calculations. There is some improvement of results for the simple targets with complexity 1 and 2. These simple 'homology modeling' targets are readily solved by many sequence based methods. For targets with complexity 3 we see improvement in length and RMSD. For the targets with complexity 4, this improvement of both parameters is most profound. In the complexes with the highest complexity 5, predictions are also improved but less than in the best case of complexity 4. This is not surprising, because the targets of complexity 5 are those for which there are no similar structures known, and HMM-SPECTR is based on comparison with known structures.

Table 4.6. Comparison of HMM-SPECTR results to CASP 4 predictions.[18] Courtesy of Oxford University Press.

Program used	HMMSPECTR prediction better than the best listed in CASP 4 results			
	Either by L or RMSD	By L	By RMSD	Complexity of a target
HMMSPECTR 1.02	64%	27%	58%	All targets
HMMSPECTR 1.02ss	79%	46%	55%	All targets
HMMSPECTR 1.02	67%	33%	67%	5
HMMSPECTR 1.02ss	67%	67%	0%	5
HMMSPECTR 1.02	88%	50%	75%	4
HMMSPECTR 1.02ss	100%	75%	88%	4
HMMSPECTR 1.02	50%	25%	50%	3
HMMSPECTR 1.02ss	75%	58%	58%	3
HMMSPECTR 1.02	60%	20%	60%	2
HMMSPECTR 1.02ss	60%	0%	60%	2
HMMSPECTR 1.02	60%	20%	60%	1
HMMSPECTR 1.02ss	80%	40%	60%	1

4.5. Prediction of Unsolved Structures

Further exploration of the value of our program was done on complex proteins for which structures are not yet available. Fig. 4.3 shows results obtained using HMM-SPECTR for structure prediction of cystic fibrosis transmembrane regulator (CFTR).

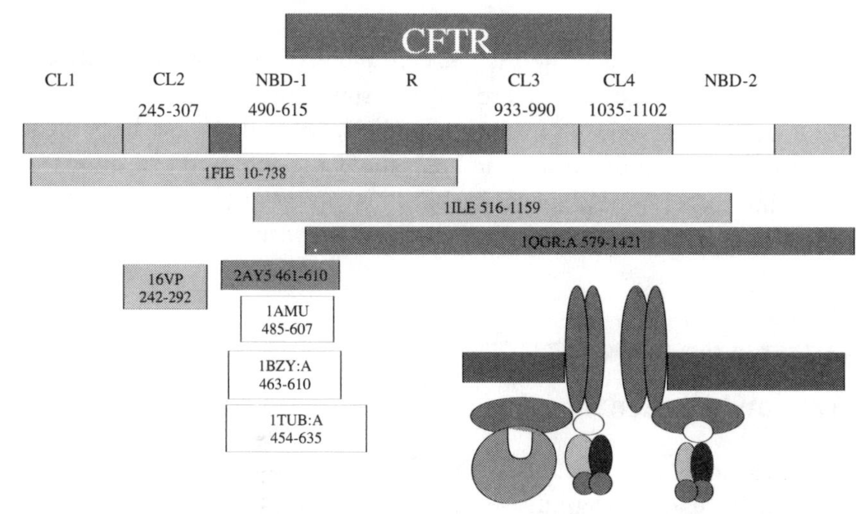

Figure 4.3. Results obtained using HMM-SPECTR for structure prediction of different domains of cystic fibrosis transmembrane regulator (CFTR).[18] The program predicted proteins consistent with the known structural domains of CFTR: 2AY5 Aromatic Amino Acid Aminotransferase, 1FIE Recombinant Human Coagulation Factor Xiii, 1ILE Isoleucyl-tRNA Synthetase, 1QGR Importin β Bound To The Ibb Domain Of Importin α, 16VP Conserved Core Of The Herpes Simplex Virus Transcriptional Regulatory Protein Vp16, 1AMU Phenylalanine Activating Domain Of Gramicidin Synthetase 1, 1BZY Human Hgprtase, 1TUB Tubulin α-β Dimer. Courtesy of Oxford University Press.

The program predicted proteins consistent with the known structural domains of CFTR. These are two of the important functional domains of CFTR—NBD-1 and NBD-2 (first and second nucleotide binding domains). HMM-SPECTR predicted a correspondence between the NBD-1 region of CFTR, and the tertiary structure of 2AY5 (aromatic amino acid aminotransferase). Following this unpublished prediction, the structure of part of ABC transporter protein (ATB-binding subunit of histidine permease) was solved in the laboratory of Sung-ho Kim at UC Berkeley.[17] This molecule has significant homology to NBD-1 of CFTR. The structure of ABC transporter was not present in the PDB and was not used in our preparation of initial HMMs. Nevertheless, when we received it directly from Dr. Kim, we constructed, on its basis, the homology model of NMD-1 of CFTR and then superimposed it with the tertiary structure of 2AY5.

Plate 4.1 (Courtesy of Oxford University Press) shows the superimposition of these two proteins,[18] 2AY5 and ABC transporter. All four helices of both molecules are nicely superimposed to each other. Moreover, three β-strands in the region between helices also have close positioning. There are inserts in each molecule (colored red and dark red) that are not superimposable, but these inserts do not compromise the striking overall correspondence of the two structures.

Acknowledgments

This research was supported by funds from the Tobacco-Related Disease Research Program (award 9RT-0057) and from the National Science Foundation (grant DBI 9911196).

References

1. Orengo,C., Jones,DT., and Thornton, JM: **Protein superfamilies and domain superfolds.** *Nature* 1994, **372**:631–634.

2. Eddy SR: **Profile hidden Markov models.** *Bioinformatics* 1998, **14**:755–763.

3. Baldi P, Chauvin Y, Hunkapiller T, McClure MA: **Hidden Markov models of biological primary sequence information.** *Proc Natl Acad Sci USA* 1994, **91**:1059–1063.

4. Krogh A, Mian IS, Haussler D. **A hidden Markov model that finds genes in *E. coli* DNA.** *Nucleic Acids Res* 1994 (Nov 11), **22**(22):4768–4778.

5. Bateman A, Birney E, Durbin R, Eddy SR, Howe KL, Sonnhammer ELL: **The Pfam contribution to the annual NAR database issue.** *Nucleic Acids Res* 2000, **28**:263–266.

6. Karplus K, Barrett C, Hughey R: **Hidden Markov models for detecting remote protein homologies.** *Bioinformatics* 1998, **14**:846–856.

7. Gough J, Karplus K, Hughey R, Chothia, C: **Assignment of homology to genome sequences using a library of Hidden Markov models that represent all proteins of known structure.** *J Mol Biol* 2001, **313**:903–919.

8. **TIGRFAM.** http://www.tigr.org/TIGRFAMs/.

9. Laurents DV, Subbiah S, and Levitt M: **Different protein sequences can give rise to highly similar folds through different stabilizing interactions.** *Protein Sci* 1994, **11**:1938–1944.

10. Bystroff C, Baker D: **Prediction of local structure in proteins using a library of sequence-structure motifs.** *J Mol Biol* 1998, **281**(3):565–577.

11. Shindyalov IN, Bourne PE: **Protein structure alignment by incremental combinatorial extension (CE) of the optimal path.** *Protein Eng* 1998, **11**:739–747.

12. Murzin AG, Brenner SE, Hubbard T, Chothia C: **SCOP: A structural classification of proteins database for the investigation of sequences and structures.** *J Mol Biol* 1995, **247**:536–540.

13. Eddy SR: **Profile hidden Markov models.** *Bioinformatics* 1998, **14**:755–763.

14. Tsigelny I, Shindyalov PE, Bourne TC, Sudhoff TC, Taylor P: **Common EF-hand motifs in cholinesterases and neuroligins suggest a role for Ca2+ binding in cell surface associations.** *Protein Sci* 2000, **9**:180–185.

15. Salamov AA, Solovyev VV: **Protein secondary structure prediction using local alignments.** *J Mol Biol* 1997, **268**:31–36.

16. *CASP 4. Fourth Meeting on the Critical Assessment on Techniques for Protein Structure Prediction, Asilomar Confrence Center, December 3–7, 2000.* Conf. Proc., Asilomar, California, 2000.

17. Hung LW, Wang IX, Nikaido K, Liu PQ, Ames GF, Kim SH: **Crystal structure of the ATP-binding subunit of an ABC transporter.** *Nature* 1998, **396**:703–707.

18. Tsigelny I, Sharikov Y, Ten Eyck LF: **Hidden Markov models based system (HMM-SPECTR) for detectiong structural homologies on the basis of sequental information.** *Prot Eng* 2002, in press.

5

The Role of Sequence Information in Protein Structure Prediction

Damien Devos, Florencio Pazos, Osvaldo Olmea, David de Juan, Osvaldo Graña, Jose M. Fernández, and Alfonso Valencia

Abstract

Protein families can be represented in the form of multiple sequence alignments, which embrace essential information about protein structure, function and evolution. Different methods extract information from multiple sequence alignments and use them as a primary source of information, often in the form of frequency of different amino acid types at each position of the alignment. Other features that could also be extracted from multiple alignments are rarely used, such as sequence conservation, the variations between sub-families, the correlation between the patterns of mutation of pairs of positions, and the distribution of apolar residues. Here we review various approaches to the systematic exploitation of these sequence-based features: in particular we focus on their combination with standard threading approaches. The initial results, in terms of remote homology detection, are very encouraging and give a glimpse of the possibilities of future approaches.

5

The Role of Sequence Information in Protein Structure Prediction

Damien Devos, Florencio Pazos, Osvaldo Olmea, David de Juan, Osvaldo Graña, Jose M. Fernández, and Alfonso Valencia

CNB-CSIC, Cantoblanco, Madrid, Spain

5.1. Introduction

5.1.1. Information Contained in Multiple Sequence Alignments in Protein Families

Multiple sequence alignments provide valuable information about many protein characteristics. They are used to detect conserved regions, which might be related with structural or functional characteristics in the three-dimensional protein structures (see for example the article).[1] Molecular phylogenetic methods use a multiple sequence alignment as starting input information. The information contained in multiple sequence alignments has also been used by different protein structure prediction methods, such as the most successful methods for predicting secondary structure, solvent accessibility and transmembrane helices.[2-8] Methods to predict two-dimensional features, that is, prediction of residue-residue relationships, also incorporate multiple sequence alignment for the extraction of pairs of positions showing a correlated mutational behavior (*correlated mutations*).[9-11] Such correlation is related to contact or closeness in the three-dimensional structure of the protein, as will be discussed later in this chapter. In the area of protein three-dimensional structures prediction, the information extracted from multiple sequence alignment has been used: in *homology modeling* to improve the fitness between target sequences and protein frameworks,[12] in *threading* either based on structural information[13-19] or purely on sequence information,[20] and in *ab initio* approaches.[21]

It is interesting to consider that the powerful information contained in multiple sequence alignments has only be used in a very simple way, such as amino acid preferences at different positions,[22] composition profiles of different alignment positions,[23] or hidden Markov model probabilistic profiles.[24] It would be interesting to incorporate other more subtle information also present in the alignments, for example the differential conservation in subfamilies of the same alignment. Here we analyze some of those possibilities following our previous line of research on this area.

5.2. Four Types of Informative Residues in Multiple Sequence Alignments

In a multiple sequence alignment, it is possible to discriminate between at least four types of informative position, depending on whether the position is totally conserved, is conserved only amongst sub-groups of sequences of the multiple alignment (also called sub-families) or displays a correlated mutational behavior with another position (Fig. 5.1), additionally we considered hydrophobic positions in this study because of their specific role in the formation of protein structure cores.

Conserved positions are those whose residue type is totally conserved for all the sequences aligned. Sequence conservation is a consequence of the evolutionary pressure at these positions to maintain a given structure and/or function.

Tree-determinant positions are determined by a more subtle way of conservation. These positions are conserved within sub-families, but the chemical type of the amino acids differs between sub-families. Their conservation suggests an evolutionary pressure on the residue type inside subfamilies. The differences between subfamilies means that this position is, with some difference between subfamilies, perhaps related

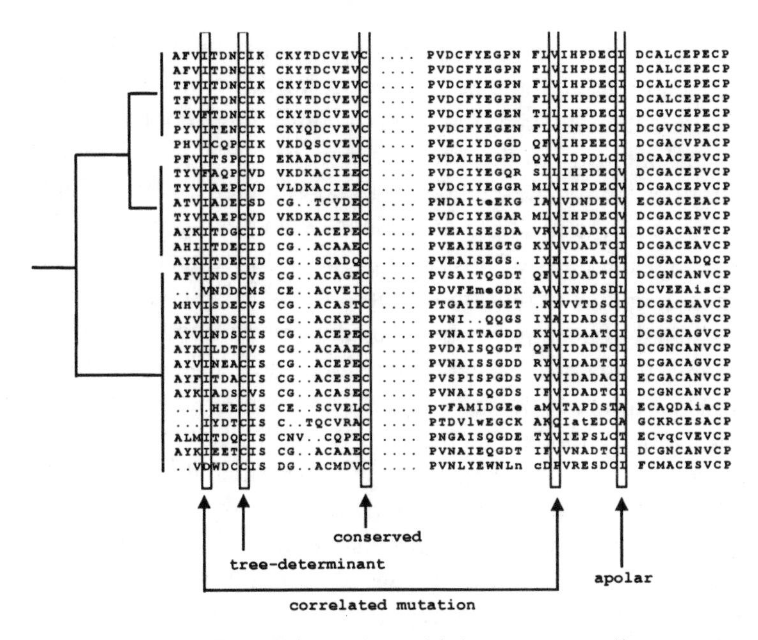

Figure 5.1. Informative positions in multiple sequences alignment.
In a typical multiple sequences alignment, four types of informative positions are highlighted. Conserved positions are totally conserved for all the sequences; tree-determinants are those whose amino acid type is conserved in sub-families but change between them; correlated positions show a concerted pattern of variation. Apolar residues are those that conserve the hydrophobic character of the amino acids.
We extracted all the sequence information from the HSSP database,[47] which compiles multiple sequence alignments for all the proteins in the PDB.
Conservation, positions with HSSP VAR<=10.
Correlation, as calculated in the article.[9] With this method, a correlation value between −1.0 and 1.0 is assigned to every pair of positions in the multiple sequence alignment. We took for the prediction a number of correlated pairs equivalent to half of the alignment length among the highly correlated residues.
Tree determinants, calculated with the "mtreedet" program described in Del Sol, *et al.* (*in prep.*). This program assigns a value between −1.0 and 1.0 to every residue in the multiple sequence alignment. We took 10% of the residues with the highest values as the predicted tree-determinants.
Apolar residues, defining Ala, Val, Leu, Ile, Pro, Phe, Trp and Met as the hydrophobic amino acids.

with differential functional specificity. A variety of algorithms have been designed to extract these positions from multiple sequence alignments.[25-30]

Correlated positions are those pairs of positions in multiple sequence alignments whose mutational patterns are related. Technical differences between different approaches have led to conflicting conclusions about the nature and intensity of this phenomenon.[9,31-38] We use the term *correlate mutations* to indicate a tendency of coordinated mutation between some positions of the multiple alignment by analyzing the correlation between changes in pairs of positions in multiple sequence alignments, with an unambiguous definition of correlation.[9,39,40] Sequence correlation is attributed to the small sequence adjustments needed to maintain protein stability against constant mutational drift. These adjustments would turn out to occur most frequently between neighbor residues in protein structures; a model that has turned out difficult to prove experimentally, with some few exceptions.[41,42,43] The calculation of correlated mutations can then be used for the prediction of three-dimensional contacts and/or physical proximity between residues.

Hydrophobic positions clustering has been traditionally considered to be an important factor of protein cores stability[44,45] and has been included here as additional information that can be directly derived from the multiple sequence alignments.

5.3. Distribution of Informative Positions in Protein Structures

The hypothesis is that those four types of informative positions tend to group together in protein structures. We quantify this tendency by comparing the distribution of inter-residue distances in the 3D structure for the informative pairs to the distribution of distances for all other pairs of positions. We have previously defined a parameter Xd that accounts for this difference between distributions[11] (Fig. 5.2). Xd is positive when the distribution of distance between the set of informative residues is shifted toward smaller distances respectively to the distribution of all other residue pairs.

We first analyze the clustering tendency of evolutionary informative positions in a representative set of protein structures. For this purpose, we determined the Xd value for a large set of protein structures deposited in the Protein Data Bank[46] and for which an informative multiple

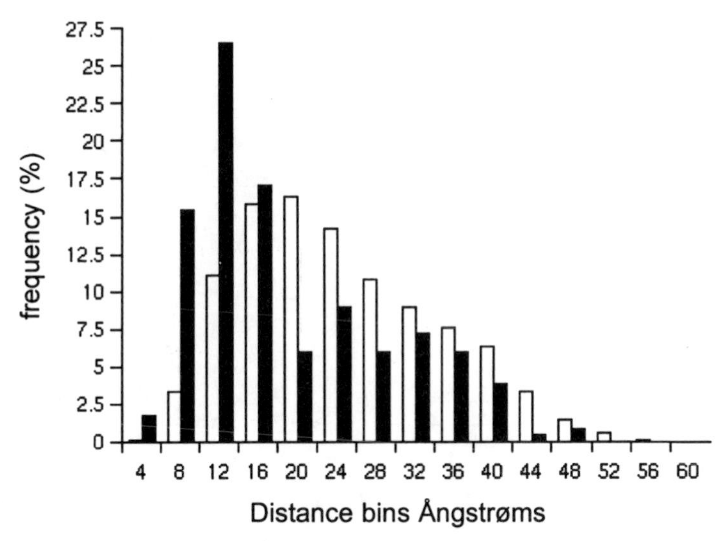

Figure 5.2. Evaluation of the distance between significant residues with the Xd parameter.
The figure represents an example of distance distribution for the 1dok structure, distances are grouped in bins of 4 Ångstrøms (*x*-axis) and the frequency reported (*y*-axis) for every residue pair or only for the pairs of conserved residues by empty or filled bars, respectively.
The Xd parameter estimates the difference between two distributions and was previously described in the article[11] with the following formula:

$$Xd = \sum_{i=1}^{i=n} \frac{P_{ic} - P_{ia}}{d_i \cdot n},$$

where n: Number of distance bins; there are 15 equally distributed bins from 4 to 60 Å; d_i: Upper limit for each bin; P_{ic}: Percentage of correlated, conserved, tree-determinant or hydrophobic pairs with distance between d_i and d_{i-1}; P_{ia}: The same percentage for every pairs of positions. Defined in this way Xd=0 indicates no separation between the two distance populations, Xd>0 indicates positive cases where the population of correlated, conserved, tree-determinant or hydrophobic pairs is shifted to smaller distances compared to the population of all pairs. Distances between residues correspond to $C_\beta - C_\beta$ distances, C_α for Glycines.

alignment could be recovered in HSSP.[47] The sequences for the analysis were chosen to cover a significant fraction of the sequence space with the minimal redundancy, that is having less than 25% similarity between any possible pair. From the multiple alignment, we determined the conserved, tree-determinant, correlated, and apolar positions (see figure legend for definitions and calculations). We then calculated the distance between each informative position and compared it to the distance between every pair of position of the sequence, expressing it in terms of Xd (Fig. 5.3). The results show that the four types of informative positions have distributions of a Poisson like shape, with a clear positive location, indicative of a tendency of those positions to cluster inside protein structures.

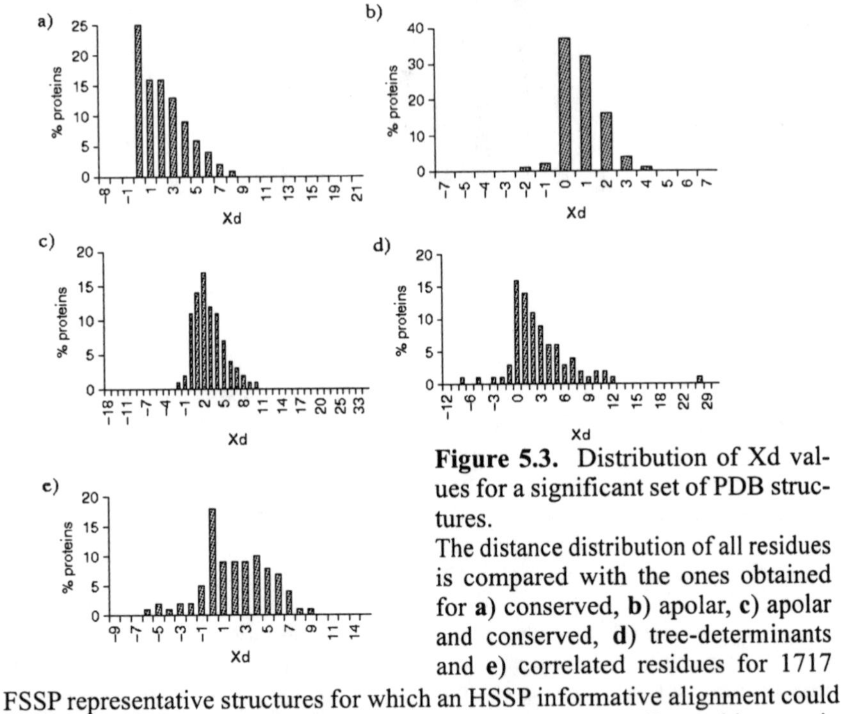

Figure 5.3. Distribution of Xd values for a significant set of PDB structures.
The distance distribution of all residues is compared with the ones obtained for **a)** conserved, **b)** apolar, **c)** apolar and conserved, **d)** tree-determinants and **e)** correlated residues for 1717 FSSP representative structures for which an HSSP informative alignment could be recovered (as at July 2001). The percentage of crystallographic protein structures (*y*-axis) at each Xd class (*x*-axis) is represented.

5.4. Informative Positions in Protein Structure Models

We proposed previously that the information contained in multiple sequence alignments could be applied to the discrimination of correctly folded proteins versus incorrectly folded models.[39] For this purpose we compared the Xd of conserved, correlated and apolar positions in real and deliberately missfolded protein structures. The latter were generated by positioning the sequence of a protein of known structure (and therefore its corresponding informative residues) into the structure of a completely different protein of the same sequence length. In this test set it was possible to observe a clear shift of the distances between pairs of conserved, correlated or apolar positions toward small distance in the case of real protein structures (Fig. 5.4); while for the corresponding incorrectly folded model, the distances were similar to the ones obtained with all the pairs of residues, indicating a random distribution of the informative residues onto the incorrectly folded structures. Overall, combining conservation, apolarity and correlation all the incorrect models could be differentiated in a large test set.[39]

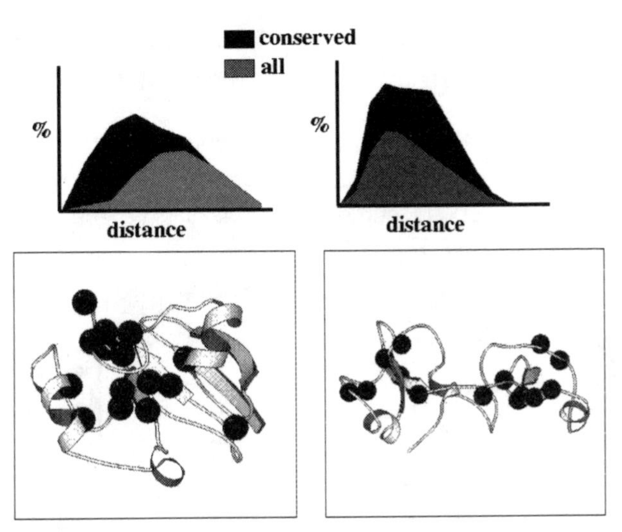

Figure 5.4. Distribution of significant residues (Xd value) in correctly and incorrectly folded proteins.
a) The proportion of residue pairs at each distance are grouped in distance bins (in Å) for conserved and all residues represented in filled and open bars, respectively. The results correspond to the representative example of the 1put protein. The correctly folded structure is represented on the left panel; the distribution has an Xd value of conserved residues of 8.79. The incorrect model generated mounting the 1put sequence in the 1pcp fold (right panel) has negative Xd value of conserved residues of −1.42, that indicates that the conserved residues can not be distinguished from other residues in the wrong model.
b) Rasmol representation of the corresponding structures, with 1put conserved residues represented as balls in their C_α atoms.

Based on this observation we proposed the use of sequence information for the selection of incorrect models generated by threading methods. In this process we analyze implicit threading models (alignments between sequences and proteins of known structure) by their distribution of correlated, conserved and apolar residues using the Xd parameter. The results show that any of these three sequence criteria could be used to improve threading results, with a clear improvement in the recognition of remote folds for different sequences by themselves or combined with the threading scores.

5.5. A Threading Server That Filters Models with Multiple Sequence Alignments Information

We have automated the analysis of sequence features extracted from multiple sequence alignments and their use to filter threading models in a web server (ThreadWithSeq server: http://www.pdg.cnb.uam.es:8081/threadwithseq.html, Pazos, *in preparation*, Fig. 5.5).

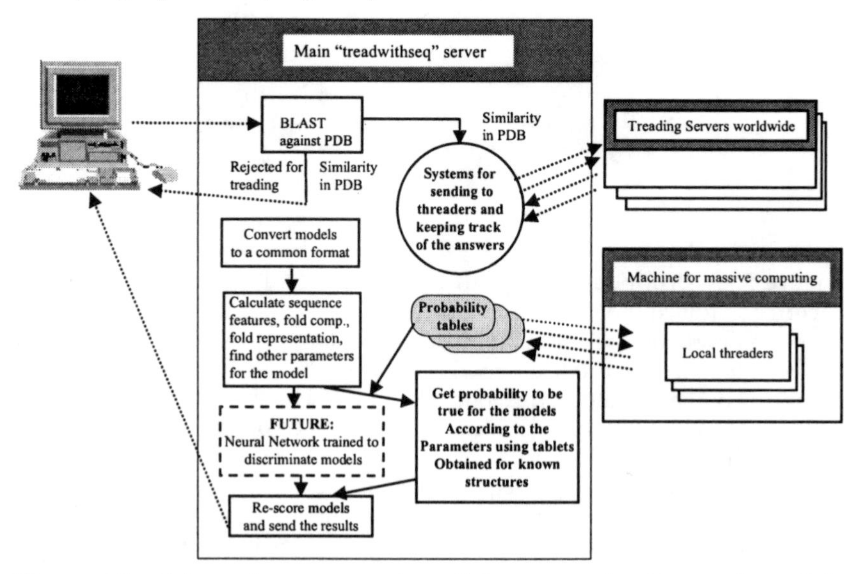

Figure 5.5. Schematic representation of the current implementation of the ThreadWithSeq server.
The server can be accessed at: http://www.pdg.cnb.uam.es:8081/threadwithseq.html.

ThreadWithSeq scans the output of various threading programs, evaluating the quality of the resulting models with a range of sequence-based information, and finally ranks the proposed folds according to the combined sequence and structure information. The threading servers/programs used in ThreadWithSeq are: TOPITS,[18] 3D-PSSM,[48] THREADER2,[49] and SAM-T99.[20]

In the current implementation, the structural similarity between the possible threading models is evaluated by retrieving their classification from the SCOP[50] and FSSP databases.[51] The characteristics used for assessing the model quality are: fold compactness, distribution of apolar residues, distribution of conserved and sub-family conserved residues (tree-determinants), and disposition of correlated mutations. The final score of the models is derived by processing all this information by a statistical analysis that was derived from a detailed study of a test set of 87 proteins of known structure (Pazos, *unpublished*).

5.6. A First Field Evaluation of the Server, the CAFASP Results

A first external evaluation of the performances of the system (in its implementation of April 2000) was performed during the second CAFASP encounter, when blind predictions of different servers were stored before the release of the crystallographic structures (http://www.cs.bgu.ac.il/~dfischer/CAFASP2/).

The *ThreadWithSeq* system is designed to recognize correct folds more than correct alignments, since it uses the alignments provided by the threading servers without further improving them with the derived sequence information. In CAFASP it indeed identified the correct fold in various cases but did not always find the best fit of the sequence in the structure. Given that *ThreadWithSeq* in that implementation does not directly use the information of the similarity between the query and target proteins, the server showed better results in the difficult cases with no recognizable similarity between the query sequence and the protein proposed as a possible fold. In CAFASP the server produced four correct models in 11 cases in which there was no sequence similarity between target and query proteins. It turned out to be the best server regarding the identification of correct folds (and not correct alignments) for this difficult class of proteins. (CAFASP evaluation, table 6: http://www.cs.bgu.ac.il/~dfischer/CAFASP2/ALevaluation/hardfr5.table).

Therefore, *ThreadWithSeq* is effective in filtering threading results with information derived from multiple sequence alignments, and it improves the fold recognition step (particularly for remotely related sequences), but it is still limited in its ability to generate correct query-target alignments (threading step). This is a limitation that is common to other threading approaches.

5.7. A CAFASP Example of the Use of Sequence Information

A good example of the results discussed above could be target T0097, a C-domain of ERp29 of 105 amino acids length (http://www.cs.bgu.ac.il/~dfischer/CAFASP2/). In this case we recovered seven homologue sequences that enabled the generation of a minimally informative alignment and the use of the different sequence based information (Plate 5.1). In Plate 5.1 CAFASP target T0097 is presented on the left, with the actual PDB code, 1g7d, between parentheses. Three of the THREADER2 hits are presented with their PDB code. The characteristics given are the THREADER2 Z-score (_z=), the ranking (between parentheses) on the first line, the Xd of conserved residues on the second, the structural similarity with target T0097 (FSSP Z-score), and ranking (between parentheses) on the third line. The structural superposition of T0097 and the second domain of 1mtyG are represented in the form of backbones on the lower right. A quasi-perfect superposition is obtained for the region with structural similarity.

The *ThreadWithSeq* server detected, as a possible correct model, the one based on the structure of the second domain of 1mtyG, since conserved residues did cluster very strongly in this model, with a Xd of 18.75. Additionally the model included approximately half of the target sequence in the 1mtyG fold. It is interesting to notice that this model was ranked in position 21 in the THREADER2 output,[49] and most probably would not have been selected by direct inspection of the THREADER2 results. Therefore, it can be said that the use of the evolutionarily informative residues "rescued" it from the background noise of the threading output. The experimental resolution of the structure confirmed that the 1mtG fold was the best available framework for half of the T0097 target sequence. Another related model was the one based on the PDB structure 1abv, ranked 4th in the THREADER2 output (FSSP Zscore of 3.4), which also has a significant distribution of conserved residues with an Xd of 1.82.

5.8. Training Neural Networks for the Discrimination of Wrong *Threading* Models Using Sequence Information

As an additional step beyond the simple combination of parameters implemented in *ThreadWithSeq*, we are training a Neural Network for the discrimination of wrong threading models by detailed information about the distribution of sequence features.

The architecture of the Neural network is based on a back-propagation model, and uses as an input vector of real numbers including three type of values:

- the score provided by different threading programs;
- values representing the sequence-derived parameters (Xd for apolar, conserved, correlated and tree-determinant residues);
- and physical parameters representing characteristics of the intrinsic models, namely fold compactness, length of the sequence-structure alignment, length of the template and query sequences and percentage of identity between the query and target sequences.

The neural network has two output units where the goodness of the model is codified as "1-0" for "good models" and "0-1" for "bad models." The internal architecture of the network contains one hidden layer with five neurons, a topology and specifications that were selected after testing the predictive power of different configurations.

In the current implementation for the training of the Network, the quality of the models was assessed by the level of structural similarity between the template and the real structure of the query sequence, as quantified with the Z-score value of the Dali program.[51] An improvement of this procedure corresponds to the direct use of the structural similarity between the explicit model and the real protein structure (for example using the MaxSub parameter).[52]

The network has been trained and tested with a non-redundant list of more than 800 proteins split in 8 groups for cross-validation. We have obtained results for two public threading servers, namely 3D-PSSM[48] and SAM-T99,[20] removing those models with sequence identity higher than 25%, which respect to the query sequence, in order to simulate a real threading prediction case. In these two cases the preliminary results are very promising, and the network is able to discriminate wrong threading models selecting the right ones, in the range of scores in which the classical threading programs are unable to point to the right model.

5.9. Conclusions

We have investigated the characteristic distribution of different types of positions that are observed in multiple sequence alignment, i.e., sequence conservation, sequence conservation in subfamilies (tree-determinants), correlated mutations and apolar residues. We have proposed a factor (Xd) for the quantification of shift in the distributions of inter-residue distances of informative residues versus the distribution of all other residues. Our results show that the informative positions tend to cluster in protein structures, with a clear tendency to be closer in space than all other residues.

We also demonstrated that this tendency could be used to distinguish correct structures from incorrectly folded models. This concept was automated in a server (*ThreadWithSeq*) that uses sequence information to filter models obtained from various threading servers. The results obtained with the first implementation of the server in the CAFASP competition were quite interesting, particularly in relation to the recognition of folds for modeling distantly related sequences.

References

1. Valencia A, Chardin P, Wittinghofer A, Sander C: **The ras protein family: evolutionary tree and role of conserved amino acids.** *Biochemistry* 1991, **30**:4637–4648.

2. Benner SA: **Patterns of divergence in homologous proteins as indicators of tertiary and quaternary structure.** *Adv Enzyme Regul* 1989, **28**:219–236.

3. Benner SA, Gerloff D: **Patterns of divergence in homologous proteins as indicators of secondary and tertiary structure: A prediction of the structure of the catalytic domain of protein kinases.** *Adv Enzyme Regul* 1991, **31**:121–181.

4. Cooperman BS, Baykov AA, Lahti R: **Evolutionary conservation of the active site of soluble inorganic pyrophosphatase.** *Trends Biochem Sci* 1992, **17**:262–266.

5. Howell N: **Evolutionary conservation of protein regions in the protonmotive cytochrome b and their possible roles in redox catalysis.** *J Mol Evol* 1989, **29**:157–169.

6. Hwang PK, Fletterick RJ: **Convergent and divergent evolution of regulatory sites in eukaryotic phosphorylases.** *Nature* 1986, **234**:80–83.

7. Rost B, Sander C: **Conservation and prediction of solvent accessibility in protein families.** *Proteins* 1994, **20**:216–226.

8. Rost B, Fariselli P, Casadio R: **Topology prediction for helical transmembrane protein at 86% accuracy.** *Protein Sci* 1996, **5**:1704–1718.

9. Göbel U, Sander C, Schneider R, Valencia A: **Correlated mutations and residue contacts in proteins.** *Proteins* 1994, **18**:309–317.

10. Olmea O, Valencia A: **Improving contact predictions by the combination of correlated mutations and other sources of sequence information.** *Folding & Design* 1997, **2**:S25–S32.

11. Pazos F, Helmer-Citterich M, Ausiello G, Valencia A: **Correlated mutations contain information about protein-protein interaction.** J *Mol Biol* 1997, **272**:1–13.

12. Overington J, Donnelly D, Johnson MS, Sali A, Blundell TL: **Environment-specific amino acid substitution tables: tertiary templates and prediction of protein folds.** *Protein Sci* 1992, **1**:216–26.

13. Shan Y, Wang G, Zhou HX: **Fold recognition and accurate query-template alignment by a combination of PSI-BLAST and threading.** *Proteins* 2001, **42**:23–37.

14. Defay TR, Cohen FE: **Multiple sequence information for threading algorithms.** *J Mol Biol* 1996, **262**:314–323.

15. Fisher D, Eisenberg D: **Protein fold recognition using sequence-derived predictions.** *Prot Sci* 1996, **5**:947–955.

16. Jones DT, Taylor WR, Thornton JM: **A new approach to protein fold recognition.** *Nature* 1992, **358**:86–89.

17. Ouzounis C, Sander C, Scharf M, Schneider R: **Prediction of protein structure by evaluation of sequence-structure fitness: Aligning sequences to contact profiles derived from three-dimensional structures.** *J Mol Biol* 1993, **232**:805–825.

18. Rost B: TOPITS: **Threading one-dimensional predictions into three-dimensional structures.** In: *Third International Conference on Intelligent Systems for Molecular Biology.* Rawlings C, Clark D, Altman R, Hunter L, Lengauer T, Wodak S, eds. Menlo Park, CA: AAAI Press, Cambridge, England 1995:314-321.

19. Rost B, Schneider R, Sander C: **Protein fold recognition by prediction-based threading.** *J Mol Biol* 1997, **270**:471–480.

20. Karplus K, Barret C, Hughey R: **Hidden Markov Models for detecting remote protein homologies.** *Bioinformatics* 1998, **14**:846–856.

21. Taylor WR: **Towards protein tertiary fold prediction using distance and motif constraints.** *Protein Eng* 1991, **4**:853–870.

22. Ouzounis C, Perez-Irratxeta C, Sander C, Valencia A: **Are Binding Residues Conserved?** In: *Proceedings of the Fifth Annual Pacific Symposium on Biocomputing.* Hunter L, eds. Hawaii, USA: World Scientist, 1998:399–410.

23. Gribskov M, McLachlan M, Eisenberg D: **Profile analysis: Detection of distantly related proteins.** *Proc Natl Acad Sci USA* 1987, **84**:4355–4358.

24. Durbin R, Eddy S, Krogh A, Mitchison G: *Biological sequence analysis: Probabilistic models of proteins and nucleic acids.* Cambridge U.K: Cambridge University Press, 1998.

25. Livingstone, CD, Barton GJ: **Protein sequence alignments: A strategy for the hierarchical analysis of residue conservation.** CABIOS 1993, **6**:645–756.

26. Casari G, Sander C, Valencia A: **A method to predict functional residues in proteins.** *Nature Struct Biol* 1995, **2**:171-178.

27. Lichtarge O, Bourne HR, Cohen FE: **An evolutionary trace method defines binding surfaces common to protein families.** *J Mol Biol* 1996, 257:342–358.

28. Andrade MA, Casari G, Sander C, Valencia A: **Classification of protein families and detection of the determinant residues with an improved self-organizing map.** *Biol Cybern* 1997, **76**:441–450.

29. Pazos F, Sanchez-Pulido L, Garcia-Ranea JA, Andrade MA, Atrian S, Valencia A: **Comparative analysis of different methods for the detection of specificity regions in protein families.** In: *Biocomputing and Emergent Computation.* Lund D, Olsson B, Narayanan A., eds. Singapore, New Jersey, London, Hong Kong: World Scientific, 1997:132–145.

30. Del Sol A, Pazos F, Valencia A: **Tree determinant residues.** 2001, Personal Communication.

31. Altschuh D, Lesk AM, Bloomer A C, Klug A: **Correlation of co-ordinated amino acid substitutions with function in viruses related to tobacco mosaic virus.** *J Mol Biol* 1987, **193**:693–707.

32. Altschuh D, Vernet T, Moras D, Nagai K: **Coordinated amino acid changes in homologous protein families.** *Protein Eng* 1988, **2**:193–199.

33. Neher E: **How frequent are correlated changes in families of protein sequences?** *Proc Natl Acad Sci USA* 1994, **91**:98–102.

34. Shindyalov IN, Kolchanov NA, Sander C: **Can three-dimensional contacts in protein structures be predicted byanalysis of correlated mutations?** *Protein Eng* 1994, **7**:349–358.

35. Taylor WR, Harrick K: **Compensating changes in protein multiple sequence alignments.** *Protein Eng* 1994, 7:342–348.

36. Chelvanayagam G, Eggenschwiler A, Knecht L, Gonnet GH, Benner SA: **An analysis of simultaneous variation in protein structures.** *Protein Eng* 1997, **10**:307–316.

37. Pollock DD, Taylor WR: **Effectiveness of correlation analysis in identifying protein residues undergoing correlated evolution.** *Protein Eng* 1997, **10**:647–657.

38. Pollock DD, Taylor WR, Goldman N: **Coevolving protein residues: maximum likelihood identification and relationship to structure.** *J Mol Biol* 1999, **287**:187–198.

39. Olmea O, Rost B, Valencia A: **Effective use of sequence correlation and conservation in fold recognition.** *J Mol Biol* 1999, **293**:1221–1239.

40. Pazos F, Olmea O, Valencia A: **A graphical interface for correlated mutations and other structure prediction methods.** *CABIOS* 997, **13**:319–321.

41. Serrano L, Kellis JT, Cann P, Matouschek A, Fersht AR: **The folding of an enzyme. II. Substructure of barnase and the contribution of different interactions to protein stability.** *J Mol Biol* 1992, **224**:783–804.

42. Gregoret LM, Sauer R: **Additivity of mutant effects assessed by binomial mutagenesis.** *Proc Natl Acad Sci USA* 1993, **90**:4246–4250.

43. Mateau M, Fersht A: **Mutually compensatory mutations during evolution of the tetramerization domain of tumor suppressor p53 lead to impaired hetero-oligomerization.** *Proc Natl Acad USA* 1999, **96**:3595–3599.

44. Huang ES, Subbiah S, Levitt M: **Recognizing native folds by the arrangement of hydrophobic and polar residues.** *J Mol Biol* 1995, **252**:709–720.

45. Huang ES, Subbiah S, Tsai J, Levitt M: **Using a hydrophobic contact potential to evaluate native and near-native folds generated by molecular simulations.** *J Mol Biol* 1996, **257**:716–725.

46. Berman HM, Westbrook J, Feng Z, Gilliland G, Bhat TN, Weissig H, Shindyalov IN, Bournem PE: **The Protein Data Bank.** *Nucleic Acids Res* 2000, **28**:235–242.

47. Sander C, Schneider R: **The HSSP data base of protein structure-sequence alignments.** *Nucleic Acids Res* 1993, **21**:3105–3109.

48. Kelley LA, MacCallum RM, Sternberg MJE: **Enhanced genome annotation using structural profiles in the program 3D-PSSM.** *J Mol Biol* 2000, **299**:501-522.

49. Jones DT, Tress M, Bryson K, Hadley C: **Succeffsul recognition of protein folds using threading methods biased by sequence similarity and predicted secondary structure.** *Proteins* 1999, **37**(S3):104–111.

50. Murzin AG, Brenner SE, Hubbard T, Chothia C: **SCOP: A structural classification of proteins database for the investigation of sequences and structures.** *J Mol Biol* 1995, **247**:536–540.

51. Holm L, Sander C: **The FSSP database: Fold classification based on structure-structure alignment of proteins.** *Nucleic Acids Res* 1996, **24**:206–210.

52. Siew N, Elofsson A, Rychlewski L, Fischer D: **MaxSub: An automated measure for the assessment of protein structure prediction quality.** *Bioinformatics* 2000, **16**:776–785.

6

Protein Fold Recognition and Comparative Modeling Using HOMSTRAD, JOY, and FUGUE

Ricardo Núñez Miguel, Jiye Shi, and Kenji Mizuguchi

Abstract

This article illustrates how tools exploiting the knowledge of protein three-dimensional structure can be used to identify homologues of known structure, generate sequence-structure alignments and assist model building. The tools described here include HOMSTRAD, a database of structure-based alignments for protein families of known structure, JOY, a program to annotate local environments in structure-based alignments, and FUGUE, a program to perform sequence-structure homology recognition. After a brief review of the whole process of homology recognition and comparative modeling, a specific example clarifies all the steps involved. This type of analysis will help obtain a better understanding of the function of many proteins whose sequences are known.

6

Protein Fold Recognition and Comparative Modeling Using HOMSTRAD, JOY, and FUGUE

Ricardo Núñez Miguel, Jiye Shi, and Kenji Mizuguchi

University of Cambridge, Cambridge, U.K.

6.1. Introduction

Divergent evolution has given rise to families of homologous proteins, where members of a family share similar but sometimes diverged amino acid sequences. Even though these distantly related members have little sequence similarity, their three-dimensional (3D) structures are very well conserved and they also share, broadly speaking, common functions. Thus, if we can somehow assign an unknown protein sequence to a known family, which has a member of known structure, we can learn about the structure and function of this unknown protein (Fig. 6.1). This is the basis of structure prediction and functional inference using sequence-structure homology recognition. This type of analysis can bridge two traditional branches of biology, sequence database searches and structural studies. With the total number of complete genomes soon to exceed 200, and a growing number of experimentally defined 3D structures, it has huge potential for providing a new type of knowledge in the post-genomic era.

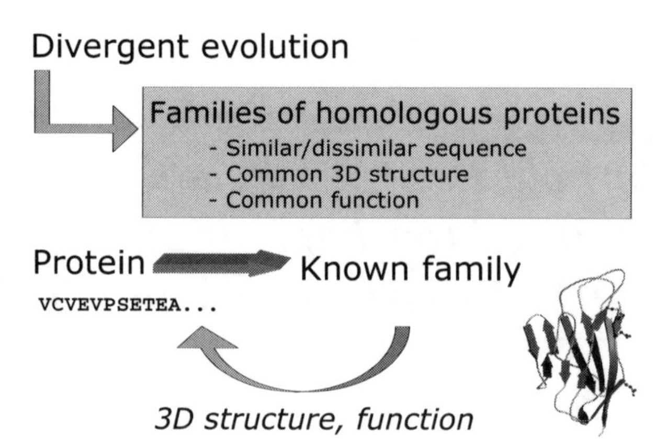

Figure 6.1. From divergent evolution to 3D protein structure and function.

We have developed various tools to facilitate many of the important steps in structure/function prediction using homology recognition. The database HOMSTRAD[1] (http://www-cryst.bioc.cam.ac.uk/homstrad/) provides information about protein families with known structure and presents a curated collection of structure-based alignments of the members of these families. We take all known protein structures, cluster them into families and align the sequences of the representative members of each family on the basis of their structures. The alignments are generated by the program COMPARER[2] to optimize the conservation of local environments and are individually checked. Because it provides structural alignments, it can be used to evaluate sequence alignments, as a standard benchmark set[3] or by direct comparison with sequence alignments in Pfam.[4] Because it is manually curated, it can even be used to benchmark automatic structure comparison methods.[5]

We use HOMSTRAD in many ways, but perhaps the most important application is the derivation of environment-specific substitution tables.[6] Each residue in a protein structure stays in a particular local environment, which can dramatically influence the amino acid substitution pattern of the residue. For instance, it is well known that residues buried in the core of the structure are more conserved than those on the surface, and residues within secondary structure elements (SSEs) are more conserved than those in coil regions. One example is illustrated in Fig. 6.2, which shows probabilities that a residue will not be substituted by any other residue type during evolution. These probabilities were calculated from the structure-based alignments in the HOMSTRAD database. Not only does a buried position have a higher conservation probability than a surface position for all 20 amino acids, the figure also shows that the increases in conservation from surface to buried residues are not uniform, i.e., the residues that undergo the largest increases in conservation are polar or charged, a typical example being asparate (represented as D). This indicates that local environments contain useful information for predicting amino acid substitution patterns.

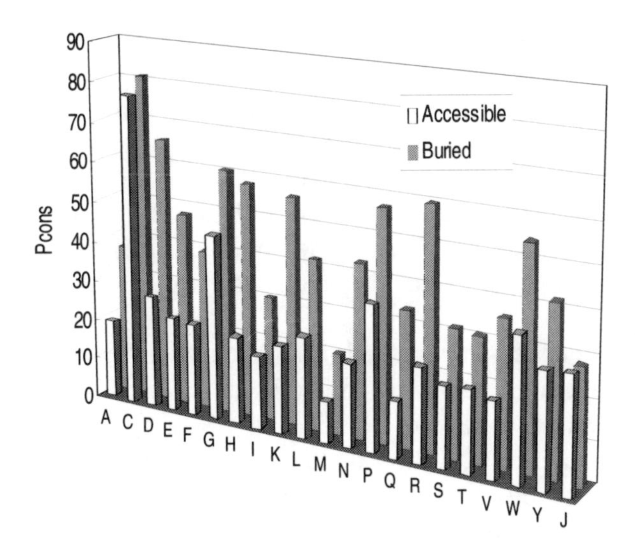

Figure 6.2. Probabilities that a particular amino acid residue will not be substituted by any other residue type during evolution. The data were calculated from selected structure-based alignments in the HOMSTRAD database. Disulphide-bonded cysteine (C) and non-disulphide-bonded cysteine (J) residues are distinguished.

The program JOY[7] (http://www-cryst.bioc.cam.ac.uk/~joy/) can define these local environments and annotate structure-based alignments. It produces formatted alignments, in which normal one-letter amino acid codes are decorated with special symbols to allow easy identification of local environments (see Fig. 6.8). The program has proved to be a useful tool in examining and optimizing sequence-structure or structure-structure alignments and identifying distant homologues.[8,9] A JOY-formatted alignment highlights unique patterns of amino acid substitutions in various environments. For example, the conservation of buried asparate residues can be easily recognized, as these are shown in bold capital letters. Thus, it helps to identify misaligned regions or residues that play important structural roles.

Using JOY and structure-based alignments in HOMSTRAD,[1] we have derived amino acid substitution tables for different environments.[6] These are one of the essential elements of the homology recognition program FUGUE.[10]

FUGUE is a tool developed to associate a query sequence with its homologues of known structure. It compares a query sequence or sequence alignment against each structural profile in a profile library derived from HOMSTRAD, and assesses the compatibility between the sequences and the structures. A structural profile consists of two matrices: a scoring matrix and a gap penalty matrix. The key feature of FUGUE is to calculate both matrices not only according to the amino acid sequence information, which is used by traditional sequence-only fold recognition methods, but also the local structural environment information.

Traditional sequence-only methods ask the question "what is the likelihood of amino acid A being substituted by amino acid B during evolution?" In contrast, when we construct the scoring matrix for the FUGUE structural profile, we ask the question, "what is the likelihood of amino acid A, within structural environment E, being substituted by amino acid B during evolution?" FUGUE uses 64 environments defined by the combination of three structural features: main-chain conformation and secondary structure (helix/strand/coil/positive phi torsion angle), solvent accessibility (accessible/inaccessible) and hydrogen bonding status (true or false for: side-chain to main-chain NH/side-chain to main-chain CO/side-chain to other side-chain). The local environment is calculated for each residue of the structure and the corresponding environment-specific amino acid substitution pattern is stored in the scoring matrix of the structural profile.

During divergent evolution, insertion/deletions occur more frequently on the surface region of the protein than in the core region and also more frequently within the coil region than within SSEs. In sequence alignments, insertions/deletions are represented as gaps. FUGUE calculates the gap penalty matrix according to the local structure information. For instance, positions in SSEs and core regions receive higher gap penalties than those in coil and surface regions and positions at the center of an SSE receive higher gap penalties than those at the termini of an SSE. These structure-dependent gap penalties are the second essential element of FUGUE.

In this article, we illustrate how these and other tools can be used to identify homologues of known structure, generate sequence-structure alignments and assist model building. After briefly reviewing the whole process of fold recognition and comparative modeling, we first describe some practical considerations in using FUGUE, which plays a key role in the whole process. We then use a specific example to illustrate all these steps, including discussions on various other tools.

6.2. Overview

Our goal is to assign an unknown sequence to a family with known structures and build an accurate model for the 3D structure of the protein, which then will allow functional inferences. There are several steps to achieve this goal. First, given a target protein sequence, one or more homologous proteins of known structure need to be identified. Second, it is important to have a good sequence alignment between the homologues and the target protein. These two steps are crucial; if the proteins identified are not true homologues, or if the two amino acid sequences are wrongly aligned, the 3D-model obtained will be wrong even if the rest of the process is perfect. In the next two sections we will discuss how FUGUE can play crucial roles in these two steps.

The third step in the process consists of obtaining the structure from the alignment by using one of the available comparative modeling programs, followed by the refinement of the obtained structure. Finally, a check of the structure is needed to avoid implausible models. If the protein structure includes some unlikely or impossible features, we go back to the alignment and try to improve it. If the alignment cannot be improved or alternative alignments do not lead to better models, it is possible that the selected homologues might be incorrect and new homologous proteins should be identified. Fig. 6.3 shows this process schematically.

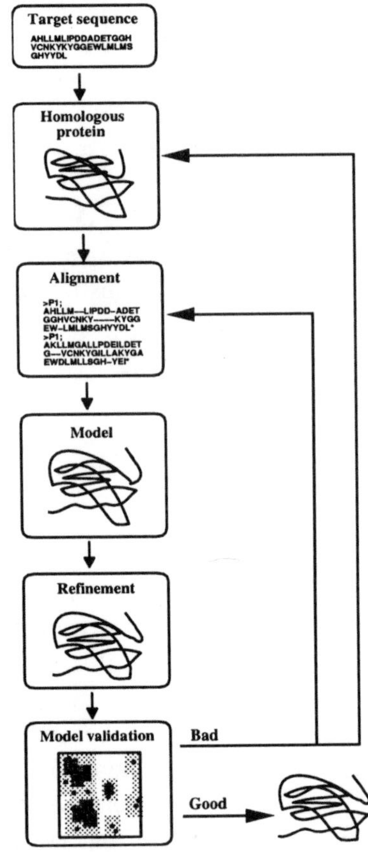

Figure 6.3. Schematic representation of the steps followed in comparative modeling.

For each step described above, there are several good tools that can be used and in some cases it is a good idea to use more than one tool and select the best results.

6.3. Identification of Homologues

FUGUE is available to the public via a web server at http://www-cryst.bioc.cam.ac.uk/fugue. Given a single query sequence, the server runs PSI-BLAST[11] to perform a search against the NCBI non-redundant sequence database and collects sequence homologues. The alignment produced by PSI-BLAST is then used to calculate a sequence profile, which describes the observed amino acid distribution at each position of the query sequence. FUGUE compares this sequence profile against each structural profile in its library derived from HOMSTRAD and assesses the compatibility between the sequences and the structures.

The sequence homologues retrieved by PSI-BLAST provide valuable information about the sequence family, which can improve the performance of FUGUE. However, in some cases non-homologous sequences (false positives) may be included in the PSI-BLAST alignment and the alignment itself may contain serious errors. Advanced users are recommended to check the PSI-BLAST alignment when receiving the FUGUE results. They can improve the alignment and resubmit it to the FUGUE server by selecting the option that tells the server to use the input alignment for sequence-structure comparison. There is also an option on the FUGUE server to skip the PSI-BLAST search and use a single input sequence for the search. This option should only be used when the user fails to obtain an alignment of reasonable quality between the query sequence and its sequence homologues.

During the database search, FUGUE aligns the query sequence profile against each structural profile using the scoring and gap penalty matrices tat are associated with the structural profile. The query sequence profile is then randomized by 100 times and an alignment score is calculated for each randomized profile. A Z-score is calculated by comparing the alignment score for the original sequence profile against the scores for the randomized ones. Higher Z-scores indicate better compatibility between the query sequence and the structure and greater probability of homology.

FUGUE was benchmarked using a test set developed by Lindahl and Elofsson.[12] In the test set, 976 proteins of known structure are clustered into families, superfamilies and folds based on the SCOP[13] classification. An all-against-all recognition test can be carried out to check how well the program being benchmarked can re-establish the correct relationships among those proteins. Fig. 6.4 shows the benchmark result for FUGUE in recognizing protein pairs that share family level similarity, together with the results for some other fold/homology recognition tools provided by Dr. Elofsson.

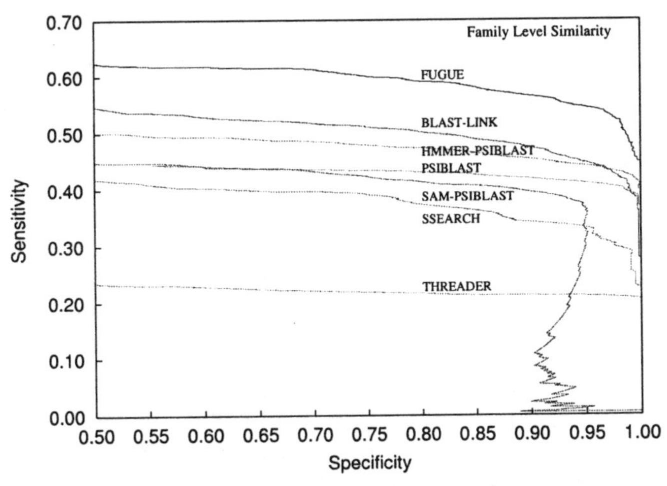

Figure 6.4. Specificity-sensitivity curves of recognition performance at the family level using the test set provided by Dr. Elofsson.[12] Data other than that of FUGUE were kindly provided by Dr. Elofsson.

FUGUE significantly outperformed other methods. For example, at 99% specificity (i.e., 1 error out of every 100 predictions of homology), FUGUE obtained a sensitivity of 49% (i.e., 49% of true homologous protein pairs were recognized), while the best performance of other methods, obtained by HMMER-PSIBLAST, hit 42% sensitivity. Z-score confidence thresholds were estimated from the benchmark result. Specificities of 99% and 95% corresponded to Z-scores of 5.6 and 4.6, respectively. In practice, we set the default Z-score thresholds at 6.0 for 99% confidence and 5.0 for 95%.

The recognition performance of FUGUE has also been benchmarked in two independent assessment exercises: CAFASP2 (http://cafasp.bioinfo.pl/) and LiveBench2 (http://bioinfo.pl/LiveBench/). FUGUE was ranked among the top servers, and was also key to the success of the Blundell group in CASP4.[14] This demonstrated the usefulness of environment-specific substitution scores and structure-dependent gap penalties in homology recognition.

6.4. Generating Sequence-Structure Alignment

FUGUE can also be used as a sequence-structure alignment program. The environment-specific substitution scores and structure-dependent gap penalties can also help to build more accurate sequence-structure alignments compared with many sequence-only alignment programs like CLUSTALW,[15] especially when the percentage sequence identity (PID) is low and the structural information becomes more significant. By using Fischer's benchmark test-set,[16] we observed that FUGUE outperformed both CLUSTALW and GenTHREADER[17] in alignment accuracy.[10]

The FUGUE homology recognition server searches for homologues in the structural profile library and automatically generates the best alignments for the top hits. This step can be shortened, however, if some homologues of known structure are already known. For example, suppose we are interested in a protein, which is known to belong to the aspartic proteinase family. Rather than submitting this sequence to the FUGUE homology recognition server, we can directly go to the asparatic proteinase page of HOMSTRAD (http://www-cryst.bioc.cam.ac.uk/cgibin/homstrad.cgi?family=asp).

This page can be reached using the search facility, either with a keyword (type in 'aspartic'), or the PDB code of a homologue if it is known (type in '5pep'). A quick BLAST search is also available (http://www-cryst.bioc.cam.ac.uk/cgi-bin/homstrad/blast.cgi). Once the aspartic proteinase page has been located, the user can simply click the blue 'ALIGN' icon at the top left corner. This will allow the submission of a user's own sequence and FUGUE will generate the optimal sequence-structure alignment.

6.5. Example

All the steps described above for homology recognition and comparative modeling will be illustrated using a particular example. NDP-4-keto-6-deoxyglucose 3,5-epimerase[18] (EvsA) from *Amycolatopsis orientalis* is involved in the production of NDP-4-*epi*-vancosamine, an L-amino-2,6-dideoxysugar needed in the biosynthesis of the heptapeptide antibiotic chloroeremomycin,[19] which is effective in combating infections from *Staphylococcus aureus.* EvsA epimerizes the positions three and five of the sugar ring (Fig. 6.5). Two molecules of NDP-4-*epi*-vancosamine are attached to the antibiotic heptapeptide backbone.

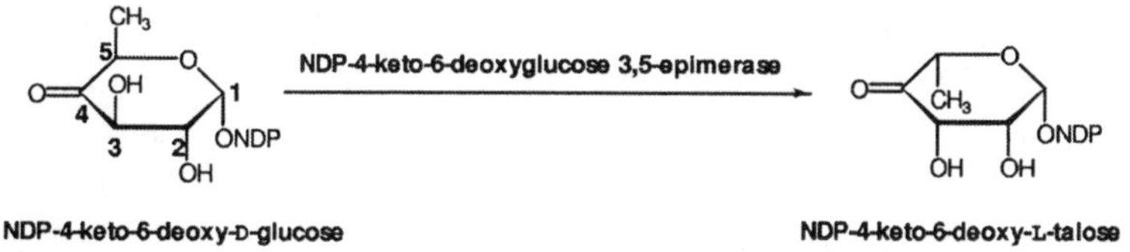

NDP-4-keto-6-deoxyglucose 3,5-epimerase

NDP-4-keto-6-deoxy-D-glucose → NDP-4-keto-6-deoxy-L-talose

Figure 6.5. Chemical reaction catalyzed by EvsA from *Amycolatopsis orientalis.*

6.5.1. Searching for Homologues

The amino acid sequence of EvsA can be obtained from SWISSPROT/TrEMBL (accesion code: O52806) and is 205 residues long. We can easily check if there are close homologues of known structure, using a BLAST[11] search against the PDB. There are numerous web servers such as the one at NCBI (http://www.ncbi.nlm.nih.gov/blast/index.html). For EvsA the BLAST search detected two statistically significant hits. Both are the same enzyme dTDP-6-Deoxy-D-Xylo-4-Hexulose 3,5-Epimerase (RmlC) from different organisms. The first hit is RmlC from *Salmonella typhimurium* (PDB code 1DZR Chain A) with an E-value of 1×10^{-22}, a PID of 33%, an alignment of 175 residues long and with 0% gaps. The second hit is RmlC from *Methanobacterium thermoautotrophicum* (PDB code 1EP0, Chain A) with an E-value of 1×10^{-23}, a PID of 33%, an alignment of 167 residues long and with 2% gaps. The structures of these proteins consist entirely of β strands and the fold is called a double-stranded β-helix. Each turn of the helix is made up of two pairs of anti-parallel strands that are linked with short turns. RmlC catalyzes the same reaction as EvsA (Fig. 6.5), the only difference being the nucleotide moiety of the substrates. It is deoxythymidine diphosphate (dTDP) in the case of RmlC whereas it is unclear in the case of EvsA.

It is not always possible to find close homologues of known structure with high PIDs. If BLAST does not detect any homologous protein in the PDB, it is necessary to perform additional analyses. Even if there are close homologues, as in our present example, it is always a good idea to perform these additional analyses, as they provide more information that can assist the alignment and model building processes.

The first, and probably most useful, piece of information can be derived from a homology search against a bigger sequence database, for example, the NCBI non-redundant database. This can be carried out, again, with the BLAST or PSI-BLAST programs, either running the program locally or using a web server. An advantage of running the program locally is that we can process the output and use various other tools. For example, the BLAST alignment can be converted into a FASTA formatted file using the program blastalign2fasta in the SEALS package.[20] This alignment can be viewed and examined, or sent directly to other programs such as FUGUE.

In the current example of EvsA, our BLAST search against the non-redundant database detected 106 homologues. The closest are the EvsA proteins from: *Streptomyces griseus, Streptomyces peucetius, Streptomyces galilaeus, Streptomyces overmitilis, Streptomyces nogalater and Saccharopolyspora erythraea.* The alignment revealed the following conserved residues: D21, R23, G24, Q48, S51, V58, R60, G61, H63, K73, V75, G80, D84, D88, R90, S93, W99, H120, F122,

Y133, Y139, D151, S167 and D170. Even though no direct structural information is available for any of these homologues, these conserved residues are likely to play important roles, by either stabilizing the structure, or being involved in catalysis.

A second piece of information, universally available, is the secondary structure prediction of the target protein. There are several programs that predict the secondary structure, for example:

- PSI-PRED[21] (http://www.psipred.net),
- PHD[22] (http://www.embl-heidelberg.de/predictprotein/predictprotein.html),
- SSPRED,[23] and
- PREDATOR.[24]

In the current example, the JPRED[25] server (http://jura.ebi.ac.uk:8888/) was used. The consensus JPRED prediction indicates three helices at sequence positions 31-38, 176-181, and 188-196, and 10 strands at positions: 11-15, 26-28, 45-49, 58-64, 73-78, 82-89, 98-104, 110-115, 122-125, and 129-135. Many secondary structure prediction programs, in fact, use the alignment obtained from a BLAST search, thus, these two analyses can be combined and performed at once.

These two analyses, sequence database searching and secondary structure prediction can be effectively combined with more sophisticated fold or homology recognition programs (see URL for LiveBench, for examples). These programs can recognize homologues of known structure, which may not be detected by BLAST or other sequence-only methods. We used FUGUE to search for homologues of EvsA.

Fig. 6.6 shows the prediction result from the FUGUE server using the sequence of EvsA as a query. The output consists of three sections: header, rank and alignment. Here we explain the first two sections in detail and the alignment section will be explained later when we discuss sequence-structure alignment.

The header section gives the version number of FUGUE, the size of the current HOMSTRAD structural profile library, the divergence of homologous sequences collected by PSI-BLAST, the confidence level of different Z-score values, and the explanations of abbreviations used in the rank section. Note that the sequence divergence is used by FUGUE internally to adjust Z-score values and should normally be ignored by the user.

The rank section lists the top 10 hits found by FUGUE, ranked by Z-score. The first column gives links to the corresponding HOMSTRAD

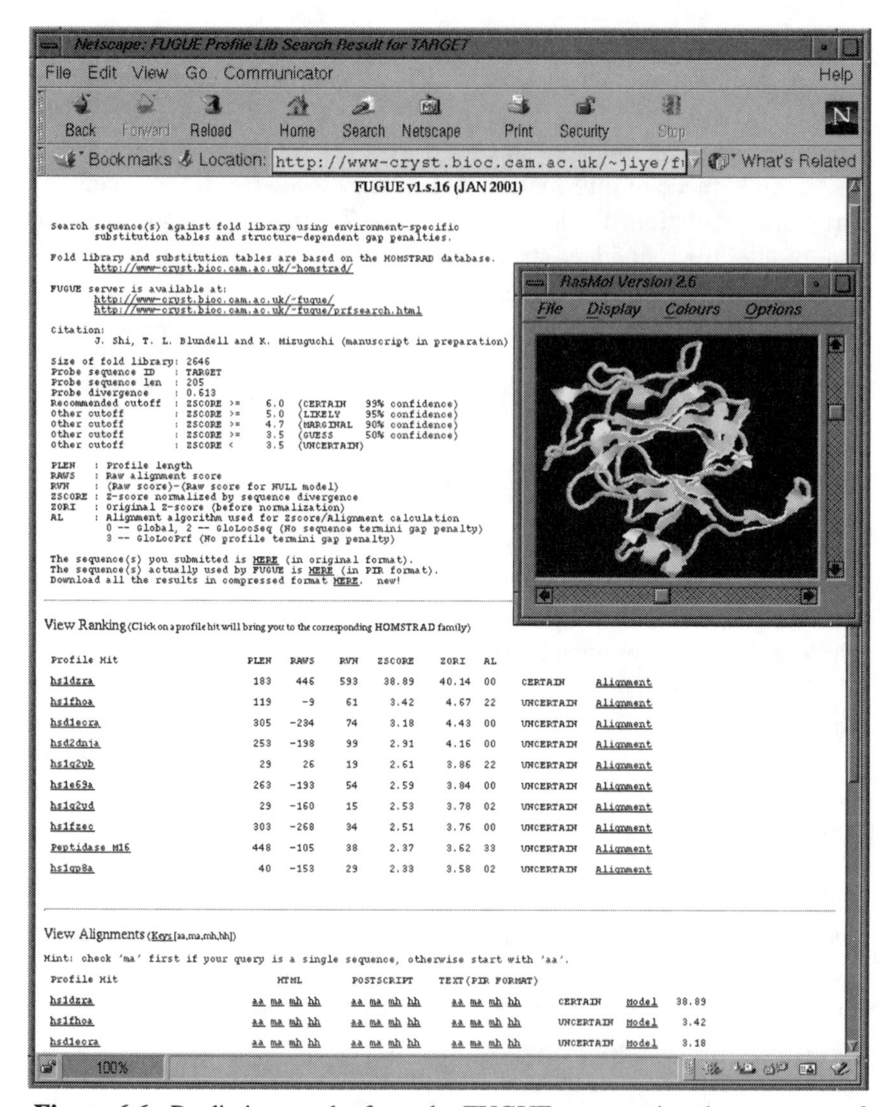

Figure 6.6. Prediction results from the FUGUE server using the sequence of EvsA as query.

family pages of the hits, where more structural/functional information can be obtained. The fifth column is the Z-score. The eighth column translates the Z-score into one of the five more understandable categories of prediction assessment: certain, likely, marginal, guess and uncertain, with decreasing confidence levels.

In our example, FUGUE searched 2646 HOMSTRAD families and predicted that the family hs1dzra (RmlC from *Salmonella typhimurium,* PDB-ID: 1dzr) is the most compatible structure of EvsA with a significant Z-score of 38.89, which corresponds to the confidence level of "certain." The second hit has a Z-score of 3.42, which indicates an "uncertain" prediction. Thus, according to the FUGUE result, we have good confidence on the prediction that the first hit is a homologue of EvsA and it can be used as a structural template for EvsA in comparative modeling.

6.5.2. Alignment

The next step, and one of the most important, is to obtain a good alignment between the query protein(s) and the database protein(s). The FUGUE server produces alignments between the query sequence and entries of the HOMSTRAD databases and annotates the alignments using JOY.[7] The annotated alignments are useful for examining whether particular local environments are compatible with the query sequence and its homologues.

The alignment section in the FUGUE output (see Fig. 6.6) gives alignments between the query sequence and each of the top 10 hits in three formats: HTML for online browsing, PostScript for printing, and pure text format for using with other programs. The HTML and PostScript versions are produced by JOY for annotations of the structure. Four types of alignments are available. The "aa" type is the alignment between the query sequence, together with its homologues collected by PSI-BLAST (all sequences), and all the structures of the corresponding HOMSTRAD family (all structures). This is the most informative type, as all the available sequence and structure information is shown. However, when there are a large number of sequence homologues collected by PSI-BLAST, this type of alignment is difficult to examine by eye. In such situations, the "ma" type, which removes the sequence homologues from the "aa" type (showing only the master sequence), can be used for visual inspection. The "mh" type is the alignment between the query sequence and the single structure, which has the highest PID to the query, in the HOMSTRAD family (master against the structure with the highest PID). It can be directly used as the input for comparative modeling software. The "hh" type represents the most similar sequence-structure pair, in terms of PID, in the "aa" type alignment.

FUGUE also builds rough models (see the "model" column in the alignment section) for the query sequence by using "mh" type alignments of the top 10 hits. For each model, backbone coordinates are copied from the template structure according to the sequence-structure alignment. For the residues in the query sequence, which do not have corresponding residues in the template structure, no coordinates are predicted. The rough models can help the user to assess the confidence of homology recognition and the alignment quality. For example, a model consisting of only short fragments suggests either a poor alignment or a non-homologous sequence-structure pair.

In Fig. 6.6, a rough model built by using the "mh" type alignment of the top hit is shown in Rasmol[26] in the upper-right corner. The model maintains most of the basic structural elements of the template structure, which, together with a significant Z-score, suggests that the FUGUE alignment between EvsA and the HOMSTRAD family hs1dzra be of reasonably good quality.

It is important to confirm that most of the conserved residues in the family alignment of the target are correctly aligned with the template. In our example all conserved residues, previously mentioned, are also conserved in the template, indicating that the FUGUE alignment is reliable. It is also important to check that the active site residues in the template protein(s) are conserved in the target. In our example the active site residues, reported in the literature,[27] in RmlC are: Phe20, Arg24, Phe27, Glu29, Gln48, Asn50, Arg60, and Tyr139. All these residues are conserved in EvsA except for Phe27, Glu29, and Asn50. These three residues in RmlC are involved in binding the nucleotide, suggesting that the nucleotide moiety of the substrate for EvsA may be different.

Another way of checking the conservation of functional residues is to go back to the template entry in HOMSTRAD. Clusters of conserved residues are often observed in the amino acid sequences of proteins with a common function. Such conserved clusters, usually called patterns, motifs or fingerprints, are catalogued in databases such as PROSITE,[28] BLOCKS,[29] PRINTS,[30] and PROF_PAT.[31] HOMSTRAD[4] has incorporated 644 PROSITE patterns, as decorations in the family alignments. Fig. 6.7 shows an example of a HOMSTRAD family "Muconate lactonizing enzyme-like" that has two PROSITE patterns, one of which is shown in the figure. The occurrence of the family pattern(s) in the query sequence is often a good indication that the homologues have been chosen correctly and that the alignment is reasonable.

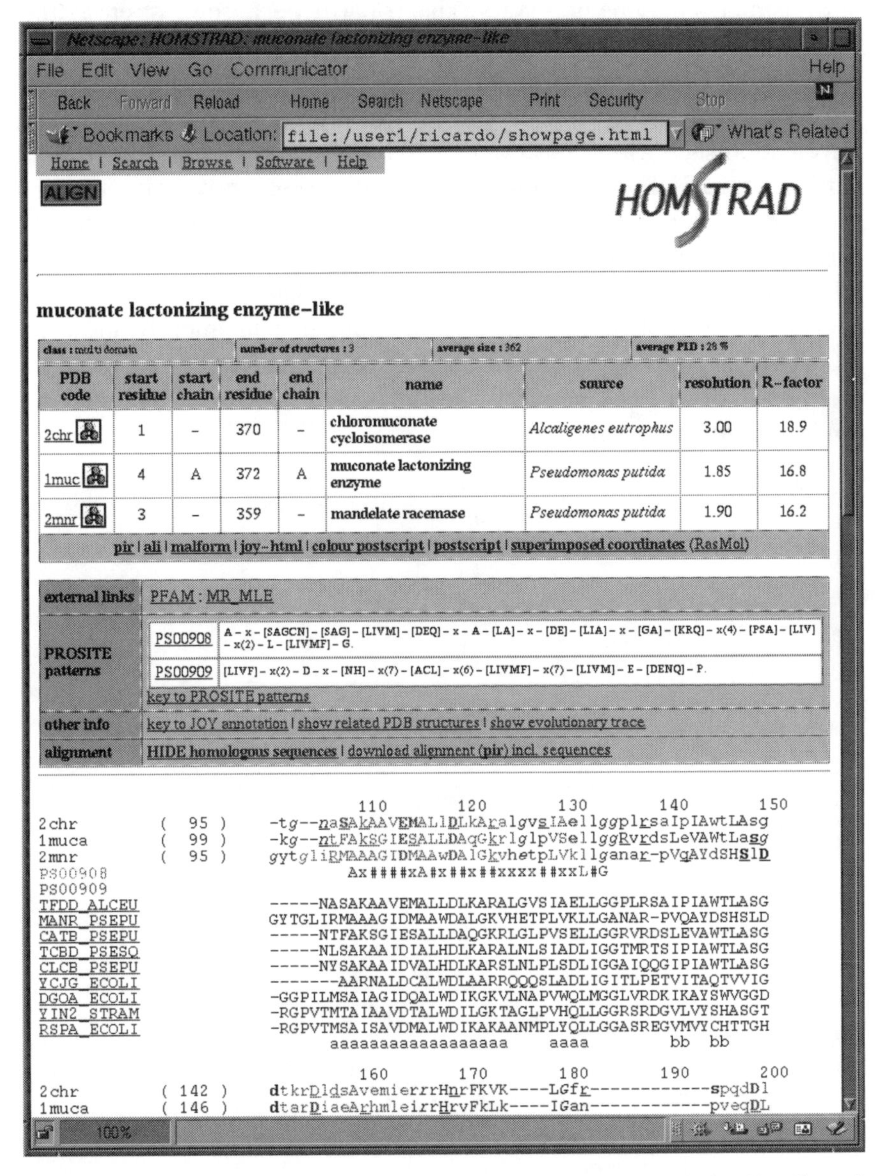

Figure 6.7. Muconate lactonizing enzyme-like HOMSTRAD family aligned with two PROSITE motifs and showing one of them.

Sequence alignment algorithms are dependent on adjustable parameters whose values determine the sequence similarity, placement of gaps and the ultimate alignment. Even using the best alignment program, manual adjustments may be necessary. This is partly because most programs do not utilize all available information. For example, FUGUE uses a PSI-BLAST alignment as input, thus takes into account the conserved residues of the query protein. It does not, however, use any functional information such as the positions of active site residues. In the present version of FUGUE, information about predicted secondary structures is not used. Therefore, it is possible to improve the alignment by altering active site residues or by maximizing the consistency between predicted secondary structures and those in the template structures. Fig. 6.8 shows the FUGUE alignment after a small manual modification, in which the last four residues in RmlC have been moved four positions to the right in order to align the two leucines in both proteins and the glutamic acid to the aspartic acid.

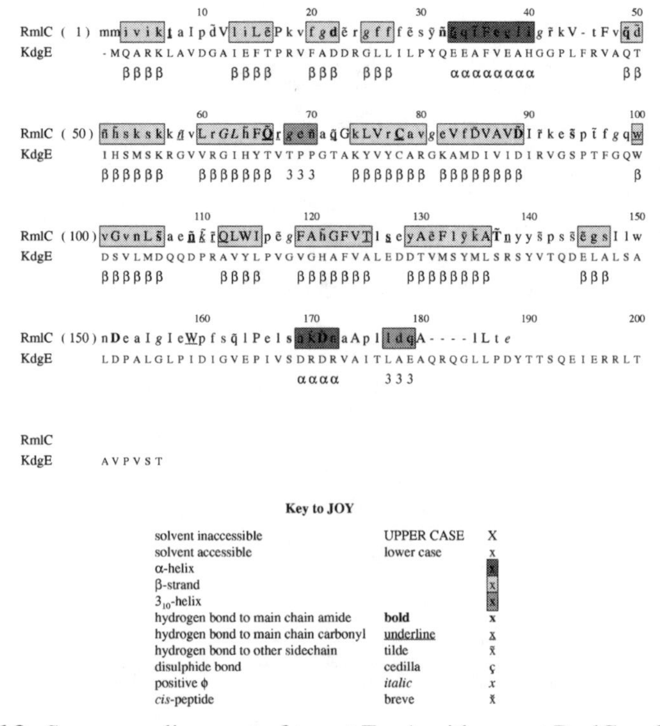

Figure 6.8. Sequence alignment of target EvsA with parent RmlC and formatted by JOY.

6.5.3. Modeling

Once a good alignment is available, a 3D-structural model of the target protein can be built. In order to obtain the model, several programs can be used including COMPOSER[32] and MODELLER.[33] These two programs produce an output file in PDB format containing the 3D coordinates of every non-hydrogen atom including all loops, the N- and C-termini and side chains. Other programs, such as SCORE,[34] produce only the backbone of the conserved parts of the protein. After the backbone, the structurally variable regions, which normally correspond to the loops, can be obtained using Sloop,[35] CODA[36] or the Loop Database of SYBYL.[37] CODA runs two programs for the prediction of the structurally variable regions of protein structures: FREAD, a knowledge-based method using a database of fragments taken from the PDB and PETRA, an *ab initio* method using a database of computer generated conformers. CODA is available on the web at http://www-cryst.bioc.cam.ac.uk/~charlotte/Coda/search_coda.html. CODA is helpful for solving the problem generated by the insertions where there is no template for these residues.

When the backbone is available the side chains may be added. To do this, programs such as SCRWL[38] or CELIAN[39] can be used. The replacement of side chain residues often results in unfavorable interactions such as steric overlaps between atoms. Relaxing these bad side chain contacts requires repositioning side chain atoms while fixing the backbone, to seek a local energy minimum. When that procedure cannot relax a local conformation from a high energy state, the original alignment may require adjustment.

6.5.4. Heteroatoms

In some cases, it is important to obtain protein-cofactor, protein-substrate or protein-cofactor-substrate complexes, especially if the aim of the work is for drug design or to study the reaction mechanism. This is difficult, however, if the template structure does not include the cofactor or substrate. Sometimes in the PDB there are multiple entries of the same protein, with and without the coordinates of the cofactors and/or substrate, and a relevant template should be chosen depending on the purpose of the study. The program MODELLER allows for the building of cofactors/substrates. The program SCORE, however, does not build cofactors/substrates automatically. In this case a coordinate superposition between the model (without cofactors) and the template (with cofactors) can be obtained using programs such as MNYFIT[40] and the coordinates for the cofactors/substrates can be transferred.

The PDB entry 1DZT is a structure of RmlC complexed with the substrate analog 3'-O-acetylthymidine-5'-diphosphate-phenyl ester. Using this entry as the template, we have built, using MODELLER, the model of EvsA complexed with this molecule (Plate 6.1a). Even though the experimental evidence to define the nucleotide moiety of the substrate is inconclusive,[41] we assume that 3'-O-acetylthymidine-5'-diphosphate-phenol is a substrate analog for EvsA. This substrate analog can be modified using programs such as SYBYL and InsightII[42] to obtain the coordinates of the real substrate. Plate 6.1b shows the result after the modification of the substrate analog with InsightII. The phenyl group has been transformed to the sugar molecule.

6.5.5. Refinements

After generating initial backbone and side chain conformations, the entire structure is energy minimized. Since energy minimization only finds a nearby local minimum for a given initial structure, molecular dynamic and Monte Carlo techniques are sometimes used to seek more energetically favorable structures.

There are several methods to evaluate what parts of the model are not well modeled. Apart from visual inspection, PROCHECK,[43] Verify3D[44] or PROSAII[45] can be used. After the energy minimization some residues can still present wrong torsion angles in the Ramachandran plot or negative values in the Verify3D output. In these cases, the templates should be checked to see whether the problem has been carried over from the template structure. If the problematic residues are placed in a loop, the entire loop can be remodeled, with CODA, for example, selecting different loops until all test results become satisfactory.

6.5.6. Model Validation

There are several methods for validating models. Programs such as Verify3D or PROSAII check whether the structure is reasonable from the perspective of sequence-structure compatibility and other programs such as PROCHECK examine the backbone and side chain stereochemistry.

Another good way to test the generated model is by realigning the model to the template structure using the structure alignment program COMPARER,[2] and annotate the alignment with JOY. This allows visual inspection of the conservation of both residues and their structural environments. A model can be validated by docking known substrates or inhibitors to the active site, if this information is known. The results should be consistent with the known specificity for substrates and inhibitors.

Finally, experimental evaluation can be performed. Site-directed mutagenesis on the active-site or binding regions can be carried out. Structurally important residues can be mutated to check if the mutation affects the folding of the protein.

6.5.7. Model

On the basis of these model validation methods, the model is either accepted, rejected, or the alignment is modified and the comparative modeling process repeated. Once the model is accepted, a large amount of useful information can be derived.

Fig. 6.9 shows cartoon representations of the model obtained for EvsA by following the previously explained steps (drawn by the programs MOLSCRIPT[46] and RASTER3D[47]). Although RmlC is known to be a dimer, it is not known whether this is also true for EvsA. We have examined amino acid sequences and structural features in the putative dimeric interface of the model of EvsA and concluded that it is likely to be a dimer. A dimer model has been obtained for EvsA, Fig. 6.9c, in which the same dimeric interactions as in RmlC have been found. In addition residues from both subunits are involved in the active site.

One of the most important pieces of information provided by a model is the knowledge of the active site. The residues that bind the cofactor and/or the substrate by hydrogen bond or salt bridge can be examined with JOY. For EvsA, the residues hydrogen-bonded to the substrate are: Gln A48, Arg A60, Tyr A139 and Arg B23. The last residue belongs to the second chain of the dimeric structure. In Fig. 6.9b all these residues as well as those involved in van der Waals contacts with the substrate are presented.

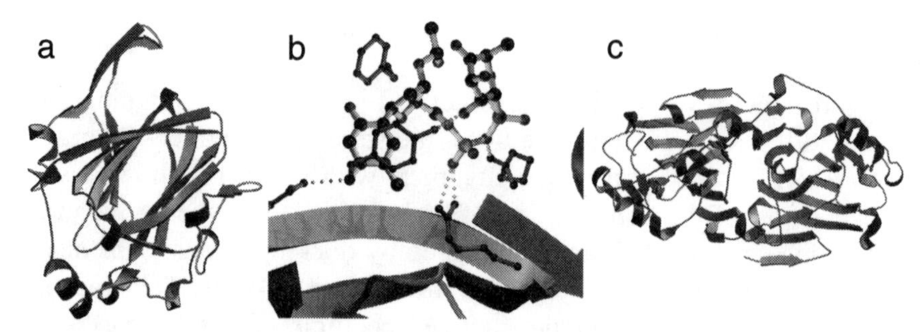

Figure 6.9. (**a**) Cartoon representation of the model of EvsA as a monomer. (**b**) Representation of the active site of EvsA. Protein side chains (dark bonds) and substrate (light bonds) are illustrated. Hydrogen bonds are represented as dashed lines. (**c**) Cartoon representation of the model of EvsA as a dimer.

Fig. 6.10 shows a schematic representation of the interactions of the substrate with the active site residues.

In this example, the main aim for modeling the EvsA enzyme is to study its active site and find some possible mutation targets that could alter the enzymatic reaction. By altering one step of the biosynthesis of the antibiotic chloroeremomycin, *Amycolatopsis orientalis* may produce a new antibiotic, which cannot be recognized by antibiotic resistant bacteria.

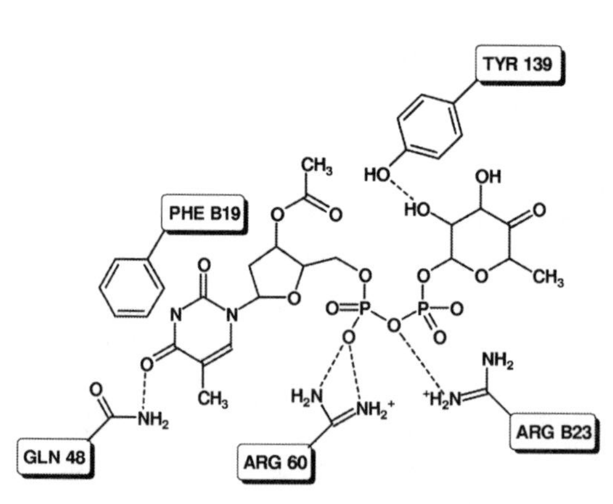

Figure 6.10. Schematic representation of the interactions at the EvsA active site. Hydrogen bonds are represented as dashed lines.

Fig. 6.11 shows part of the active site of EvsA (the figure has been produced by Rasmol[26]). Tyr 139 is hydrogen bonded to the hydroxyl group in the C2 of the sugar, while the hydroxyl group in the C3 has a closer residue, Thr 141, but the distance (4.2 Å) is still too large for hydrogen bond formation. Probably Thr 141 is involved in the isomerization of the C3 carbon of EvsA. If mutations of T141Y and Y139T are made, it is possible that the situation could be reversed.

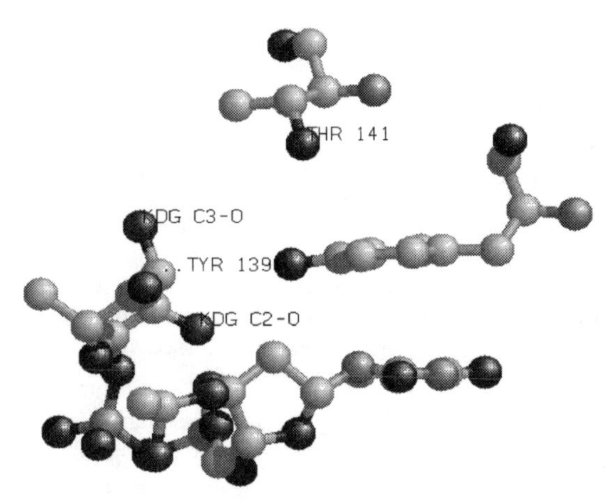

Figure 6.11. Substrate and two residues from the active site of EvsA.

Tyr 141 could be hydrogen bonded to the hydroxyl group in the C3, thus preventing its isomerization, while Thr 139 would be at the correct distance to help the isomerization of the hydroxyl group in the C2. This means that the mutant might catalyze a 2,5-epimerization instead of a 3,5-epimerization.

6.6. Conclusions

We have discussed some of the key issues in predicting the structure and functions of proteins by homology recognition. The new tools described here are particularly useful in improving the identification of homologues and sequence-structure alignments. By exploiting information about protein 3D structure we can obtain a better understanding of the function of many proteins whose sequence are known.

Acknowledgments

We thank Simon Lovell and Lucy Stebbings for reading the manuscript and Oliver W. Choroba for supplying useful information. RNM thanks the *Ministerio Español de Educación y Cultura* for financial support. KM is a Wellcome Trust Research Career Development Fellow.

References

1. Mizuguchi K, Deane CM, Blundell TL, Overington JP: **HOMSTRAD: A database of protein structure alignments for homologous families.** *Protein Sci* 1998, **7**:2469–2471.

2. Sali A, Blundell TL: **Definition of general topological equivalence in protein structures. A procedure involving comparison of properties and relationships through simulated annealing and dynamic programming.** *J Mol Biol* 1990, **212**:403–428.

3. **(a)** Thompson JD, Plewniak F, Poch O: **BAliBASE: A benchmark alignment database for the evaluation of multiple alignment programs.** *Bioinformatics* 1999, **15**:87–88; **(b)** Bahr A, Thompson JD, Thierry JC, Poch O: **BAliBASE (Benchmark Alignment data BASE): Enhancements for repeats, transmembrane sequences and circular permutations.** *Nucleic Acids Res* 2001, **29**:323–326.

4. de Bakker PIW, Bateman A, Burke DF, Miguel RN, Mizuguchi K, Shi J, Shirai H, Blundell TL: **HOMSTRAD: Adding sequence information to structure-based alignments of homologous protein families.** *Bioinformatics* 2001, **17**: 748–749.

5. Guda C, Scheeff ED, Bourne PE, Shindyalov IN: **A new algorithm for the alignment of multiple proteinstructures using Monte Carlo optimization.** *Pac Symp Biocomput* 2001:275–286.

6. Overington JP, Donnelly D, Johnson MS, Sali A, Blundell TL: **Environment specific amino acid substitution tables: Tertiary templates and prediction of protein folds.** *Protein Sci* 1991, **1**:216–226.

7. Mizuguchi K, Deane CM, Blundell TL, Johnson MS, Overington JP: **JOY: Protein sequence-structure representation and analysis.** *Bioinformatics* 1998, **14**:617–623.

8. Burke DF, Deane CM, Nagarajaram HA, Campillo N, Martin-Martinez M, Mendes J, Molina F, Perry J, Reddy BVB, Soares CM, Steward RE, Williams M, Carrondo MA, Blundell TL, Mizuguchi K: **An iterative structure-assisted approach to sequence alignment and comparative modelling.** *Proteins* 1999, **Suppl 3**:55–60.

9. Parker JS, Mizuguchi K, Gay NJ: A family of proteins related to **Spaetzle, the Toll receptor ligand, are encoded in the Drosophila genome.** *Proteins* 2001, **45**:71–80.

10. Shi J, Blundell TL, Mizuguchi K: **FUGUE: Sequence-structure homology recognition using environment-specific substitution tables and structure-dependent gap penalties.** *J Mol Biol* 2001, **310**:243–257.

11. Altschul SF, Madden TL, Schaffer AA, Zhang J, Zhang Z, Miller W, Lipman DJ: **Gapped BLAST and PSI-BLAST: A new generation of protein database search programs.** *Nucleic Acids Res* 1997, **25**:3389–3402.

12. Lindahl E, Elofsson A: **Identification of related proteins on family, superfamily and fold level.** *J Mol Biol* 2000, **295**:613–625.

13. Murzin AG, Brenner SE, Hubbard T, Chothia C: **SCOP: A structural classification of proteins database for the investigation of sequences and structures.** *J Mol. Biol* 1995, **247**:536–540.

14. Williams MG, Shirai H, Shi J, Nagendra HG, Mueller J, Mizuguchi K, Miguel RN, Lovell SC, Innis CA, Deane CM, Chen L, Campillo N, Burke DF, Blundell TL, de Bakker PIW: **Sequence-structure homology recognition by iterative alignment refinement and comparative modelling.** *Proteins* 2001, **Suppl 5**:92–97.

15. Thompson JD, Higgins DG, Gibson TJ: **CLUSTALW: Improving the sensitivity of progressive multiple sequence alignment through sequence weighting, position-specific gap penalties and weight matrix choice.** *Nucleic Acids Res* 1994, **22**:4673–4680.

16. Fischer D, Elofsson A, Rice D, Eisenberg D: **Assessing the performance of fold recognition methods by means of a comprehensive benchmark.** *Pac Symp Biocomput* 1996:300-318.

17. Jones DT: **GenTHREADER: An efficient and reliable protein fold recognition method for genomic sequences.** *J Mol Biol* 1999, **287**:797–815.

18. Wageningen AMA, Kirkpatrick PN, Williams DH, Harris BR, Kershaw JK, Lennard NL, Jones M, Jones SJM, Solenberg PJ: **Sequencing and analysis of genes involved in the biosynthesis of a vancomycin group antibiotic.** *Chem Biol* 1998, **5**:155–162.

19. Williams DH, Bardsley B: **The vancomycin group of antibiotics and the fight against resistant bacteria.** *Angew Chem Int Ed* 1999, **38**:1172–1193.

20. Walker DR, Koonin EV: **SEALS: A System for easy analysis of lots of sequences.** *ISMB* 1997, **5**:333–339.

21. Jones DT: **Protein secondary structure prediction based on position-specific scoring matrices.** *J Mol Biol* 1999, **292**:195–202.

22. Rost B, Sander C: **Combining evolutionary information and neural networks to predict protein secondary structure.** *Proteins* 1994, **19**:55–72.

23. Mehta P, Heringa J, Argos P: **A simple and fast approach to prediction of protein secondary structure from multiple aligned sequences with accuracy above 70%.** *Protein Sci* 1995, **4**:2517–2525.

24. Frishman D, Argos P: **Knowledge-based secondary structure assignment.** *Proteins* 1995, **23**:566–579.

25. Cuff JA, Clamp ME, Siddiqui AS, Finlay M, Barton GJ: **Jpred: A consensus secondary structure prediction server.** *Bioinformatics* 1998, **14**:892–893.

26. Sayle RA, Milner-White EJ: **RasMol—Biomolecular graphics for all.** *Trends Biochem Sci* 1995, **20**:374–376.

27. Giraud M-F, Leonard GA, Field RA, Berlind C, Naismith JH: **RmlC, the third enzyme of dTDP-L—rhamnose pathway, is a new class of epimerase.** *Nature Struct Biol* 2000, **7**:398–402.

28. (a) Bairoch A: **The PROSITE dictionary of sites and patterns in proteins, its current status.** *Nucleic Acids Res* 1993, **21**:3097–3103; (b) Hofmann K, Bucher P, Falquet L, Bairoch A: **The PROSITE database, its status in 1999.** *Nucleic Acids Res* 1999, **27**:215–219.

29. **(a)** Henikoff S,Henikoff JG: **Automated assembly of proteins block for database searching.** *Nucl Acids Res* 1991, **19**:6565–6572; **(b)** Henikoff JG, Henikoff S, Pietrokovski S: **New features of the Vlocks Database servers.** *Nucleic Acids Res* 1999, **27**:226–228.

30. Attwood TK, Flower DR, Lewis AP, Mabey JE, Morgan SR, Scordis P, Selley JN, Wright W: **PRINTS prepares for the new millennium.** *Nucleic Acids Res* 1999, **27**: 220–225.

31. Bachinsky AG, Frolov AS, Naumochkin AN, Nizolenko LPh, Yarigin AA: **PROF_PAT 1.3: Updated database of patterns used to detect local similarities.** *Bioinformatics* 2000, **16**:358–366.

32. Srinivasan BN, Blundell TL: **An evaluation of the performance of an automated procedure for comparative modellingof protein tertiary structure.** *Protein Eng* 1993, **6**:501–512.

33. Sali A, Blundell TL: **Comparative protein modelling by satisfaction of spatial restraints.** *J Mol Biol* 1993, **234**:779–815.

34. Deane CM, Kaas Q, Blundell TL: **SCORE: Predicting the core of protein models.** *Bioinformatics* 2001, **17**:541–550.

35. **(a)** Donate LE, Rufino SD, Canard LHJ, Blundell TL: **Conformational analysis and clustering of short and medium size loops connecting regular secondary structures. A database for modelling and prediction.** *Protein Sci* 1996, **5**:2600–2616; **(b)** Burke DF, Dean CM, Blundell TL: **A browsable and searchable web interface to the database of structurally based classification of loops—Sloop.** *Bioinformatics* 2000, **16**:513–519.

36. Deane CM, Blundell TL: **CODA: A combined algorithm for predicting the structurally variable regions of protein models.** *Protein Sci* 2001, **10**:599–612.

37. *SYBYL.* Tripos, Inc., St. Louis, Missouri.

38. Bower MJ, Cohen FE, Dunbrack RLJr: **Prediction of protein side-chain rotamers from a backbone-dependent rotamer library: A new homology modeling tool.** *J Mol Biol* 1997, **267**:1268–1282.

39. Chen L. *CELIAN*: *Personal Communication.*

40. Sutcliffe MJ, Haneef I, Carney D, Blundell TL: **Knowledge based modelling of homologous proteins, part I: Three-dimensional frameworks derived from the simultaneous superposition of multiple structures.** *Protein Eng* 1987, **1**:377–384.

41. **(a)** Kirkpatrick PN, Scaife W, Hallis TM, Liu H, Spencer JB, Williams DH: **Characterization of a sugar epimerase enzyme involved in the biosynthesis of a vancomycin-group antibiotic.** *Chem Commun* 2000, 1565–1566; (b) Chen H,

Thomas MG, Hubbard BK, Losey HC, Walsh CT, Burkart MD: **Deoxysugars in glycopeptide antibiotics: Enzymatic synthesis of TDP-L-epivancosamine in chloroeremomycin biosynthesis.** Proc Natl Acad Sci USA 2000, **97**:11942–11947.

42. *InsightII.* Accelrys, Ins (former MSI), San Diego, California.

43. Laskowski RA, MacArthur MW, Moss DS, Thornton JM: **Procheck—A program to check the stereochemical quality of protein structures.** *J Appl Cryst* 1993, **26**:283–291.

44. Luthy R, Bowie JU, Eisenberg D: **Assessment of protein models with 3-dimensional profiles.** *Nature* 1992, **356**:83–85.

45. Sippl MJ: **Recognition of errors in the three-dimensional structure of proteins.** *Proteins* 1993, **17**:355–362.

46. Kraulis PJ: **MOLSCRIPT—A Program to produce both detailed and schematic plots of protein.** *J Appl Cryst* 1991, **24**:946–950.

47. Merritt EA, Murph M: **RASTER3D version-2.0—A program for photorealistic molecular graphics.** *Acta Cryst D* 1994, **50**:869–873.

48. Nicholls A, Sharp K, Honig B: **Protein folding and association: Insights from the interfacial and thermodynamic properties of hydrocarbons.** *Proteins* 1991, **11**:281–296.

7

Fully Automated Protein Tertiary Structure Prediction Using Fourier Transform Spectral Methods

Carlos Adriel Del Carpio Muñoz
and Atsushi Yoshimori

Abstract

Divergence in sequence through evolution precludes sequence alignment based homology methodologies for protein folding prediction from detecting structural and folding similarities for distantly related proteins. Homolog coverage of actual data bases is also a factor playing a critical role in the performance of those methodologies, the factor being conspicuously apparent in what is called the twilight zone of sequence homology in which proteins of high degree of similarity in both biological function and structure are found but for which the amino acid sequence homology ranges from about 20% to less than 30%. In contrast to these methodologies a strategy is proposed here based on a different concept of sequence homology. This concept is derived from a periodicity analysis of the physicochemical properties of the residues constituting proteins primary structures. The analysis is performed using a front-end processing technique in automatic speech recognition by means of which the cepstrum (measure of the periodic wiggliness of a frequency response) is computed that leads to a spectral envelope that depicts the subtle periodicity in physicochemical characteristics of the sequence. Homology in sequences is then derived by alignment of the spectral envelopes. Proteins sharing common folding patterns and biological function but low sequence homology can then be detected by the similarity in spectral dimension. The methodology applied to protein folding recognition underscores in many cases other methodologies in the twilight zone.

7

Fully Automated Protein Tertiary Structure Prediction Using Fourier Transform Spectral Methods

Carlos Adriel Del Carpio Muñoz and Atsushi Yoshimori

Toyohashi University of Technology, Tempaku, Toyohashi, Japan

7.1. Sequence Alignment and Protein Structure Modeling

Many attempts have been made to elucidate the principles ruling the coding of protein function into its sequence of amino acids. Since the function of protein and its structure are intimately related the rules leading to the formation of specific high level structures in the biopolymer have to govern also the creation of specific function-related regions of the sturucture. Typical approaches to finding this type of rule deal especially with the characterization of the homology of several proteins both in function and amino acid composition. The concept underlying these methodologies is that proteins with similar structural characteristics and biological function share similarities in amino acid composition. This is a natural consequence of molecular evolution, which has resulted in a dense network of specific inter protein interaction relationships that has provided biologists with a practical method to infer protein shape or biological function based on structure or function of its relatives. Accordingly a vast number of methodologies have been devised to extract protein similarities at the primary structure level and predict secondary and higher level structural features for amino acid sequences. These methodologies lead to assignment of tendencies in secondary structure, and eventually tertiary structure, to segments of the amino acid sequence constituting a determined protein as well as the extraction of structural motifs, domains, or regions involved in expression of the biological function for the bio-macromolecule.

Homology in protein sequences is retrieved by alignment of the sequences of amino acids. The method most frequently used for this purpose is based on the dynamic programming algorithm,[1] which allows not only two sequence alignments but also multi-sequential alignments. Other methodologies are based on genetic algorithms,[2-8] simulated annealing, neural networks,[9,10] etc.

Comparative modeling of protein structure based on homology search consists in selecting putative homologs by alignment of structures in a data base of sequences using several strategies based on sequence information, available functional information, and different types of mutation matrices. Regions of high confidence alignment—result of a consensus of all the used strategies—are then taken as templates on which insertions, gaps and other low confidence fragments are mapped. Combination of database search and other molecular mechanics, molecular dynamics, and *ab initio* strategies result in improved performance of the modeling process.[11-13]

Incompleteness of homolog coverage resulting from both data base insufficiency or the failure to recognize many proteins that are known to be homologous—result of the divergence in sequence through evolution—by current alignment programs play a critical role in the potential of this type of methodologies. Several other strategies, characterized by a higher sensitivity in sequence alignment have been developed in order to detect distant homology in protein sequence and thus overcome this problem. Algorithmic procedures have been devised based on comparison of mutation patterns expressed as profiles of allowed mutations at various positions along the sequence; the allowed mutation patterns being derived from substitution patterns in highly homologous sequences. Gribskov *et al.*, reported on a program to compare single sequences to families of profiles.[14] The position specific iterated (PSI) BLAST (Atschul *et al.*)[15] and Hidden Markov Models (HMMs) (Eddy S. R.,[16,17] Karplus *et al.*[18], and Krogh *et al.*[19]) constitute sensitive alignment techniques together with BLOCKS of Henikoff *et al.*,[20] and FFAS of Rychlewski *et al.*[21] which are oriented to compare two profiles to each other. A further approach, often called threading, is to use the structure of one of the proteins being compared to assess whether the second could possibly have similar structure.[22] Another technique that explores sequence homology among distant proteins without constructing the explicit multiple sequence alignment or profile is the Intermediate Sequence Search (ISS) developed by Karplus *et al.*,[18] Park *et al.*,[23] and Salamov *et al.*[24] This alignment strategy makes use of the transitive nature of homology relationships among proteins. The strategy can be described in terms of proteins A, B, and C, where if A is homologous to both B and C, then a homology relationship can be established among B and C. Construction of the SCOP database[25] took advantage of this technique to recognize remote evolutionarily related sequence pairs, an increment of 70% in the detection of related proteins being reported by the authors as compared to FASTA[26] while error-yield rate remains unchanged. Salamov *et al.*[24] improved the methodology further by extending the number of intermediates to more than one in a scheme they called Multiple Intermediates Sequence Search (MISS). The strategy behind MISS consists in using connections of the type A-I_1-I_2-I_n-B, where $I_1,I_2,\ldots,I_n$ are intermediates in the similarity search path relating A and B. They show the potentiality of the method as compared with the simple ISS and other methodologies since many distantly homologous proteins missed by those methodologies can be identified by MISS. ISS and MISS do not rely on sometimes misleading and ambiguous multiple alignments therefore the information provided by these two methodologies is superior in terms of evolutionary relationships than other homology search strategies. The latter comprehend methodologies based on Hidden Markov Models (HMM) and similar profile comparison oriented ones. A drawback of the MISS strategy is, however, the difficultness involved in its fully automated implementation since results obtained by the methodology often contain bogus due to the intrinsic diverging tendency of the procedure. To deal with this inherent instability of the MISS, Li *et al.*[26] introduce a series of filters in the intermediate searches to avoid spurious seeds that make the method diverge. Using the BLAST and PSI-BLAST program families and combining them with the filtering procedure they confine MISS searches into single converging directions, a strategy they call Saturated BLAST.

Nevertheless, problems still remain with alignment techniques especially those associated with the length of the sequence of amino acids, the unambiguous differentiation of domains within the sequence, the number of gaps to increase the number of matching residues and other factors the determination of which plays a critical role in the relevance of the information obtained. Furthermore, there is still a tendency in all the methods mentioned above towards establishing rather high threshold values for homology searches as well as assign arbitrary gap-weights to the alignment processes. This fact is in clear incongruity with the widely accepted principle in molecular biology that structure and not sequence composition is conserved through evolution; the latter giving rise to the so called twilight zone of structure and sequence relationship. This is the zone in which proteins of high degree of similarity in both biological function and structure are found but for which the amino acid sequence homology ranges from about 20% to less than 30%.[27]

7.2. Protein Function and Structure Elucidation by Spectral Analysis

In contrast to the previously mentioned methodologies based on sequence alignment and profile comparison a strategy based on periodicity analysis of the physicochemical properties of the amino acids constituting the sequence has been developed recently,[28-31] the aim of which is the identification of the fold for a determined sequence of residues.

Periodicities in the polar and/or apolar character of the amino acid sequence of a protein has been examined by several types of techniques aiming at the elucidation of protein structure and biofunction; the rationale behind them being the fact that evaluation of proteins of known three dimensional structure reveals a strong tendency towards periodicity in physicochemical properties of the constituting residues leading to the adoption of specific secondary and tertiary structures as well as biological function. From the structural point of view, for instance, a strong periodicity in the hydrophobicity of the sequence is found when the structure is folded to form α-helices; the period of the sequence in average being about 3.6 residues. A similar periodicity is found in the portions of sequence that fold into strands of β-sheets, the typical period in hydrophobic residues being 2.3 in this case.[32] In general, these observations suggest that periodicity in the hydrophobicity of a protein primary structure is a factor in the formation of its secondary structure. Similarly it has also been observed that many protein sequences tend to form segments of maximum amphiphilicity[32] suggesting that segments of secondary structure fold at hydrophobic surfaces resulting from cooperative folding of other parts of the protein. Accordingly the amphiphilicity of protein segments plays a major role in the structure adopted by that particular segment within the folding of the protein.[32] A well studied measure of amphiphilicity and, thus, hydrophobic periodicity in primary protein structures is the hydrophobic moment,[32] which is actually the modulus of the Fourier transform of the one-dimensional hydrophobicity function expressed by:

$$\mu = \left\{ \left[\sum_{n=1}^{N} H_n \sin(\delta n) \right]^2 + \left[\sum_{n=1}^{N} H_n \cos(\delta n) \right]^2 \right\}^{1/2} , \qquad (7.1)$$

where H_n is the numerical hydrophobicity of the nth residue and $\mu(\delta)$, the hydrophobic moment, is the component of the periodicity having the frequency δ, in other words, the angle expressed in radians at which successive side chains emerge from the backbone, when the periodic segment is viewed down its axis (for α-helices $\delta = 100°$and for β-sheets $\delta = 160{\sim}180°$).[32]

Periodicity and spectral analyses have also been applied to predict hydrophobic cores of proteins,[33] domains of transmembrane proteins.[34] Similarly a study by Mandell *et al.*[35] based on wavelet analysis revealed that imaginary parts of the coefficients discriminated among proteins; the tertiary structure of which were essentially dominated by helices, sheets and structural combinations of both.

Biological function has also been related to periodicity of physicochemical parameters of the residues constituting the primary structure of proteins; and several methodologies based on spectral analysis have been devised to extract this type of information. The work of Cosic *et al.*,[28-31] which is summarized in their Resonant Recognition Model (RRM), constructed on the hypothesis that certain periodicities exist within the distribution of energies of delocalized electrons along a protein molecule which are critical for the expression of their biological function.

Firstly using a Fourier transform, numerical series representing the electron-ion interaction potential values for each residue constituting a protein primary structure are transformed to the frequency domain in the RRM. Assuming equidistant positions for the amino acids in the numerical series, a maximum frequency of 0.5 for the spectrum is established making the number of points exert influence only on the resolution of the spectrum. This makes that for a sequence of N amino acids the resolution of the spectra be $1/N$, while the nth point in the spectral function corresponds to the frequency $f = n/N$.

To extract common characteristics from this type of spectra for two proteins known to have similar biological function a cross-spectral function is defined which is composed of real and complex conjugate values of the spectra of both proteins respectively.

The method is extended to the analysis of groups of proteins sharing bio-function similarities for which a consensus spectrum is obtained from the analysis of the multiple pair wise cross-spectral functions.

Particular segments of primary structure which are most affected by the change of amplitude of the frequency in the spectral analysis are defined as Hot Spots, and they have shown that these segments or hot spots are directly related to regions that are involved in bio-function expression (binding regions, clefts, etc.).[28-31]

The methodology was further improved by introducing wavelet transforms to recognize and characterize protein structural and functional properties.[31]

Accordingly from this overview it can be concluded that functional and structural properties of proteins can be related to special functions derived from the spectral analysis of the distribution of physical and physicochemical parameters of the amino acids constituting the primary structures of a protein.

7.3. Spectral Analysis and Folding Pattern Recognition

Del Carpio and Yoshimori[36,37] have developed a new methodology that relates this intrinsic, though, subtle periodicity of the physicochemical characteristics of the amino acids constituting the primary sequence of a protein with the folding characteristics of its three dimensional structure. They exploited the classification of several protein primary structures into families and super-families in the SCOP database[25] and carried out a systematic analysis of the periodicities characterizing each folding pattern in every group of sequences in the database together with the physicochemical factors on which these periodicities were the most remarkable. The hypothesis underlying the methodology was the assumption that each class, family and super-family of proteins were characterized by some type of periodicity over a particular set of physicochemical parameters of the constituting residues. This characteristic set of parameters is called the set of dominant coefficients and is constituted by a number of features that are selected through a statistical procedure among several hundreds of reported amino acid physicochemical parameters.

The main feature that characterizes the methodology is the spectral analysis—the decomposition of a signal into its frequency components—using a series of Fourier direct and inverse transforms to express any underlying periodicity that is not apparent in the simple profile of physicochemical parameters of a sequence. This analysis leads to a spectral description of the primary structures by means of which similar 3D structures can be detected when even the homology in composition is very small. Accordingly a methodology has been adopted and improved to align primary structures of proteins based on alignment of the spectra representing the structures in the frequency domain.

The methodology has proved efficient especially at recognizing folding characteristics of protein structures in the twilight zone,[27] that is, similarities in tertiary structure of proteins with less than 25% of homology in primary structure. Using the sets of dominant coefficients for each type of folding, automatic assignment of an unknown sequence to its putative class, family, and superfamily is performed with high probability. This constitutes the basis of a new fully automated system for protein folding recognition, and a detailed description of the methodology as well as the results obtained is given in what follows.[37]

7.3.1 Spectral Representation of Protein Primary Structures

Digital spectral analysis—decomposition of a signal into its frequency components—is a very useful technique in many engineering fields, applied sciences and data processing. The conventional algorithm to perform this analysis is a Fourier transform (FT); the assumption underlying a FT analysis being that a frequency-domain description of a signal reveals important information unperceivable in the time domain signal.

One major problem in a mere FT when used for this task is, however, the shape of the profile in the frequency domain which contains abrupt changes that overwhelm subtle details of an underlying and more general characteristic function of periodicity of the physicochemical parameter sequence. Consequently, the authors adopted a well known technique of front-end processing in robust automatic speech recognition (ASR), the objective of which is to preserve critical linguistic information while suppressing irrelevant information such as speaker-specific characteristics, channel characteristics, and noise.[38] This analysis-synthesis technique is based on the transformation of a signal into its *cepstrum* which is a measure of the periodic wiggliness of a frequency response plot. This is a *homomorphic* transformation, and the concept of the cepstrum is a fundamental part of the theory of *homomorphic* systems for processing signals that have been combined by convolution. The cepstrum is calculated as the logarithm of the power spectrum of a signal and leads to a logarithmic periodgram for which the spectral envelope is obtained as a smooth curve depicted by connecting the main local peaks of the minute structure of the frequency spectrum.

The technique applied to the analysis of the sequence of physicochemical parameters representing the primary structure of a protein allows extraction of information in the form of the spectral envelop of the cepstrum which is used to model the relationship between the tertiary structure of a protein and the distribution of physicochemical parameters of the residues in the primary sequence.

The model allows the direct comparison of two or more residue sequences and therefore is oriented to express similarities among amino acid sequences at these two levels of protein structure. The comparison process is reduced to an alignment of spectral envelops representing the primary structures of the proteins to be compared as described later.

The first stage thus consists in the representation of the protein primary structure as a numerical series by assigning to each amino acid a physicochemical parameter value that may be related to inter-residue interaction (hydrophobic, coulomb, polarities, ionization, etc.) or any physical property related to the conformation of the residues within the protein structure (volume, surface, accessibilities, etc.). Amino acids are assumed to be equidistant.[28] This numerical series or profile of physicochemical characteristics is manipulated as a waveform in a way similar to a discrete-time signal (digital signal). Since a digital signal generally represents some underlying analog function, and because of the aperiodic nature of the profile, and in order to extract information that is not possible to obtain in the time domain, the profile is converted to the frequency domain by applying to it a Fourier transform; the result being a discrete Fourier spectrum.

If the physicochemical profile for a sequence of N amino acids is represented by x_n, the discrete Fourier transform (DFT) X_k is computed by:

$$X_k = \sum_{n=0}^{N-1} x_n \exp\left(-j\frac{2\pi kn}{N}\right), \quad (0 \le k \le N-1). \tag{7.2}$$

Fig. 7.1 illustrates the profile of physicochemical characteristics for the sequence:

"TGVTLFVALYDYEARTEDDLSFHKGEKFQILNSSEGDWWEARSLTTGETGYIPSNYVAPVDSIQAEE"

of PDB code 1nyf for which the Eisenberg hydrophobic[39] index is used.

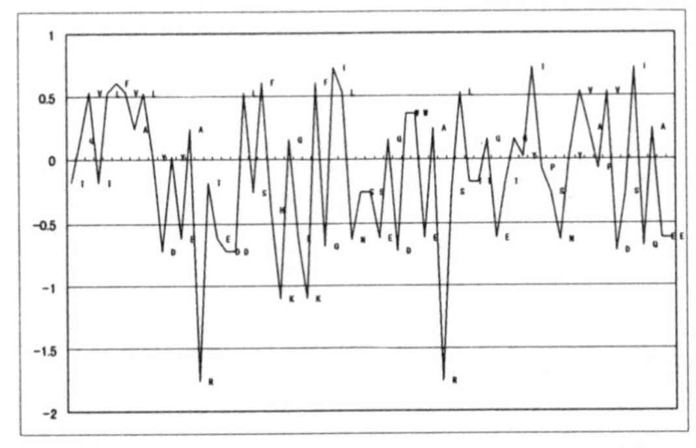

Figure 7.1. Profile of Eisenberg Hydrophobic Indices for the primary structure of 1NYF.

Applying the FFT to the profile of physicochemical properties leads to a set of spectral coefficients or harmonics that can be regarded as samples of the underlying ideal continuous spectral function. Each component occupies a definite harmonic frequency, corresponding to a single spectral coefficient as illustrated in Figs. 7.2 and 7.3 for the real and imaginary parts of the transform.

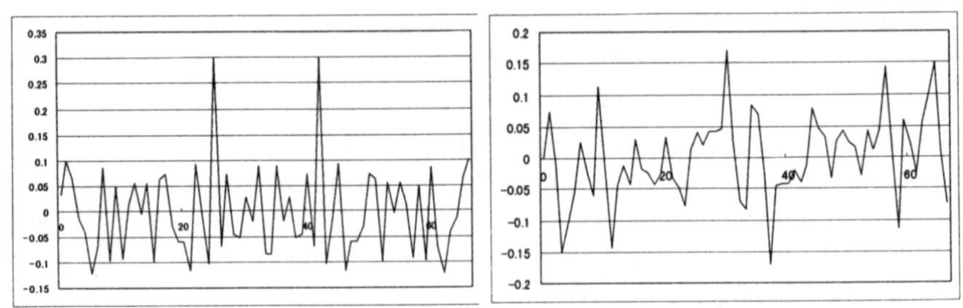

Figure 7.2. Real Part of the FT for the profile of characteristics in Fig. 7.1.

Figure 7.3. Imaginary Part of the FT for the profile of characteristics in Fig. 7.1.

Considering all these facts and from Figs. 7.2 and 7.3 it is evident that an important component of the fine structure of the transform is expressed and represented by the spectral envelope. This component or tendency underlying the profile of physicochemical characteristics may express a particular relationship among the primary sequence and the folding patterns of the three dimensional structure.

The process to compute the cepstrum—leading to the spectral envelope—for any spectra is illustrated in the flowchart of Fig. 7.4; each step marked with a number being summarized in Table 7.1.

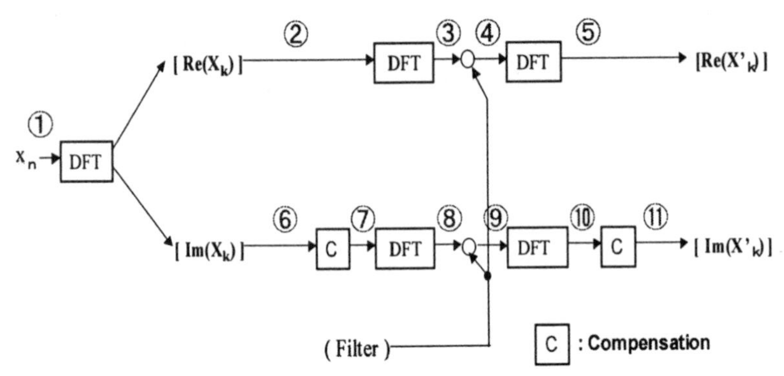

Figure 7.4. Flowchart for the computation of the Spectral Envelope.

Extraction of the spectral envelope is performed for both the real and the imaginary parts of the initial FT of the profile of physicochemical parameters. These two calculations differ in that the imaginary part has point symmetry as compared to the real part, the symmetry of which is linear. Thus a compensation operation is performed with the imaginary part of the spectra. In Table 7.1, steps 1 to 5 describe the steps involved in the computation of the real part of the spectral envelope while steps 6 to 11 describe the procedure to obtain the imaginary part of the envelope.

Table 7.1. Steps in the computation of the Spectral Envelope

Step #	Step Description
1.	Application of the FFT to obtain the DFT of the profile of physico-chemical characteristics x_n.
2.	Application of the inverse discrete Fourier transform (IDFT) to the real part of the spectrum obtained in Equation (7.2), replacing with 0 the coefficients of the imaginary part.
3.	After the IDFT analysis the cepstrum of the real part is taken.
4.	Filtering the cepstrum, i.e., removing peaks below an arbitrary cepstrum dimension.
5.	The real part of the spectral envelope is obtained by performing a DFT analysis of the filtered cepstrum in Step 4.
6.	Extraction of the imaginary part of the FT of the initial profile of physicochemical characteristics.
7.	To overcome the problem of the point symmetry of the imaginary part as compared to the linear symmetry of the real part, a compensation operation is performed. Then perform an inverse FT (IFT) of the spectrum after replacement of the coefficients of the imaginary part of the new complemented spectrum with zero's and).
8.	Extract the cepstrum after the IFT.
9.	Filtering of the IFT by removing peaks below a threshold cepstrum dimension.
10.	Taking a new DFT of the filtered spectrum leads to a spectral envelope of the compensated imaginary part of the initial spectrum.
11.	To restore the point symmetry from the linear symmetry of the spectral envelop of the imaginary part in Step 10, a new compensation operation is performed, and the imaginary part of the spectral envelope is obtained.

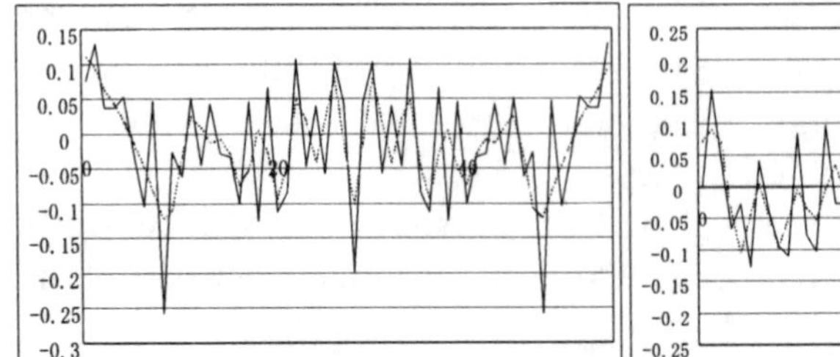

Figure 7.5. Real Part of the spectral envelop for the FT of the profile of characteristics in Fig. 7.1.

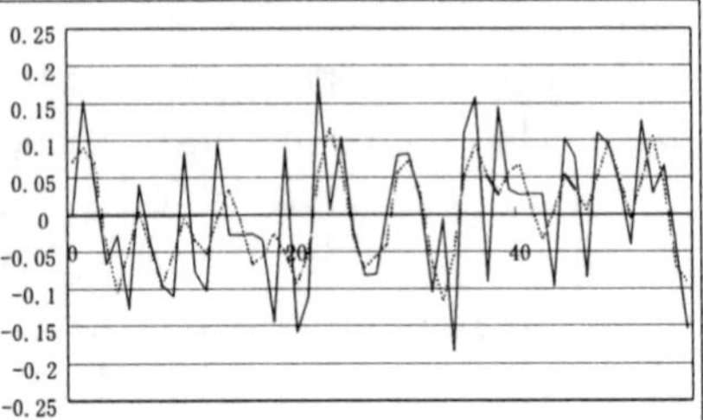

Figure 7.6. Imaginary Part of the spectral envelop for the FT of the profile of characteristics in Fig. 7.1.

Observation of both the real and imaginary parts of the spectral envelopes derived by the automatic procedure reveals their averaging character of the wiggling in the original spectra. Accordingly, the spectral envelope provides a way of expressing the underlying subtle relationship of the physicochemical parameter sequence by which a protein primary structure is represented.

This method of expressing the sequence of amino acids allows its comparison by alignment of the spectral envelops, a metric of similarity being derived straightforwardly as described in the next section.

7.3.2 Spectral Alignment and Protein Structure Similarity

Representation of protein primary structures as the spectral envelopes of the FT of physicochemical feature profiles for the amino acid sequences allows a straightforward analysis of the similarities of two or more sequences that can be reflected in the 3D structural similarity.

Because the final objective of the methodology is the automatic recognition of protein folding, it must be able to deal with sequences of different lengths, yet, possessing similar folding patterns. Accordingly a procedure for flexible spectral matching becomes necessary in order to process and compare any two sequences with different number of amino acids represented by the spectral envelopes. The methodology selected to perform this flexible matching is the alignment of the spectra, in order to find the best spectral pattern fit among the spectra being compared. Since the optimal fit is required, a procedure based on a dynamic programming algorithm (DP) is suitable. DP is an optimization algorithm frequently used in alignment of amino acid sequences by directly comparing the letters representing the residues or values extracted from amino acid relational tables such as mutation matrices.[40]

The idea behind the methodology is that patterns bearing similarity in spectral space, originated by similar segments of residue physicochemical features, are translated in particular folding patterns for the proteins.

Direct application of the DP algorithm would lead, however, to a spectral matching of different characteristics than that applied to a sequence of characters (character pattern match). In the usual DP a penalty is imposed when no match (gap) occurs between sequences and the search can continue in both the vertical and horizontal directions, a number of continuous gaps being allowed without interfering with the search process. This can not be done with spectral matching since it would lead to an unlimited flexibility of the match operation. To avoid this negative effect when using DP for spectral comparison, a gradient is imposed to the search so that the match can continue smoothly in the diagonal direction as shown in Fig. 7.7.

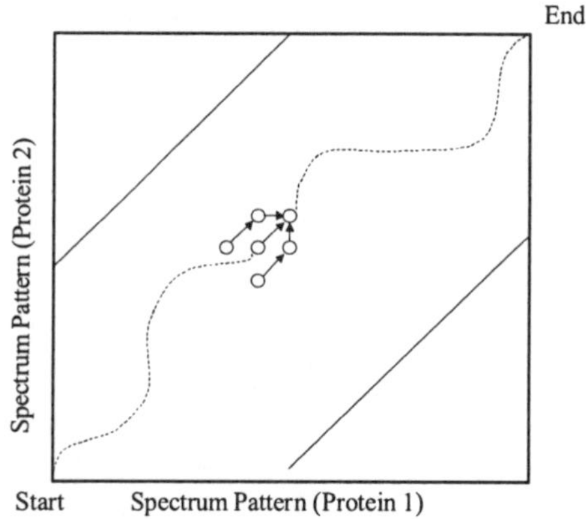

Figure 7.7. DP for spectral comparison.

This equates to reduce the number of gaps in the DP process, since a horizontal and vertical advance is performed and allowed only once, and recurrent gaps are inexistent.

This gradient constrained DP for flexible spectral alignment is expressed by:

$$g(i_k,j_k)=\min\begin{Bmatrix} d(i_k,j_k)+\min\{2d(i_k-1,j_k)+g(i_k-2,j_k-1),2d(i_k,j_k-1)+g(i_k-1,j_k-2)\}, \\ 2d(i_k,j_k)+g(i_k-1,j_k-1) \end{Bmatrix}, \tag{7.3}$$

where $g(i_k,j_k)$ shows the best match at a step.

$d(i_k,j_k)$ is the similarity metric used in the process and is expressed as:

$$dist(i,j)=\left\{\mathrm{Re}\left(X'_{Ai}\right)-\mathrm{Re}\left(X'_{Bj}\right)\right\}^2+\left\{\mathrm{Im}\left(X'_{Ai}\right)-\mathrm{Im}\left(X'_{Bj}\right)\right\}^2 . \tag{7.4}$$

Here X'_{Ai} is the ith harmonic of spectral envelop for spectrum A and X'_{Bj} is the jth harmonic of spectral envelop for spectrum B.

Values of $d(i_k, j_k)$ close to zero stand for closeness in spectral dimension and thus high similarity in physicochemical feature distribution while higher values stand for increased dissimilarity.

7.3.3 Automatic Protein Folding Pattern Recognition

Applying the procedure described above to the automatic folding recognition process consists in expressing the primary sequences of proteins and follow the steps illustrated in the flow chart of Fig. 7.8 for any two arbitrary proteins A and B.

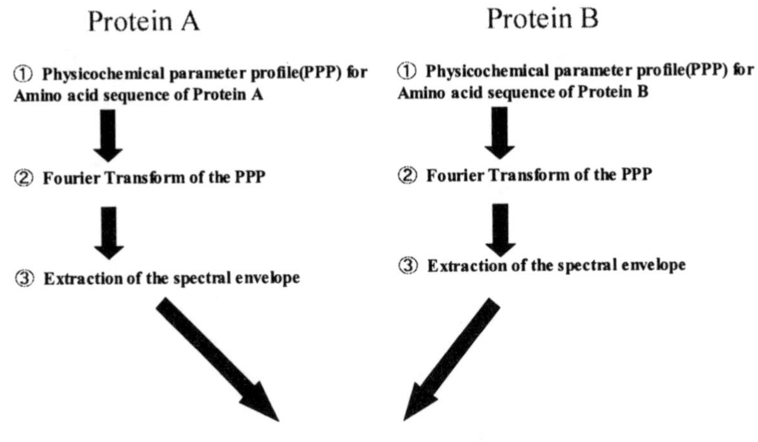

Figure 7.8. Flowchart for Evaluation of Protein Similarity using Spectral Analysis.

The results at each step of procedure in Fig. 7.8 can be illustrated comparing spectral characteristics for target T099 in CASP4,[41] with that of protein 1nyf which is found to be the protein with the highest homology in the PDB. Results obtained using the methodology described here can be then compared with those obtained using the conventional alignment procedures.

Accordingly, in the first place using the Eisenberg hydrophobic coefficients, the PPP for both sequences, are shown in Fig. 7.9.1a, and 7.9.1b respectively. Applying the Fourier transform leads to the spectra for both sequences the real and imaginary parts of which are shown in Fig. 7.9.2a and 7.9.2b for T099 and Fig. 7.9.2c and 7.9.2d for NYF respectively. Spectral envelopes are computed applying the procedure in Fig. 7.4 and are illustrated together with the FT for the sequences of T099 and NYF in Fig. 7.9.2a and 7.9.2b for the first sequence and 7.9.2c and 7.9.3d for the latter.

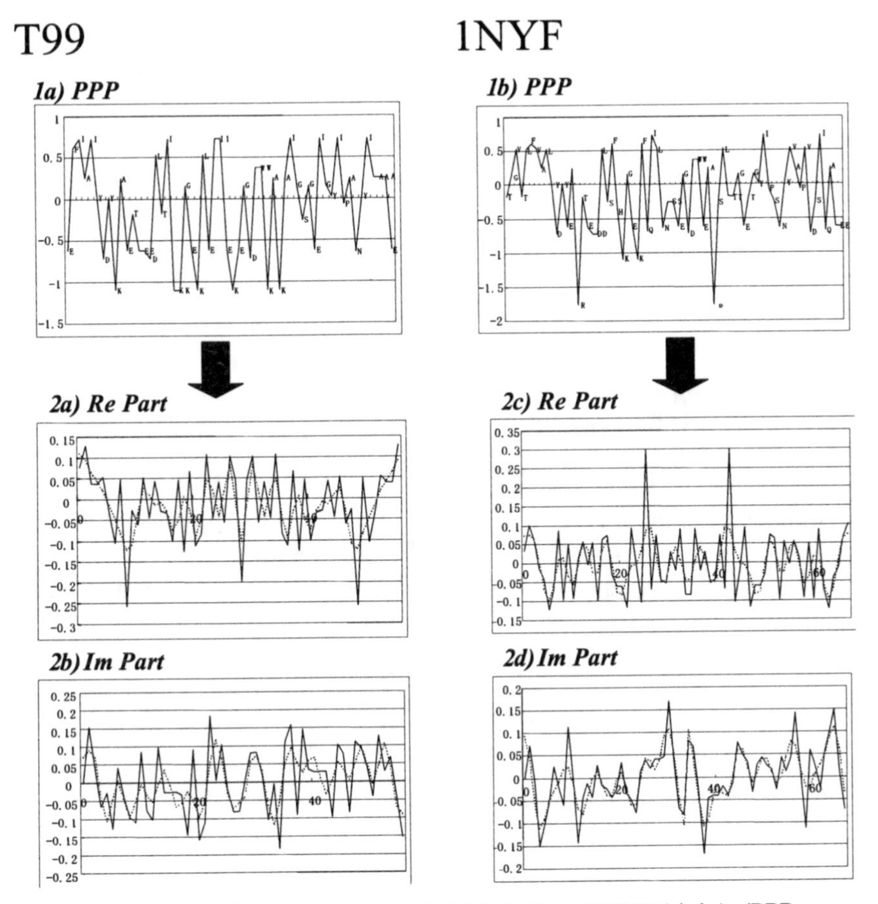

Figure 7.9. Spectral representation of T099 (left) and NYF (right). (PPP: profile of physicochemical properties, Re.Part: spectrum real part, Im.Part: spectral imaginary part).

Finally, dynamic alignment of the spectral envelopes by the procedure in Fig.7.4 are calculated and are shown in Figs. 7.10a and 7.10b for alignment of the real part and imaginary part of the spectral envelopes.

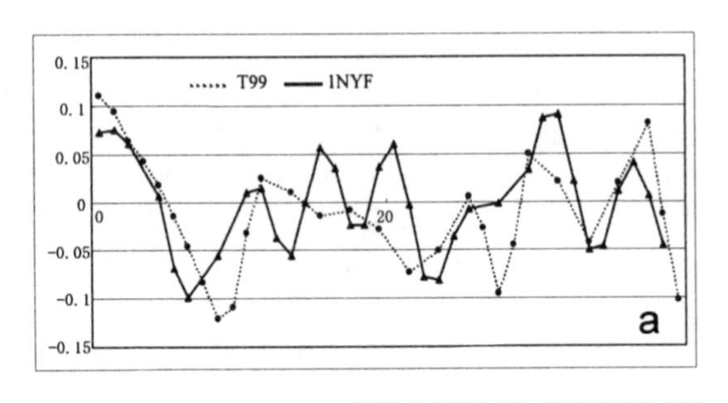

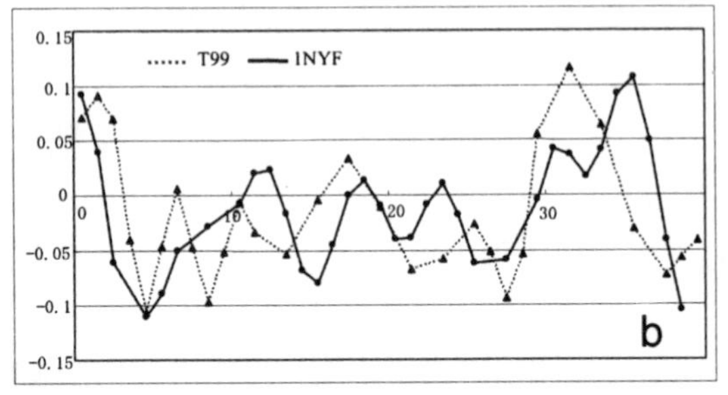

Figure 7.10. Spectral Envelope Alignment using DP for T0099 (dotted) and NYF (solid). (**a**) Real part and (**b**) Imaginary part.

The results by alignment using the CLUSTAL system[42] for multiple sequence alignment (Fig. 7.11) are in high agreement with the folding recognition performed by the spectral methodology described above although only one physicochemical parameter (the hydrophobic index) is used in the example.

```
T0099 ----EFIAIYDYKAETEEDLTIKKGEKLEIIEK-EGDWWKAKAIGSGEIGYIPANYIAAAE------
1NYF  TGVTLFVALYDYEARTEDDLSFHKGEKFQILNSSEGDWWEARSLTTGETGYIPSNYVAPVDSIQAEE
           *:*:***:*.**:**:::****::*::. *****:*::: :** ****:**:*..
```

Figure 7.11. Alignment of sequences for T0099 and 1NYF using CLUSTAL W (1.81)[42].

7.4. Automatic Classification of Protein Foldings

7.4.1. Dominant Physicochemical Parameters

As previously stated the aim of the spectral methodology introduced so far consists in finding the relationship between protein folding patterns and the distribution of physicochemical properties of the sequence of amino acid residues. The hypothesis underlying the methodology is based on the fact that most of the proteins have structural similarities with other proteins and, in some of these cases, share a common evolutionary origin. On the other hand, unique spectral patterns for sequences can be obtained using different amino acid physicochemical characteristics and a correspondence can be established among these particular spectral patterns and the main factors that make proteins adopt a specific folding pattern. Accordingly, it can be assumed that proteins possessing common folding characteristics, i.e., those belonging to the same folding pattern category may also possess homologous spectral characteristics that can be extracted by analysis of spectral information. Extraction of spectral information is carried out by the spectral alignment procedure described above. However, since protein folding may obey different physicochemical and physical characteristics, finding the parameters for which the alignments are optimal (within a protein folding category) leads to the determination of the dominant physicochemical properties for any particular class of proteins.

This evaluation is performed for the proteins recorded in the SCOP database.[26,47] The SCOP database, created by manual inspection aims to provide a detailed and comprehensive description of the structural and evolutionary relationships between all proteins whose structures are known. In this way, it provides a broad survey of all known protein folds, detailed information about the close relatives of any particular protein, and a framework for research and classification. The analysis introduced here is carried out after reprocessing the structural information found in each family and superfamily of the data base. This preprocessing operation consists in constructing sets of proteins with amino acid residue homologies of less than 80% in order to maintain diversity in residue sequence.

Physicochemical parameters are obtained from the AAINDEX[43-46] database which is a compilation of a list of 402 amino acid indices for the twenty naturally occurring residues.

Dominant physicochemical parameters for each class of proteins are obtained by alignment of the spectra for all the pairs of proteins constituting the class and using the 402 amino acid indices. Five indices are selected as the dominant physicochemical parameters for which the spectral alignment scores are the highest among the 402 amino acid

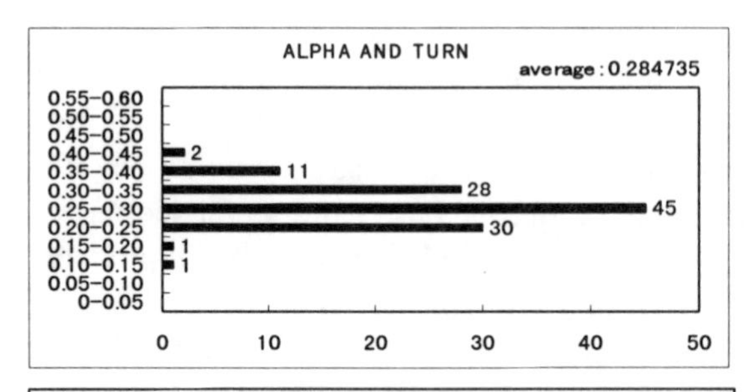

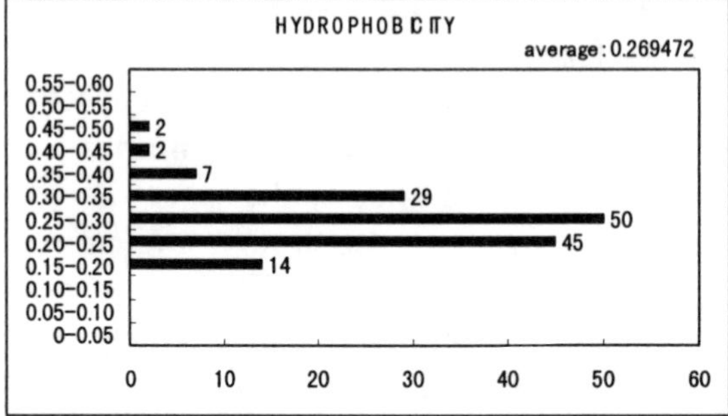

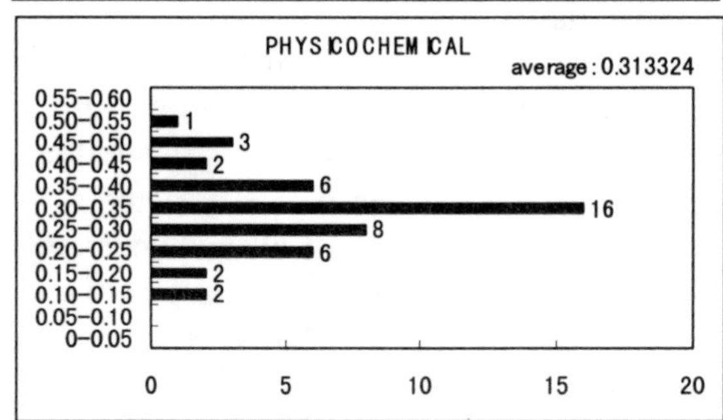

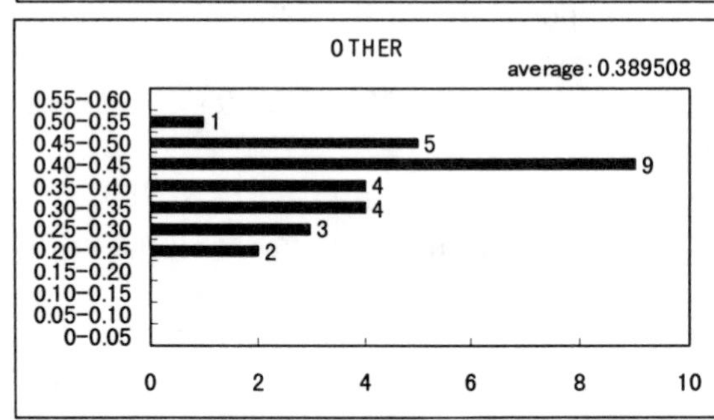

Figure 7.12. Analysis of similarity discrimination power for several types of amino acid physicochemical indices.

indices. Fig. 7.12 summarizes the results of the statistical analysis leading to the dominant physicochemical parameters for class 1.004.001.001. The ordinate indicates the spectral similarity after alignment as computed using Eq. (7.2) and using indices belonging to the category shown on the top of the histogram. The number of indices for which similarity at a determined level occurs is indicated at the right side of each bar.

From the graph it is possible to ascertain that the spectra for this class computed using indices related to the alpha and turn tendencies and hydrophobicity show a strong similarity among their member proteins. It is possible to find sequences with meaningful similarity (distance in spectral dimension of less than 0.20) for several indices related to this category. Two of these indices can be used to get similarities of less than 0.20 using alpha and turn propensity type coefficients. This number of indices is 14 when hydrophobic type of physicochemical parameters are considered. In general only 19 indices can be used to discriminate this particular superfamily of proteins. Fig. 7.13 summarizes the entire gamma of indices and their incidence in the evaluation of sequence similarities. Only six indices may be considered when a discrimination value under 0.20 may be desired.

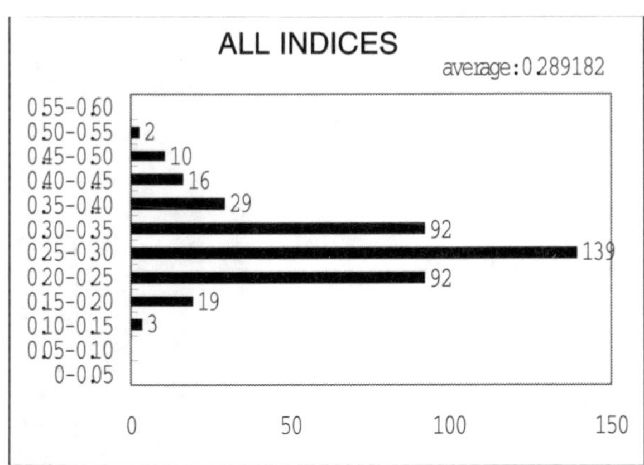

Figure 7.13. Similarities for protein 1KO using 403 amino acid indices.

7.4.2. Classification of Protein Folding by Spectral Analysis

Applying the procedure described in the former section to every superfamily in the SCOP database leads to the definition of the dominant physicochemical parameters for each superfamily. Table 7.2 illustrates some superfamilies together with the corresponding five dominant parameters and the best alignment scores obtained for every one. Furthermore, the main folding characteristics for the particular group are illustrated using a ribbon model for the representative molecule. Meticulous observation of this table shows the level of spectral alignment at which the indices are obtained for every superfamily as well as the fact that several indices of high discriminative power belong to the same physicochemical index category. For instance, four out of the five dominant indices for superfamily 1.001.002.002 belong to the category of coefficients of hydrophobicity. This fact illustrates that in many of the superfamilies analyzed less than 2 parameters are enough to discriminate protein folding patterns. In contrast, some superfamilies can not be described sufficiently well with five of the indices selected out of the 402 utilized so far. Accordingly, specification of other type of physicochemical parameters for these superfamilies is required.

Table 7.2. Dominant Physicochemical Properties for Folding Pattern Discrimination. (For index description see Table 7.3.)

SuperFamily		Ranking				
		1	2	3	4	5
1.001.001.001	Index	O303	H311	A093	A016	A369
	Score	0.084	0.092	0.095	0.096	0.106
1.001.002.002	Index	H212	P390	H113	H240	H180
	Score	0.045	0.047	0.048	0.048	0.049
1.001.016.001	Index	P096	P219	A018	O303	O385
	Score	0.145	0.215	0.218	0.235	0.258
1.001.025.006	Index	A016	A093	P096	H311	A369
	Score	0.005	0.006	0.045	0.047	0.069
1.001.030.001	Index	P219	P096	A018	O303	A338
	Score	0.018	0.023	0.028	0.039	0.041
1.001.046.001	Index	A016	A093	H396	H395	H311
	Score	0.003	0.004	0.006	0.006	0.016
1.001.080.001	Index	P096	P216	P159	P383	H271
	Score	0.050	0.109	0.113	0.114	0.117

Continued on next page

SuperFamily	Ranking					
		1	2	3	4	5
1.001.100.001	Index	P096	O303	P383	P219	H088
	Score	0.031	0.035	0.036	0.038	0.039
1.001.111.008	Index	O303	P096	H110	P397	A306
	Score	0.031	0.036	0.045	0.046	0.047
1.001.120.001	Index	P096	P219	O303	A338	H110
	Score	0.015	0.017	0.019	0.025	0.025
1.002.077.003	Index	O303	A306	A331	C190	H147
	Score	0.025	0.039	0.042	0.045	0.045
1.002.081.001	Index	C191	C193	H003	H004	A335
	Score	0.042	0.043	0.044	0.044	0.045
1.002.085.001	Index	O303	A023	A269	A018	H013
	Score	0.022	0.045	0.049	0.051	0.052
1.003.001.002	Index	P096	A018	P219	P383	H311
	Score	0.449	0.458	0.459	0.466	0.475
1.003.001.004	Index	P096	O303	P219	P383	H147
	Score	0.032	0.044	0.046	0.046	0.047

Continued on next page

SuperFamily	Ranking					
		1	2	3	4	5
1. 004. 001. 001	Index	A018	P096	P219	C144	H332
	Score	0. 103	0. 107	0. 149	0. 155	0. 159
1. 004. 020. 001	Index	P096	P219	H327	H057	H202
	Score	0. 033	0. 044	0. 060	0. 062	0. 063
1. 004. 036. 002	Index	C116	P219	A304	P096	P383
	Score	0. 034	0. 041	0. 046	0. 046	0. 047
1. 004. 110. 001	Index	H395	H396	H070	H088	H133
	Score	0. 004	0. 004	0. 025	0. 034	0. 038
1. 004. 120. 001	Index	H378	H401	A335	A023	H088
	Score	0. 018	0. 019	0. 020	0. 020	0. 020

Table 7.3. Dominant indices for protein folding pattern discrimination (Table 7.2.)

Index	Property	Authors*
	A. Alpha and turn propensities	
A016 BUNA790101	Alpha-NH chemical shifts	(Bundi Wuthrich, 1979)
A018 BUNA790103	Spin spin coupling constants 3JHalpha NH	(Bundi Wuthrich, 1979)
A023 CHAM830102	A parameter defined from the residuals obtained from the best correlation of the Chou Fasman parameter of beta sheet	(Charton-Charton, 1983)
A093 FINA910102	Helix initiation parameter at posision i i+1 i+2	(Finkelstein et al.1991)
A269 QIAN880112	Weights for alpha helix at the window position of 5	(Qian Sejnowski, 1988)
A304 RACS820108	Average relative fractional occurrence in AR(i 1)	(Rackovsky Scheraga, 1982)
A306 RACS820110	Average relative fractional occurrence in EL(i 1)	(Rackovsky Scheraga, 1982)
A331 RICJ880110	Relative preference value at C5	(Richardson Richardson, 1988)
A335 RICJ880114	Relative preference value at C1	(Richardson Richardson, 1988)
A338 RICJ880117	Relative preference value at C"	(Richardson Richardson, 1988)
A369 TANS770104	Normalized frequency of chain reversal R	(Tanaka Scheraga, 1977)

Continued on next page

Index	Property	Authors*
	C. Composition	
C116 HUTJ700101	Heat capacity	(Hutchens 1970)
C144 KARP850103	Flexibility parameter for two rigid neighbors	(Karplus Schulz, 1985)
C189 NAKH920101	AA composition of CYT of single spanning proteins	(Nakashima Nishikawa, 1992)
C190 NAKH920102	AA composition of CYT2 of single-spanning proteins	(Nakashima Nishikawa, 1992)
C191 NAKH920103	AA composition of EXT of single spanning proteins	(Nakashima Nishikawa, 1992)
C193 NAKH920105	AA composition of MEM of single spanning proteins	(Nakashima Nishikawa, 1992)
	H. Hydrophobicity	
H003 ARGP820102	Signal sequence helical potential	(Argos et al.1982)
H004 ARGP820103	Membrane buried preference parameters	(Argos et al., 1982)
H013 BROC820102	Retention coefficient in HFBA	(Browne et al., 1982)
H057 CIDH920104	Normalized hydrophobicity scales for alpha/beta proteins	(Cid et al., 1992)
H070 EISD860102	Atom based hydrophobic moment	(Eisenberg-McLachlan, 1986)
H088 FAUJ880111	Positive charge	(Fauchere et al., 1988)
H110 GRAR740101	Composition	(Grantham, 1974)
H113 GUYH850101	Partition energy	(Guy, 1985)
H133 JOND750102	pK (-COOH)	(Jones, 1975)
H147 KRIW790101	Side chain interaction parameter	(Krigbaum-Komoriya, 1979)
H180 MEEJ810101	Retention coefficient in NaClO4	(Meek-Rossetti, 1981)
H202 NAKH900106	Normalized composition from animal	(Nakashima et al., 1990)
H212 NOZY710101	Transfer energy, organic solvent/water	(Nozaki-Tanford, 1971)
H240 PLIV810101	Partition coefficient	(Pliska et al., 1981)
H311 RACS770101	Average reduced distance for Ca	(Rackovsky-Scheraga, 1977)
H327 RICJ880106	Relative preference value at N3	(Richardson-Richardson, 1988)
H332 RICJ880111	Relative preference value at C4	(Richardson-Richardson, 1988)
H378 VASM830103	Relative population of conformational state E	(Vasquez et al., 1983)
H401 ZIMJ680104	Isoelectric point	(Zimmerman et al., 1968)
	P. Physicochemical properties	
P096 GARJ730101	Partition coefficient	(Garel et al., 1973)
P159 LEVM760107	van der Waals parameter e	(Levitt, 1976)
P216 OOBM770104	Average non-bonded energy per residue	(Oobatake-Ooi, 1977)
P219 OOBM850102	Optimized propensity to form reverse turn	(Oobatake et al., 1985)
P383 WEBA780101	RF value in high salt chromatography	(Weber-Lacey, 1978)
P390 WOLS870102	Principal property value z2	(Wold et al., 1987)
P397 ZASB820101	Dependence of partition coefficient on ionic strength	(Zaslavsky et al., 1982)
	O. Other properties	
O303 RACS820107	Average relative fractional occurrence in A0(i-1)	(Rackovsky-Scheraga, 1982)
O385 WERD780102	Free energy change of e(i) to e(ex)	(Wertz-Scheraga, 1978)

* References to all abovementioned indices can be found in sources[45,46] of this chapter.

7.5 Protein Folding Pattern Recognition by Spectral Analysis

Protein folding prediction using the spectral methodology introduced in this chapter proceeds by comparing the spectra for the unknown or target sequence with the spectra of every member of a superfamily and for all the superfamilies, the dominant parameters of which have been calculated. These dominant parameters are used for constructing the spectra (for both the target molecule and the known proteins).

Spectral alignment scores for each of the five spectra are calculated. The protein folding pattern is assigned to the target the alignment scores for which are the highest in average; the representative folding being, therefore, the one with the highest similarity to the target within the group, both in spectral superimposition and number of amino acid residues.

Since the number of amino acids in the sequences constituting the group of proteins used in derivation of the dominant parameters is, however, different to that of the target molecule the spectral analysis methodology developed leads to characterization of folding patterns rather than a structure that can be superimposed to the target molecule. Consequently, the structure derived by the spectral analysis performed consists in assigning a set of folding characteristics to the target molecule from which the real folding can be built by threading or any other similar protein three dimensional modeling method.

Plate 7.1 illustrates a set of folding recognition experiments using the present methodology. The set of target molecules corresponds to CASP4 and is summarized in Table 7.4.

Structural information for each target molecule consists in its native fold together with the most similar structure found by spectral alignment. Furthermore, the result of the alignment of the real and imaginary parts of the spectral envelopes for both molecules is also presented. This figure illustrates the potentiality of the methodology in modeling protein structures.

Table 7.4 lists the name of the targets in Plate 7.1 together with the class to which the parent structure belongs as well as the dominant index by means of which the highest alignment score is obtained.

Table 7.4. Summary of the Folding Recognition Experiment using Spectral Analysis

TARGET	PARENT	Rank	CLASS	Dominant Index
T94	1b87_A	3	1.4.89.1	H240
T110	1c52	5	1.1.3.1	H14
T111	1pgb_A	2	1.3.1.7	H318
T112	1lcf_A	5	1.3.89.1	P214
T125	1cqx_A	5	1.1.1.1	O303
T128	1mrg	5	1.4.143.1	H317

Since the system outputs a number of structures ranked by spectral alignment and similarity score, the table reveals that the parent structure output by the system is within the first five folds that the system recognizes.

The methodology thus proves effective in recognizing protein fold patterns searching for similarities among proteins of low primary structure homology since for the targets listed in Table 7.3 it amounts to less than 30% in amino acid composition.

This proves to certain extent the underlying hypothesis on which the methodology was developed. It proves concretely the fact that characteristics in physicochemical properties distributions along the sequence of amino acids are reflected in particular protein folds that natural proteins adopt.

Although the methodology can be improved in many ways especially in those aspects concerning the type of properties used to represent protein similarities at the primary structure level, the hypothesis on which the methodology is based may not undergo any substantial change.

References

1. Smith TF, Waterman MS: **Identification of common molecular subsequences.** *J Mol Biol* 1981, **147**:195–197.
2. Anbarasu LA, Narayanasamy P, Sundararajan V: **Multiple molecular sequence alignment by island parallel genetic algorithm.** *Current Science—New Delhi* 2000, **78**(7):858–863.
3. Szustakowski JD, Weng Z: **Protein structure alignment using a genetic algorithm.** *Proteins* 2000, **38**(4):428-440.
4. Zhang C, Wong AKC: **A genetic algorithm for multiple molecular sequence alignment.** *Bioinformatics* 1997, **13**(6):565–582.
5. Zhang C, Wong AKC: **Toward efficient multiple molecular sequence alignment: A system of genetic algorithm and dynamic programming.** *IEEE Trans Systems Man Cybern—Part B—Cybernetics* 1997, **27**(6):918–932.
6. Kwong S, Chau CW, Halang WA: **Genetic algorithm for optimizing the nonlinear time alignment of automatic speech recognition systems.** *IEEE Trans Industrial Electronics* 1996, **43**(5):559–566.
7. Notredame C, Higgins DG: SAGA: **Sequence alignment by genetic algorithm.** *Nucleic Acids Res* 1996, **24**(8):1515–1524
8. Wild DJ, Willett P: **Similarity searching in files of three-dimensional chemical structures. Alignment of molecular electrostatic potential fields with a genetic algorithm.** *J Chem Inf Comput Sci* 1996, **36**(2):159–167.

9. Pancoska P, Blazek M, Keiderling TA: **Relationships between secondary structure fractions for globular proteins. Neural network analyses of crystallographic data sets.** *Biochemistry* 1992, **31**:10250–10257

10. Fredholm H, Bohr J, Bohr S, Brunak RMJ, Cotterill B, Lautrup SB: **Neural Networks applied to the study of Protein Sequences and Protein Structures.** In: *Site-Directed Mutagenesis and Protein Engineering.* El-Gewely MR, ed., Amsterdam: Elsevier Science, 1991:41–46.

11. Mackay DHJ, Cross AJ, Hagker AT: **The role of energy minimization in simulation strategies of biomolecular systems.** In: *Prediction of protein structure and the principles of protein conformation.* Fasman GD, ed., New York: Plenum Press, 1989:317–358.

12. Maple J, Dinar U, Hagler AT: **Derivation of force field for molecular mechanics and dynamics from *ab initio* energy surfaces.** *Proc Natl Acad USA* 1988, **85**:5350–5354.

13. Karplus M, Mc Cammon JA: **Dynamics of proteins: Elements and function.** *Ann Rev Biochem* 1983, **53**:263–300.

14. Gribskov M, McLachlan AD, Eisenberg D: **Profile analysis: Detection of distantly related proteins.** *Proc Natl Acad Sci USA* 1987, **84**:4355–4358.

15. Altschul SF, Madden TL, Schaffer AA, Zhang Z, Miller W, Lipman DJ: **Gapped BLAST and PSI-BLAST: A new generation of protein database search programs.** *Nucleic Acids Res* 1997, **25**:3389–3402.

16. Eddy SR: **Multiple alignment using hidden Markov models.** *ISMB* 1995, **3**:114–120.

17. Eddy SR: **Hidden Markov models.** *Curr Opin Struct Biol* 1996, **6**:361–365.

18. Karplus K, Barrett C, Hughey R: **Hidden Markov models for detecting remote protein homologies.** *Bioinformatics* 1998, **14**:846–856.

19. Krogh A, Brown M, Mian IS, Sjolander K, Haussler D: **Hidden Markov models in computational biology. Applications to protein modeling.** *J Mol Biol* 1994, **235**:1501–1531.

20. Henikoff S, Pietrokovski S, Henikoff JG: **Superior performance in protein homology detection with the blocks database servers.** *Nucleics Acids Res* 1998, **26**:309–312.

21. Rychlewski L, Jaroszewski L, Li W, Godzik A: **Comparison of sequence profiles. Strategies for structural predictions using sequence information.** *Protein Sci* 2000, **9**:232–241.

22. Moult J: **Predicting protein three-dimensional structure.** *Curr Opin Biotechnol* 1999, **10**:583–588.

23. Park K, Teichmann SA, Hubbard T, Chothia C: **Intermediate sequences increase the detection of homology between sequences.** *J Mol Biol* 1997, **273**:349–354.

24. Salamov AA, Suwa M, Orengo CA, Swindells,M.B: **Combining sensitive database searches with multiple intermediates to detect distant homologues.** *Protein Sci* 1999, **9**:232–241.

25. Pearson WR, Lipmann DJ: **Improved tools for biological sequence comparison.** *Proc Natl Acad Sci USA* 1988, **85**:2444–2448.

26. Li W, Pio F, Pawlowski K, Godzik A: **Saturated BLAST: An automated multimple intermediates sequences search used to detect distant homology.** *Bioinformatics* 2000. **16**:1105–1110.

27. Doolittle RF: *Of Urfs and Orfs:A Primer on How to Analyze Derived Amino Acid Sequences.* Mill Valley, California: University Science Books, 1986.

28. Cosic I, Hodder AN, Aguilar MI, Hearn MTW: **Resonant recognition model and protein topography. Model studies with myoglobin, hemoglobin, and lysozyme.** *Eur J Biochem* 1991, **198**:113–119.

29. Cosic I: **Macromolecular bioactivity: Is it resonant interaction between macromolecules?—Theory and applications.** *IEEE Trans Biomed Eng* 1994, **41**:1101–1114.

30. Cosic I: *The Resonant Recognition Model of Macromolecular Bioactivity: Theory and Applications.* Basel, Switzerland: Birkhouser Verlag, 1997.

31. Fang Q, Cosic I: **Protein structure analysis using the resonant recognition model and wavelet transforms.** *Aust Phys Eng Sci Med* 1998, **21**:179–185.

32. Eisenberg D, Weiss RM, Terwillinger TC: **The hydrophobic moment detects periodicity in protein hydrophobicity.** *Proc Natl Acad Sci USA* 1984, **81**:140–144.

33. Hirakawa H, Muta S, Kuhara S: **The hydrophobic cores of proteins predicted by wavelet analysis.** *Bioinformatics* 1999, **15**(2):141–148.

34. Lio P, Vannucci M: **Wavelet change-point prediction of transmembrane proteins.** *Bioinformatics* 2000, **16**(4):376–382.

35. Mandell AJ, Selz KA, Shlesinger MF:**Wavelet transformation of protein hydrophobicity sequences suggests their memberships in structural families.** *Physica A* 1997, **244**:254–262.

36. Del Carpio CA, Yoshimori A: **Spectral analysis for protein folding recognition using dominant amino acid biochemical properties.** *CASP4 (Fourth Meeting on the Critical Assessment of Techniques for Protein Structure Prediction), Asilomar Conference Center, December 3–7, 2000.* Asilomar, California, 2000:A-17.

37. Del Carpio CA, Yoshimori A: **Protein folding pattern recognition using signal processing theory.** Personal Communication.

38. Rabiner LR, Gold B: *Theory and Application of Digital Signal Processing.* Englewood Cliffs, New Jersey: Prentice-Hall, Inc., 1975.

39. Eisenberg, D: **Three-dimensional structure of membrane and surface proteins.** *J Ann Rev Biochem* 1984, **53**:595–623.

40. Needleman SB, Wunsch CD: **A general method applicable to the search for similarities in the amino acid sequences of two proteins.** *J Mol Biol* 1970, **48**:443–453.

41. *CASP4.* http://predictioncenter.llnl.gov/casp4/.

42. Thompson JD, Higgins DG, Gibson TJ: **CLUSTALW: Improving the sensitivity of progressive multiple sequence alignment through sequence weighting, positions-specific gap penalties and weight matrix choice.** *Nucleic Acids Res* 1994, **22**:4673–4680.

43. Nakai K, Kidera A, Kanehisa M: **Cluster analysis of amino acid indices for prediction of protein structure and function.** *Protein Eng* 1988, **2**:93–100.

44. Tomii K, Kanehisa M: **Analysis of amino acid indices and mutation matrices for sequence comparison and structure prediction of proteins.** *Protein Eng* 1996, **9**:27–36.

45. Kawashima S, Ogata H, Kanehisa M: **AAindex: Amino acid index database.** *Nucleic Acids Res* 1999, **27**:368–369.

46. *AAINDEX.* http://www.genome.ad.jp/dbget/aaindex.html.

47. *SCOP.* http://scop.mrc-lmb.cam.ac.uk/scop/.

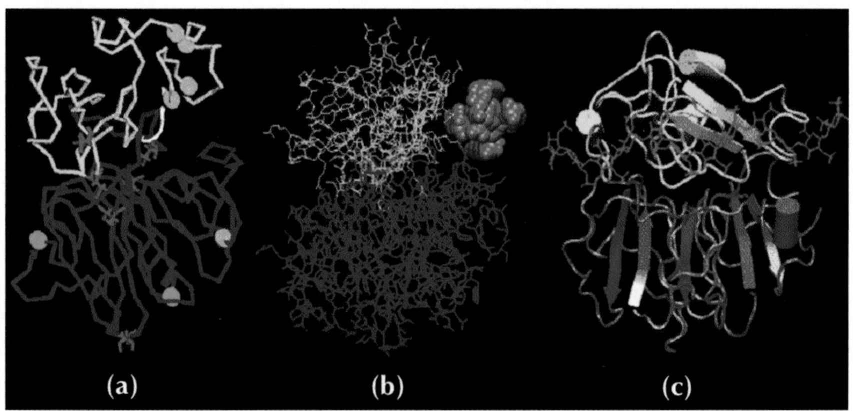

Plate 1.1. Structural model of vitronectin. (**a**) the docking structure between the central (*blue*) and C-terminal (*yellow*) domains, with cysteines (*red lines*), presumed heparin-binding residues (354-363 in *white* ribbons), sites susceptible to protease (residues 305, 361, 370, 379 and 383 with *light blue spheres*), and N-linked glycosylation sites (residues 150 and 223 with *green* spheres). (**b**) and (**c**) show the predicted docking conformation between the heparin (in *red*) and the combined structure of the central (*blue*) and C-terminal (*yellow*) domains from two different perspectives.

Protein structure

Structural elements { helix, strand, loop }

Plate 2.1. The protein structure is represented as a ribbon with the secondary structure elements identified by different colors. *Green*, *red*, and *gray* ribbons represent the protein basic structural elements: helices, strands, and loops, respectively.

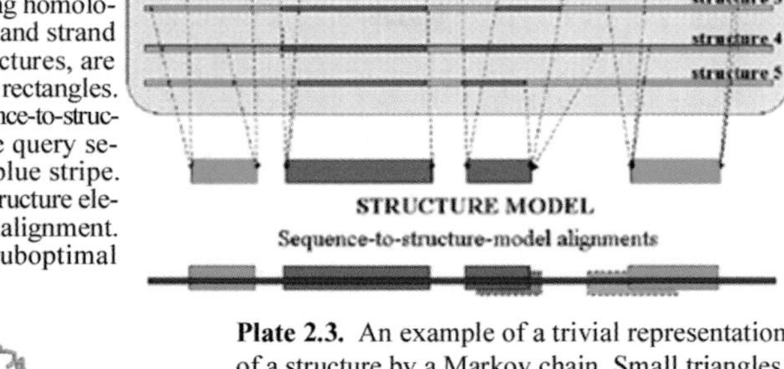

Plate 2.2. Top: Representation of related structures by a structure model. The structure model represents a set of common structural features among homologous proteins. Here, helical and strand regions, common to all structures, are represented as *red* and *green* rectangles. **Bottom:** Example of the sequence-to-structure-model alignments. The query sequence is represented as a blue stripe. Brightly colored secondary structure elements represent the optimal alignment. The faded colors depict a suboptimal alignment.

Plate 2.3. An example of a trivial representation of a structure by a Markov chain. Small triangles, circles, and rectangles represent hidden states in a DSM. Lines connecting them represent transitions from state to state. Here, all transition probabilities are equal to one and thus there are no numbers associated with the lines connecting the states. Each state corresponds a residue position in the structure.

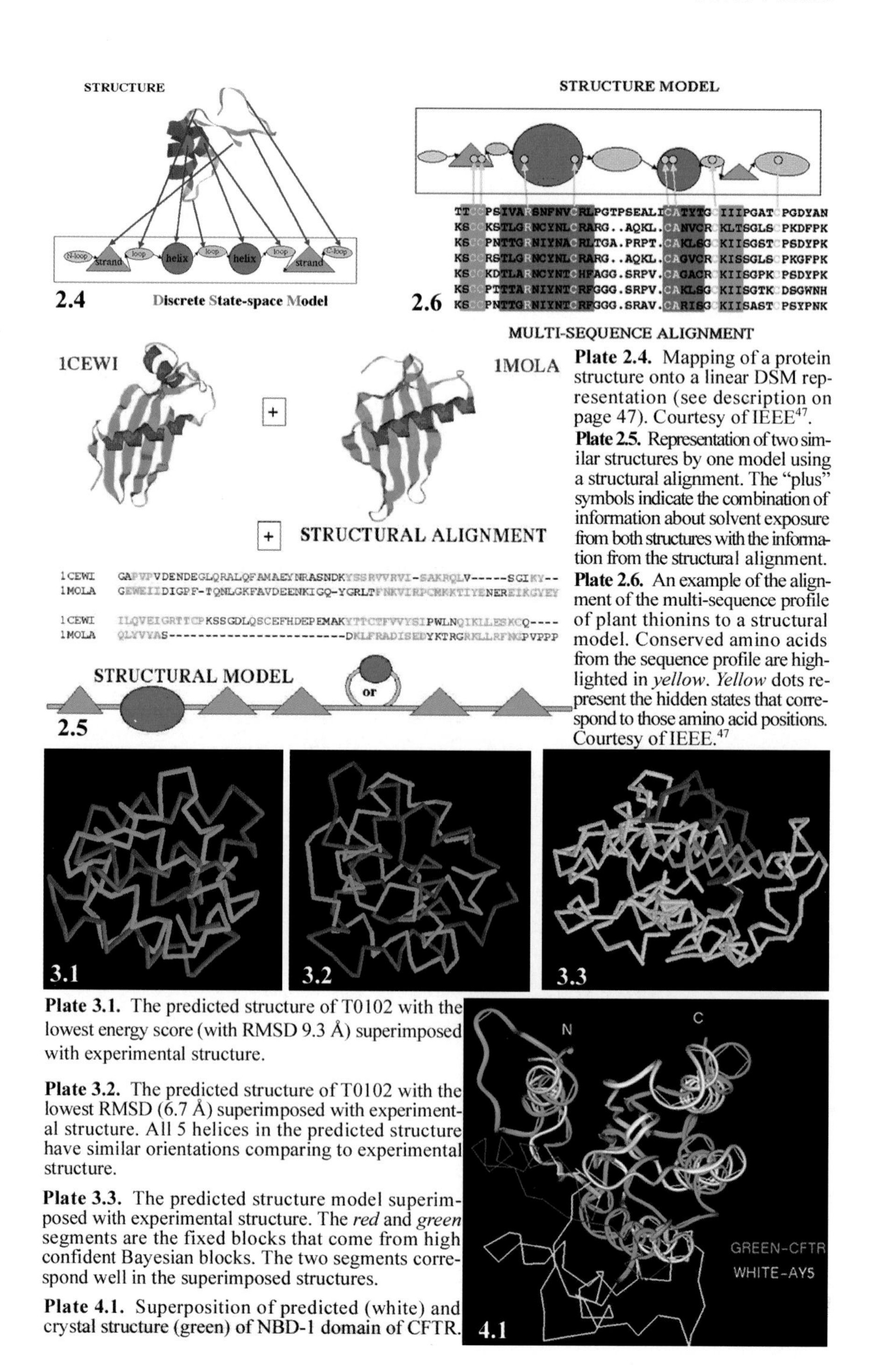

Plate 2.4. Mapping of a protein structure onto a linear DSM representation (see description on page 47). Courtesy of IEEE[47].

Plate 2.5. Representation of two similar structures by one model using a structural alignment. The "plus" symbols indicate the combination of information about solvent exposure from both structures with the information from the structural alignment.

Plate 2.6. An example of the alignment of the multi-sequence profile of plant thionins to a structural model. Conserved amino acids from the sequence profile are highlighted in *yellow*. *Yellow* dots represent the hidden states that correspond to those amino acid positions. Courtesy of IEEE.[47]

Plate 3.1. The predicted structure of T0102 with the lowest energy score (with RMSD 9.3 Å) superimposed with experimental structure.

Plate 3.2. The predicted structure of T0102 with the lowest RMSD (6.7 Å) superimposed with experimental structure. All 5 helices in the predicted structure have similar orientations comparing to experimental structure.

Plate 3.3. The predicted structure model superimposed with experimental structure. The *red* and *green* segments are the fixed blocks that come from high confident Bayesian blocks. The two segments correspond well in the superimposed structures.

Plate 4.1. Superposition of predicted (white) and crystal structure (green) of NBD-1 domain of CFTR.

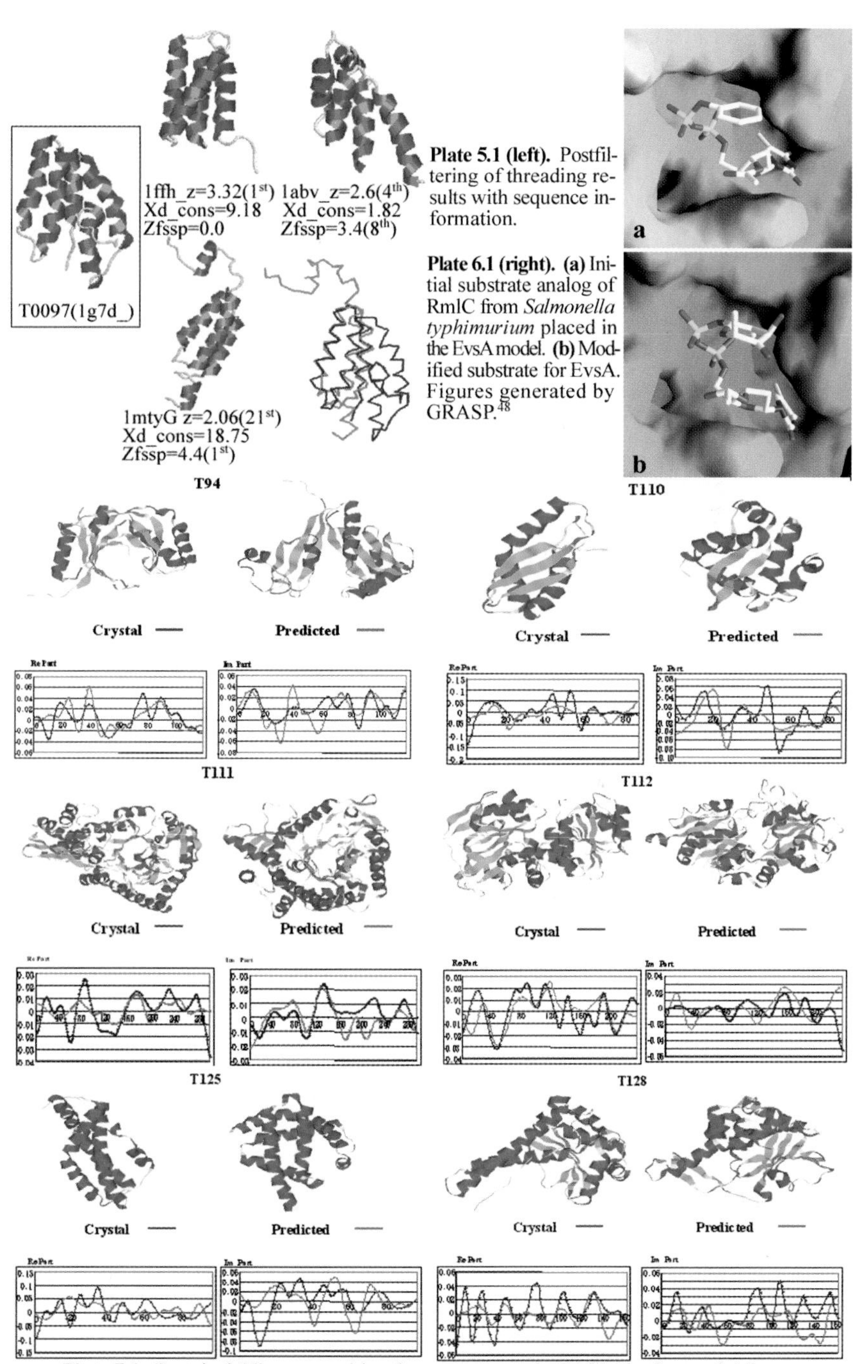

Plate 5.1 (left). Postfiltering of threading results with sequence information.

Plate 6.1 (right). **(a)** Initial substrate analog of RmlC from *Salmonella typhimurium* placed in the EvsA model. **(b)** Modified substrate for EvsA. Figures generated by GRASP.[48]

Plate 7.1. Protein folding recognition for some targets in CASP4 by Spectral Alignment.

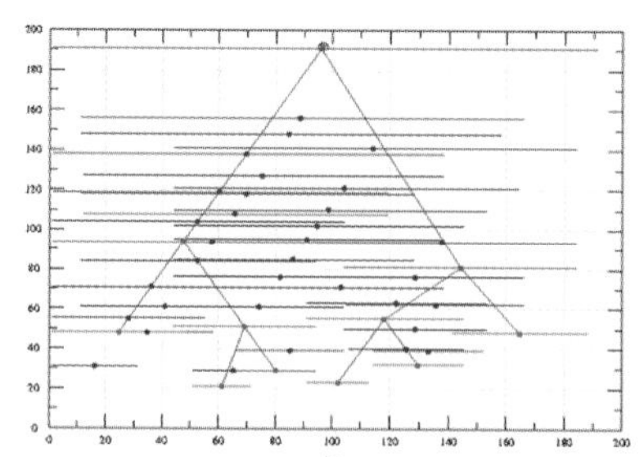

Plate 8.1. The fragment map of human hemopexin (pdb code 1pex), a sequentially folding protein. The building blocks are marked in red, while local minima, which are not assigned as building blocks, are blue.

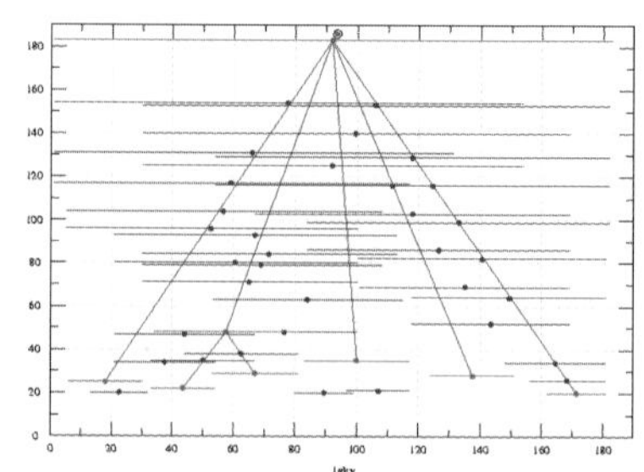

Plate 8.2. The fragment map of guanylate kinase (pdb code 1gky), a complex folding protein. The building blocks are marked in red, while local minima, which are not assigned as building blocks, are blue.

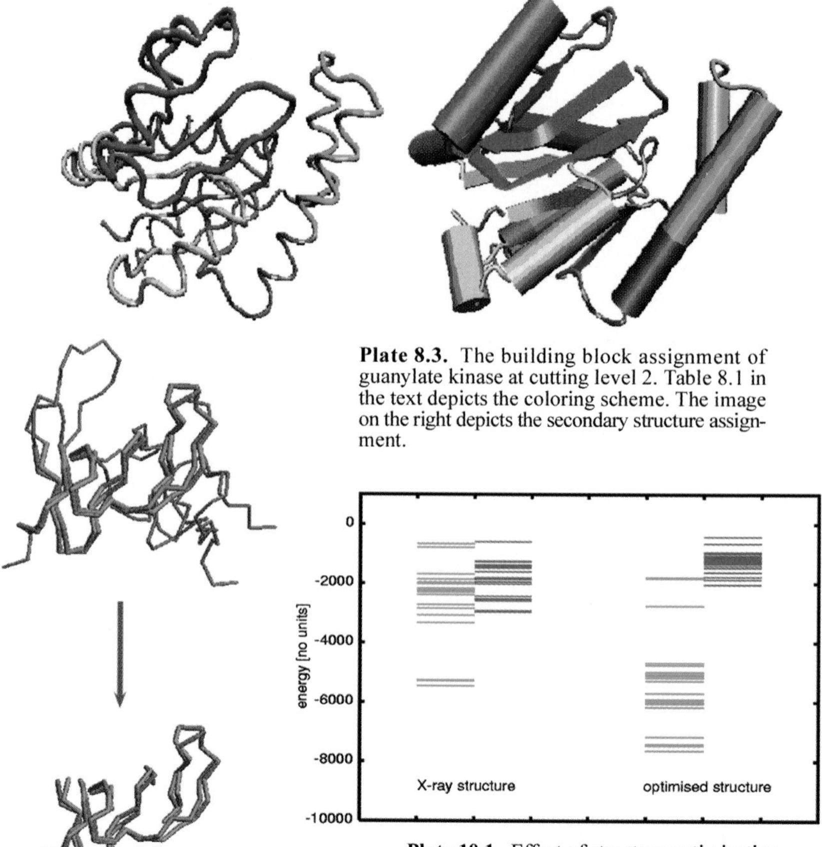

Plate 8.3. The building block assignment of guanylate kinase at cutting level 2. Table 8.1 in the text depicts the coloring scheme. The image on the right depicts the secondary structure assignment.

Plate 10.1. Effect of structure optimization on energy levels for lysozyme (153l).

Plate 9.1. Structural alignment of mamba intestinal toxin (PDB code 1IMT, in *blue*) and porcine pancreatic procolipase B (PDB code 1PCN, in *gold*).

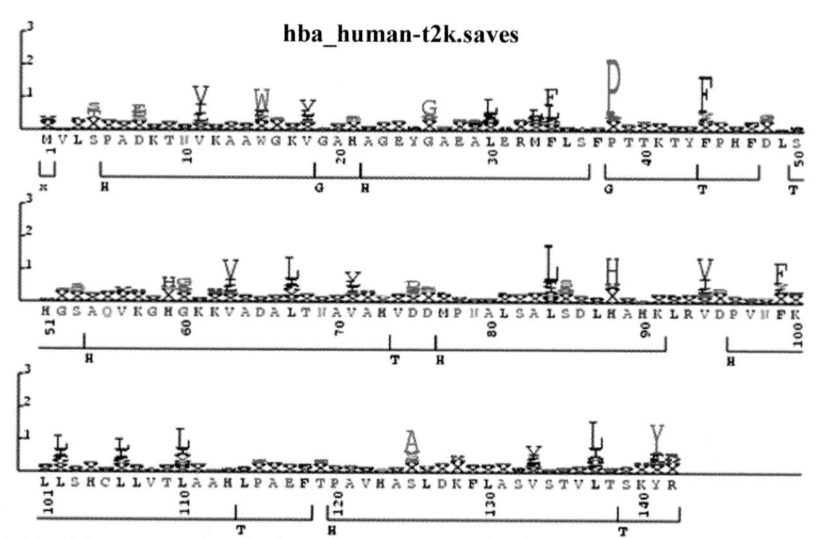

Plate 11.1. Sequence logo for the q300 sequences found by SAM-T2K using HBA_HUMAN as a seed (See details on page 309.)

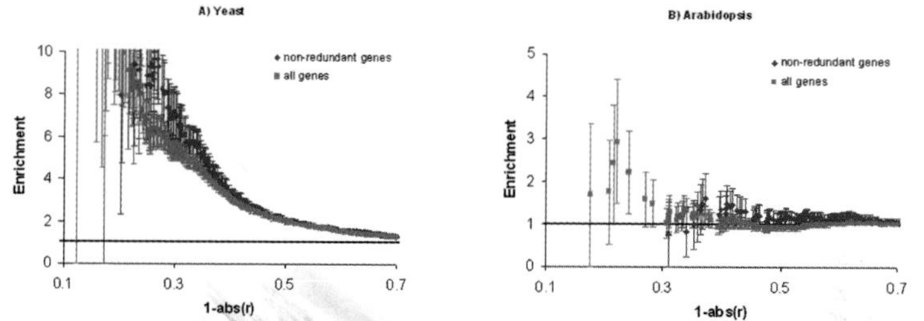

Plate 12.1. (**a**) The fold enrichment in yeast as a function of *r*. (**b**) The fold enrichment in *Arabidopsis* as a function of *r*. Courtesy of World Scientific Co. Pte. Ltd. (Singapore.)[24]

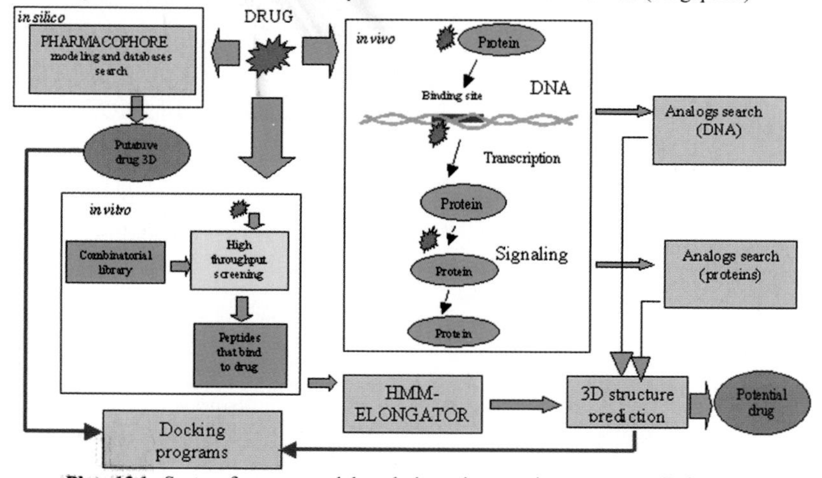

Plate 13.1. System for automated drug design using protein structure prediction.

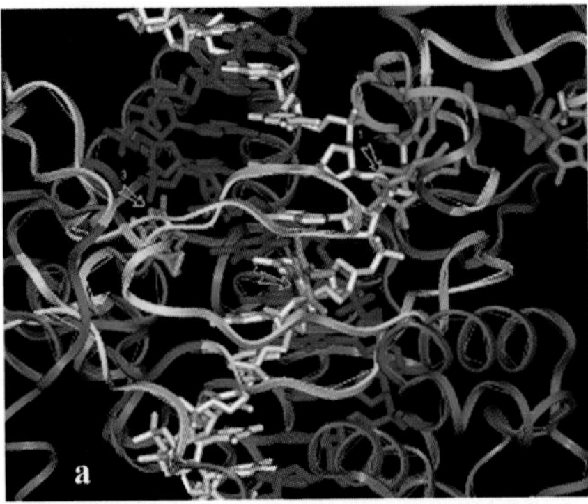

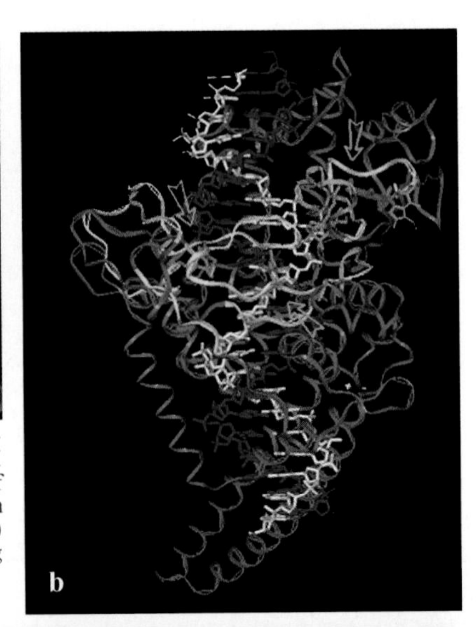

Plate 13.2. The most probable positions of docking the drag molecule *(pink)* to Topoisomerase I *(red)*. Arrows show the places where positions of the docked drug molecule interfere with position of DNA *(blue* and *white* strands). The sites *(yellow)* were predicted by HMM-ELONGATOR for drug binding. (**a**) Enlarged view of binding sites. (**b**) General view.

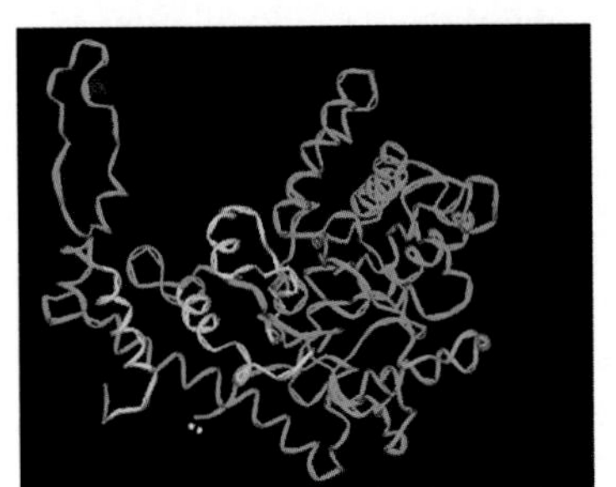

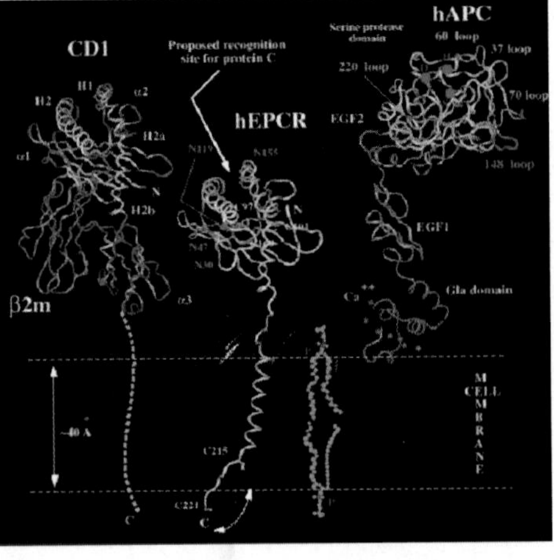

Plate 13.3. (Left) The predicted by HMM-SPECTR 3D structure of c-Myc protein, which was proposed by PROTEIN-ELONGATOR: *yellow*—the regions of c-Myc corresponded to library peptides.

Plate 14.1. (Right) The 3D model of Human EPCR.[21] (See details on page 362.) Courtesy of B. Villoutreix and Oxford University Press.

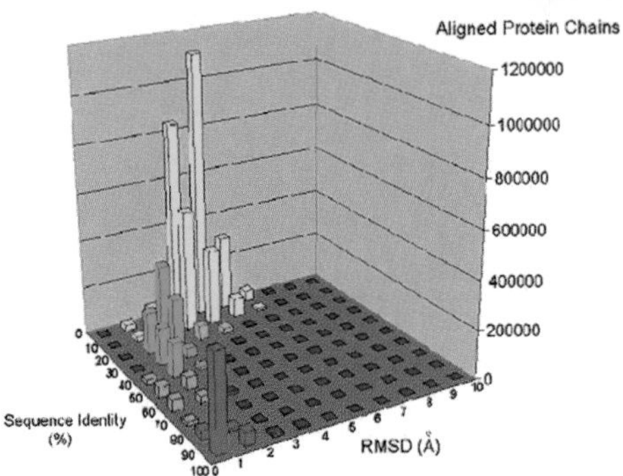

Plate 17.1. CE fold space (distribution of significant structure similarities in 'Sequence identity'—'RMSD' coordinates.) *Yellow*—analogous fold (low sequence identity, moderate to high RMSD.) *Green*—homologous folds, superfamilies and families (moderate sequence identity, low RMSD.) *Blue*—homologous and analogous folds. *Orange*—homologous folds, close family members, orthologs (high sequence identity, low RMSD.) *Magenta*—mutants, alternative experiments for the same structure (very high sequence identity, low RMSD.)

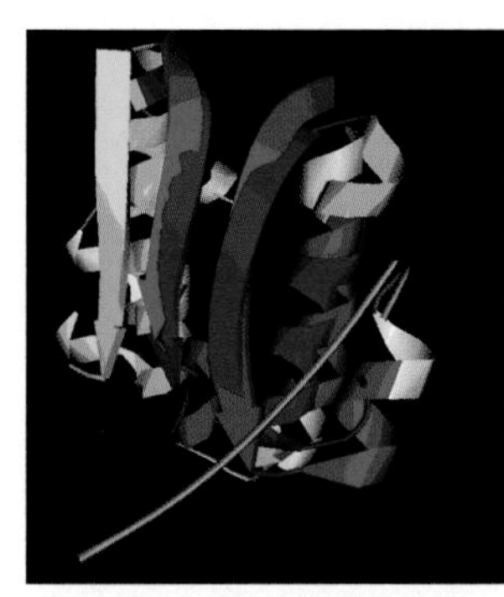

Plate 18.2. (Right) The alignment ribbon image corresponding to the second domain in Table 18.3.2. The target (1G29:A) is *white*, the probe (1B0U:A)—*yellow*. The aligned fragments are shown in *red* (target) and *blue* (probe). Note that this domain comprises the α-helices on the left and two β-strands, whereas the first domain (see Plate 18.1) covers the same β-strands, but the α-helices on the right.

Plate 18.1. The alignment ribbon image corresponding to the first domain in Table 18.3.2. The target (1G29:A) is *white*, the probe (1B0U:A)—*yellow*. The aligned fragments are shown in *red* (target) and *blue* (probe).

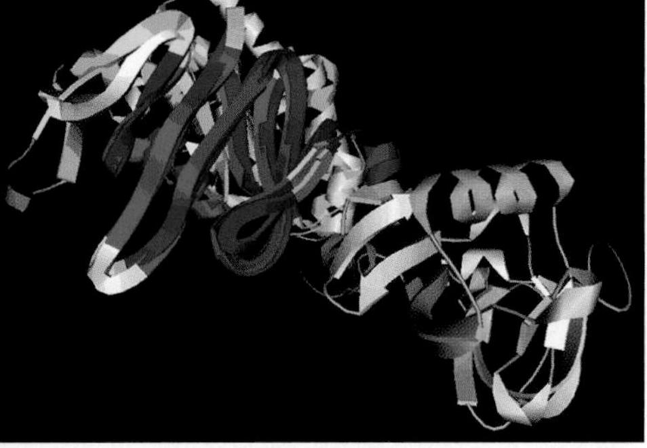

Plate 18.3. (Up) The alignment ribbon image representing the first domain in Table 18.1. (see also Table 18.2). The aligned fragments shown in Plates18.1 and 18.2 (piece 3, 4, and 5 in Table 18.2) are hidden behind the β-sheet twist near the center of the Plate.

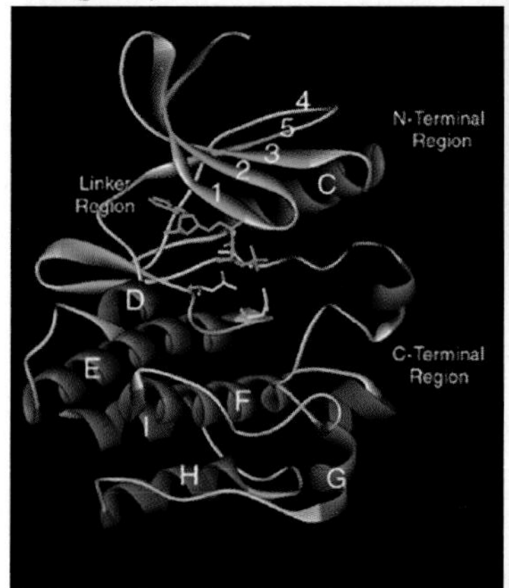

Plate 20.1. (Left) The shared structure of the protein kinase catalytic domain. See page 468 for detailed description. Plates for this chapter were created with WebLab Viewer™.

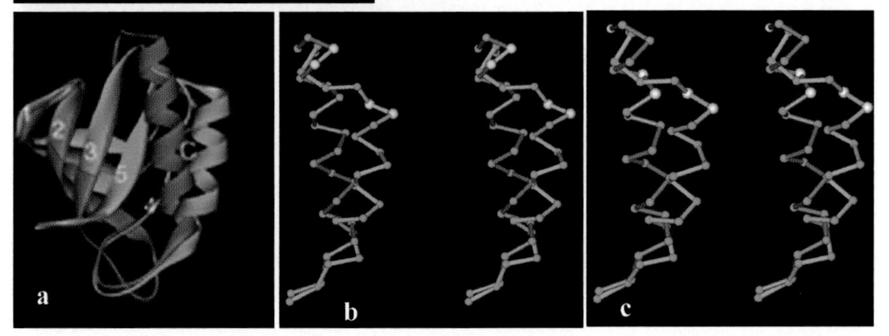

Plate 20.2. (**a**) Displacement of the C-helix in an alignment of two tyrosine kinases, Insulin Receptor Tyrosine Kinase (*orange*, PDB id 1IR3:A) and c-Src (*light blue*, PDB id 2SRC). See page 472 for detailed description. (**b**) Helix C as aligned by CE, shown in stereo. This element is shown alone for clarity, but is in the same orientation as in **a**. See page 472 for detailed description of color coding. (**c**) Helix C from our manual alignment, shown in stereo. All conventions are the same as in **b**. See pages 472–473 for detailed description.

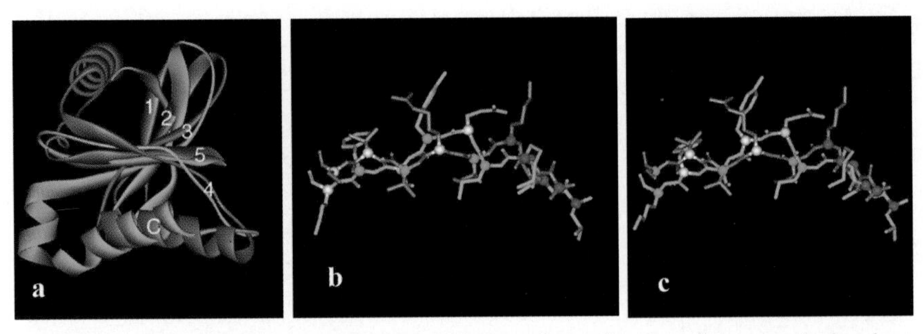

Plate 20.3. (**a**) Strand 4 of the protein kinase fold, shown in an alignment of the n-terminal regions of the Actin-Fragmin kinase (*orange*, PDB id 1CJA:A) and cAMP-Dependent Protein Kinase (*light blue*, PDB id 1CDK:A). See page 473 for detailed description. (**b**) Hypothetical alignment of strand 4 of 1CDK:A and 1CJA:A. Strands are color keyed in same way as in **a**. See pages 473–474 for detailed description of color coding. (**c**) Alignment of strand 4 as generated by CE and our manual alignment. All conventions are the same as in **b** (see page 474.)

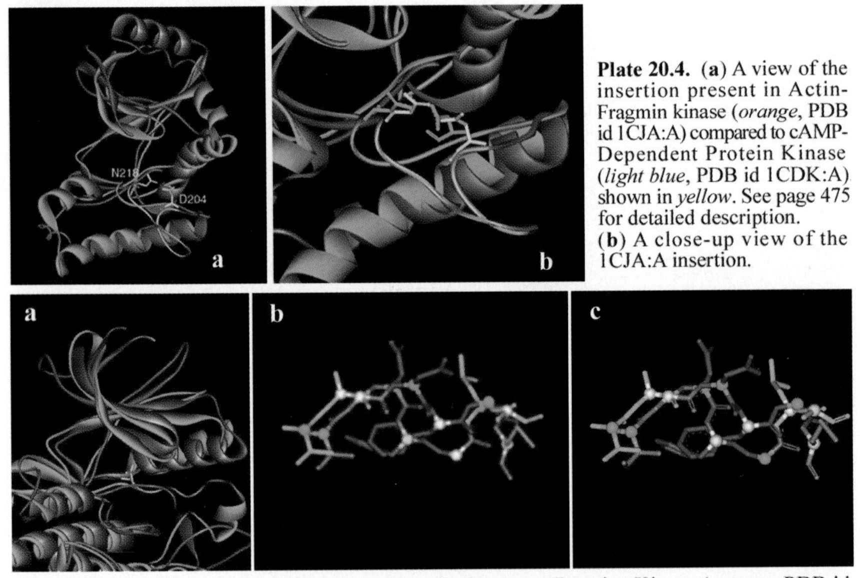

Plate 20.4. (**a**) A view of the insertion present in Actin-Fragmin kinase (*orange*, PDB id 1CJA:A) compared to cAMP-Dependent Protein Kinase (*light blue*, PDB id 1CDK:A) shown in *yellow*. See page 475 for detailed description. (**b**) A close-up view of the 1CJA:A insertion.

Plate 20.5. (**a**) A view of an alignment of Insulin Receptor Tyrosine Kinase (*orange*, PDB id 1IR3:A) and c-Src (*light blue*, PDB id 2SRC), shown with the DFG motif in *yellow* in 2SRC. The Asp residue of this structure is shown. See page 476 for detailed description. (**b**) The DFG motif and surrounding sequence as aligned by CE. The two chains are colored as in **a**, with the DFG motif in each in a darker color (*red* for 1IR3:A, *blue* for 2SRC). See page 476 for detailed description of color coding. (c) The DFG motif as aligned by our manual alignment. Color coding and all other aspects are identical to b, with the exception of the alignment.

8

From the Building Blocks Folding Model to Protein Structure Prediction

Nurit Haspel, Chung-Jung Tsai, Haim Wolfson, and Ruth Nussinov

Abstract

To fold an amino acid chain into its three dimensional structure is a computationally huge effort, requiring an enormous amount of time. To reduce the time required in such an endeavor, here we propose to mimic the folding pathway undertaken in nature. To carry out such a scheme, we make use of our recently developed building block protein folding model. This model has been shown to be consistent with experiment and theory.

The model postulates that protein folding is a hierarchical process. Folding initiates by local building block fragments, which constitute local minima along the protein sequence. These conformationally fluctuating building blocks combinatorially hierarchically assemble to finally yield the native fold. The model further postulates that folding is guided by the native interactions. We have applied this model to the entire protein structural data bank. The resulting building blocks have been clustered. Via a building block assigning algorithm, sequence comparisons and weighting scheme, building blocks are assigned to a target protein sequence. In the next step their structures are simulated and finally, in the most computationally intensive step, combinatorially assembled. The combinatorial assembly step is not available yet. Here we illustrate the first stages of this proposed computational protein folding scheme.

8

From the Building Blocks Folding Model to Protein Structure Prediction

Nurit Haspel[†], Chung-Jung Tsai[‡], Haim Wolfson[†], and Ruth Nussinov[†‡]

[†]*Tel Aviv University, Tel Aviv, Israel,*
[‡]*NCI-Frederick, Frederick, Maryland*

8.1 Introduction

The enormity and the challenge of the problem of computationally predicting the structure of a protein from its amino acid sequence is only too well recognized. On the other hand, the importance of addressing it is equally well understood. The scientific approaches and concepts to the so-called protein folding problem have changed over the years. About 35 years ago, Levinthal postulated what came to be known as the Levinthal paradox. Given the huge number of possible intermediates, Levinthal wondered how the native protein conformation could be achieved on a biological time-scale of a couple of minutes. At that time it was commonly accepted that each amino acid chain goes through some given pathway, and consequently the same intermediates. The early '90s have seen a revolution of concepts. The problem of protein folding was viewed in the context of populations and ensembles. It was realized that the different intermediates were sampled by the population, not by a given individual chain. The so-called "new view"[1-12] proposed that there are multiple pathways going down the energy landscape. There are no obligatory intermediates that each chain has to sample. While this approach suggested that the Levinthal paradox simply does not exist, it built no simplifying inroads into the immense computational problem. On the practical side, the way to computationally predict protein structure still addressed the huge conformational space, which needs to be sampled by effective search algorithms. This is particularly awesome for long chains. However, even relatively short ones, on the order of tens of residues are still practically impossible.

The last few years have again seen some conceptual shift. Now popular belief holds that while in principle many pathways exist for the chain to undertake as it rolls down the funnel slope, in practice only very few of these are highly populated. Currently we accept that there is a major folding pathway which most molecules follow. This pathway is guided by the native interactions, where local elements fold and hierarchically bind to yield the native fold.[13,14] Hierarchical folding is implicated by the funnel-shaped landscape. Hence, in principle, we should be able to progressively dissect the protein structure to reveal its folding pathways.[15-20] The building block computational procedure follows such a rationale. The building block folding model postulates that protein folding is a hierarchical process. It initiates with the local conformationally fluctuating building blocks and proceeds with their hierarchical combinatorial assembly into stable, independently folding hydrophobic units (HFUs), domains, and finally the entire protein. This process is guided by the native state.

Such a scheme enables, not only deriving folding pathways, but reducing the complexity of computational structural prediction. First, we cut a protein sequence into its likely building block fragments. Second, rather than simulate the conformation of the entire chain, we simulate the fragments. Third, we combinatorially dock the building blocks, seeking "optimal" conformations. This last combinatorial assembly step is the bottle-neck of the calculation. Hence, our goal is at each stage to cut the sequence into as few fragments as possible. The second major hurdle of this computational scheme is how to identify building blocks on the sequence, so we cut the sequence into the "correct" fragments. To address this sequence-cutting or building block assignment problem, we carry out a systematic study of the likely building blocks we already have, namely, those derived from protein structures available in the protein data bank (PDB).[21] In parallel, we design a graph theoretic algorithm to use this information in the cutting of the chain. The graph algorithm is a computationally efficient way to do the assignment.

Below, we first describe the building block folding model and its consistency with current experimental and theoretical results.[17,18,22] We follow with a description of the algorithm. We outline the procedure of cutting the native structure to obtain both its building block components, and the pathway through which they associate. This folding pathway can be straightforwardly read from the anatomy tree, which is automatically created by the algorithm. We further give two examples. While these illustrate the major folding pathway predicted by the algorithm, it is important to realize that the pathway actually followed is an outcome of the physical conditions. Under different sets of conditions, different pathways may be preferred. Finally, we next describe how the rich dataset of building blocks, which has been created from all proteins in the PDB can be utilized in the cutting of sequences into their candidate building block fragments. The second aspect of the problem, that of combinatorially putting them together to create a tertiary fold, is beyond the scope of this chapter.

8.2 Protein Folding: A Process of Intra-Molecular Building Block Recognition

Protein folding may be viewed as a process of intramolecular recognition.[15,23-25] Such recognition implies that a native structure is the outcome of mutual stabilization between its favorable constituent structural units. Hence, protein folding, and protein-protein association, which involves intermolecular recognition, are similar processes governed by similar principles. In both. the driving force is the hydrophobic effect,[26] although charge or polar complementarity are also important contributors.[27] This analogy between folding and binding is consistent with viewing protein folding as a hierarchical process,[13,14] involving a combinatorial assembly of a set of conformationally fluctuating building blocks.[15,18] Such a favorable assembly leads to compact, stable, independently folding hydrophobic units. However, unlike the case of protein-protein association, in protein folding the structural units are chain-linked. Associations between sequentially connected structural elements, whether building blocks, or independent stable folding units. are entropically and kinetically favorable.

While the independently folding hydrophobic unit is stable even when separated from the remainder of the chain, this is not the case for a building block. A building block is of variable size, and is determined by the local coding of the sequence, and hence by local interactions. It may consist of a single secondary structure element, or a contiguous fragment of interacting elements. In particular, a building block may be defined as a fragment with transient, highly populated conformation. A given building block fragment may have several favorable alternate conformations. A building block may twist or may open up during cooperative hydrophobic collapse, losing its intra-segment interactions. If isolated, the building block is unstable. It is the cooperative, mutual interactions between building blocks that stabilize both the building block, and the entire protein conformation. According to the building block folding model, folding is a hierarchical process: formation of fluctuating local building blocks, their association into hydrophobic folding units; association of these units to form domains; and subsequently, entire proteins. Such a definition of the building blocks also implies that not all fragments of the sequence can be classified as building blocks. Some sequence segments do not show preferred populated conformations. In our scheme, such segments are considered as "unassigned" blocks, along with short flexible linkers between the building blocks. Such a description is consistent with the occurrence of a stable structure containing both native and non-native interactions in the denatured state.[28]

8.3. Experimental and Theoretical Support for the Building Block Concept

Building blocks have been either explicitly observed or implied in both experiment and theory. In solution, peptides frequently co-exist in a wide range of conformations. Short peptides such as α-helices or β-strands, have been observed in solution with substantial population times.[29-31] Peptides have been shown to have stable structures at temperatures below the living temperature.[32] Secondary structures have been observed during the very early stages of the folding process.[33,34] Peptides have also been considered as model systems for initiation of protein folding.[35-37] Recurring supersecondary structures have been reported in a number of database structure analyses.[38-40] The description of the initial formation of "microdomains" in protein folding in the collision-diffusion model[41] is also characterized by salient features similar to the formation of building blocks. The "foldon" approach,[42,43] where a protein is built from an assembly of foldons, also corresponds well to the building blocks concept described here. Additionally, compact structures of short peptide fragments have been detected in many three dimensional structures of proteins.[44,45] Nevertheless, when identifying a building block directly from the native structure we face a problem: What we see is the final, static conformation.

Viewing the protein structure as composed of building block units suggests that protein folding is a combinatorial assembly of these blocks, rather than of single amino acid residues. This is in contrast to the proposed tertiary nucleation protein folding mechanism where single, isolated residues uniquely form critical interactions regardless of their sequence separation and local neighborhoods. A combinatorial assembly of building block fragments, each of which consists of a contiguous stretch of residues, implicitly reduces the search space of the chain during the folding process.

Considering the protein structure as being made of building blocks rationalizes a range of observations and descriptions: First, building blocks are consistent with both a two-state kinetics and folding with detectable, "molten globule" like, intermediates.[13,14] According to the building block model, both types of mechanisms may be viewed as essentially being identical. The difference is in the barrier heights. In very stable proteins showing a two-state kinetics, such as in chymotrypsin inhibitor 2 (CI2)[46] and in cold-shock protein B (Csp B),[47] folding is fast, as the barriers which are encountered are low. On the other hand, in proteins showing a three-state kinetics, such as in α-lactalbumin (α LA), apomyoglobin (apoMb), Rnase H, barnase, and cytochrome c (cyt c),[13] the secondary structures of the building blocks may partially form first. Here, during the combinatorial assembly of the building blocks, they twist and open. These larger conformational changes lead to higher barriers, slower folding rates, and consequently detectable intermediates.

Second, for some cases, such as in β-lactoglobulin,[48] a non-native α-helical intermediate has been observed during the refolding of this predominantly β-sheet protein. Such conformational change of a fragment of the sequence fits the building block model. In this case, the β-sheet conformation has a higher population time, and therefore is the one observed in the final, native structure. During folding, the initial, transient α-helical building block conformation opens up and refolds into the more stable β-structure.

Third, the building block folding model is in agreement with the diffusion-collision model.[41] In the building block model, the description of the building blocks is in terms of population times. Only if we look at building blocks from this point of view, can their hierarchical assembly be consistent with both the diffusion-collision model and with peptide populations. Thus, for example, NMR studies showing high population times of hydrophobic β-hairpins[49-52] are related to building block conformational populations.

Fourth, the barrier heights relate to the population times of the building block fragments, as compared to their conformations in native protein structures. If the conformation of the fragment has a high population time, and that same conformation is observed in the final native state, the barrier will be low. In contrast, if there is a conformational change and the conformation which is observed in the native state is different from the conformation of the fragment in solution, a high barrier will be observed. Thus, if both the conformation that has a high population time in solution and the native protein structure are available, the kinetics of the protein may be predicted. Heat capacity measurements yield estimates of population times.

Fifth, the building blocks model is consistent with hydrophobic collapse.[53-56] Hydrophobic collapse may involve cooperative formation of interactions between building blocks, rather than between isolated residues. The critical issue is the barrier heights. If, however, the conformations of the mutually stabilizing interacting building blocks in the collapsed state are similar to the ones in the native state with only minor alterations, the barriers are low, and a fast two state-like folding is observed.

Additionally, sixth, the general definition of a transition state is not a single obligatory step through which all molecules pass.[57] It has been shown that transition state ensembles. at least in β-sheet proteins, can be well defined and conformationally restricted.[58,59] Some native contacts are more probable than others. Recent studies of β-sheet proteins[58-60] suggest that the formation of a β-hairpin along with some compact tertiary contacts may be obligatory steps in the folding of these proteins. In our terminology, a β-hairpin is cut as a building block. In such cases, a stable hairpin with a high population time[49-52] is retained in the final, native structure. Data in support of obligatory on-pathway intermediates in the folding of the Greek key protein apo-pseudoazurin have also been presented.[61] Further support for the importance of local interactions, and hierarchy in folding (and binding) may be derived from the encouraging results obtained by a recent protein structure prediction algorithm, which is based on enumeration of local structures in proteins, and the use of a library of short sequence-structure motifs.[62]

8.4. The Building Block Cutting Algorithm

The stability measurement of an isolated contiguous fragment in a particular conformation should reflect the population time of the conformation during the folding process.[18] The larger the measurement, the larger the time that the fragment spends in this conformation. To locate building blocks from a native protein 3-D structure, we find a set of non-overlapping contiguous fragments that have the highest conformational stability among all other possible candidate combinations. In practice, we allow few residue-overlaps between the fragments.

For a polypeptide chain with a size of N_e residues, and with a size limit of a building block set to N_s residues, the total number of candidate fragments is $N_{total} = \Sigma(N_e - N_i + 1)$, where N_i is the sum from N_s to N_e. A scoring function measures the conformational stability of each fragment, regardless of its size. The "fragment map" is a two-coordinate system giving the position of the fragment (by its central residue) versus the fragment size. The local minima depicted on the map are the locations of the building blocks of a given protein.

8.5. The Scoring Function

The scoring function is fragment size-independent. It is based on a previous scoring function to locate hydrophobic folding units.[24] The HFU scoring function has four ingredients: compactness, hydrophobicity, degree of isolatedness, and number of segments. A building block has only one segment. The function is a linear combination of the three measurements. Each is a quantity calculated as the deviation from the averaged value of known protein structures. The building block scoring function, $Score^{BB}$, is

$$Score^{BB}(Z,H,I) = \frac{(Z^1_{Avg}-Z)}{Z^1_{avg}} + \frac{(H-H^1_{Avg})}{H^1_{avg}} + \frac{(I^1_{Avg}-I)}{I^1_{avg}} + \frac{(Z^2_{Avg}-Z)}{Z^2_{avg}} + \frac{(H-H^2_{Avg})}{H^2_{avg}} + \frac{(I^2_{Avg}-I)}{I^2_{avg}} .$$

Here Z, H, and I are respectively, the compactness, the hydrophobicity, and the degree of isolatedness of a candidate fragment. The corresponding arithmetic average, X_{avg}, and standard deviation, X_{dev}, were determined from a non-redundant dataset of 930 representative single-chain proteins. Average and standard deviation with superscript 1 are calculated with respect to fragment size. Average and standard deviation with superscript 2 are calculated as a function of the fraction of the fragment size to the whole protein.

Compactness, Z. Z is the solvent accessible surface area of a fragment ASA_{surf}, divided by its minimum possible surface area. The minimum possible surface area is the surface area of a sphere with an equal volume as that of the fragment, *Vol*. It is calculated by an integration of all individual exposed solvent accessible surface areas. Z is:

$$Z = \frac{ASA_{surf}}{\left(36\pi * Vol^2\right)^{\frac{1}{3}}} .$$

Hydrophobicity, H. The hydrophobicity is defined as:

$$H = \frac{ASA^{non}_{buried}}{ASA^{non}_{buried} + ASA^{non}_{surf}} ,$$

where ASA^{non}_{buried} and ASA^{non}_{surf} are the buried and the exposed non-polar ASA, respectively. The hydrophobicity is calculated as the fraction of the buried non-polar area.

Isolation, I. The degree of isolation is:

$$I = \frac{ASA^{Non}_{B \to E}}{ASA_{frag}},$$

where $ASA^{Non}_{B \to E}$ is the non-polar ASA that was originally buried in the interior of a protein but became exposed after cutting and ASA_{frag} is the solvent accessible surface area of the isolated fragment. The size-independent stability form of the function assumes that fragments of different sizes have equal averaged conformational stability. Thus, instead of using statistical values, the H_{avg} and I_{avg} in the scoring function are calculated from a fitted straight line, to reflect the relative size-dependent stability.[18]

8.6. The Cutting Procedure

Step 1: Locating building blocks. For a given protein 3-D structure, we calculate the stabilities of all fragment candidates. To locate all local minima on the fragment map, we look for the highest stability in a defined local region. A local region is a quasi-circle, with a radius of 7.5% of the size of a candidate fragment. To locate all of the building blocks, every fragment candidate is examined to see if it is the highest scoring within its local region as defined above. If a candidate is the highest scoring one, we register it in the list of building blocks.

Step 2: An iterative top-down cutting process to construct the anatomy tree. The folding process does not follow a single pathway. The anatomy tree should straightforwardly yield the most likely folding pathway(s). It should also identify the set of the most likely building blocks which, via combinatorial assembly, form the native protein conformation. The anatomy of the protein structure is organized as a tree which grows upside down, with the starting native protein node at the top. Each node represents a fragment. A node can sprout multiple branches to create child nodes. These are generated via a multi-cutting procedure. If a new node does not produce a child, it is an end node. The level of a node is determined by counting the number of steps that are needed to trace back to the root node. Tree growth stops when no new children nodes can be generated. At the end of the process, the collection of end nodes is the set of the most likely building blocks. And, the tree organization depicts the most likely folding pathway.

The recursive top-down multi-cutting procedure allows any number of branches at any level. Starting with a node fragment and the building blocks created in the first step, we search for a set of fragment that constitutes the entire node fragment. Here, a short 7 residues overlap between building blocks is allowed. If an unassigned segment is less than 15 residues, it is left unassigned. Otherwise, it becomes a low-score building block, not listed among the original building blocks. A node is an end node if there are no two building blocks with scores above a threshold value. The sum of the first two fragment scores is used to rank all building blocks. At each branching level, we locate the HFUs by combinatorially assembling the collection of building blocks. This procedure is straightforwardly implied from the building block folding model.[18] Fig. 8.1 and Plate 8.1 present an example of the building block cuttings and the anatomy tree of a simple, sequentially folding algorithm, hemopexin (PDB code-1pex). In sequential folding, building blocks that are sequentially connected interact in the three-dimensional fold. Inspection of the fragment map and the anatomy tree immediately illustrates that, just by looking at the tree, we can immediately see the sequential folding pathway undertaken by the protein.

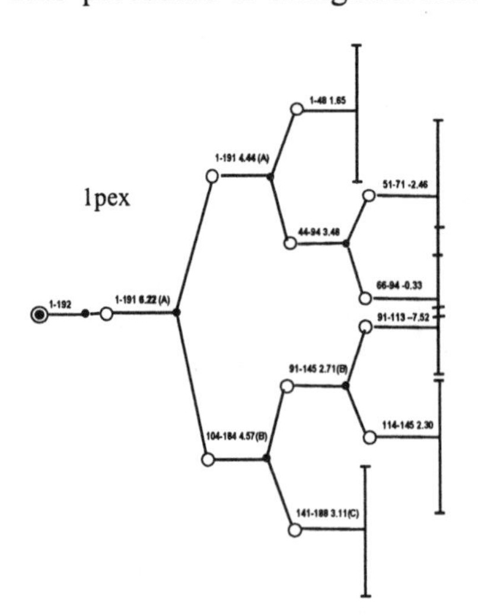

Figure 8.1. The anatomy tree of human hemopexin (PDB code 1pex), a sequentially folding protein.

8.7. Critical Building Blocks

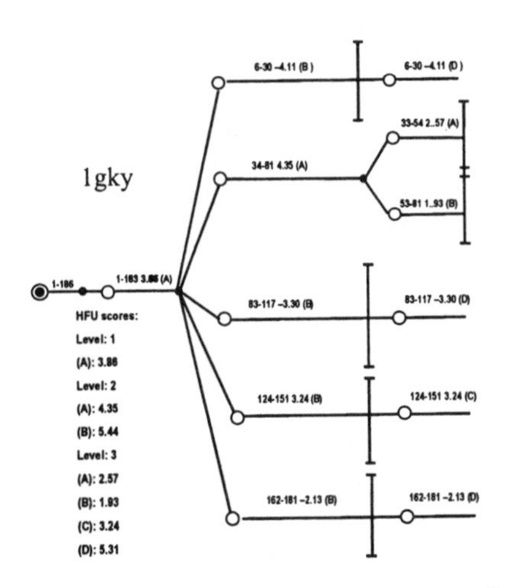

Figure 8.2. The anatomy tree of guanylate kinase (PDB code 1gky), a complex folding protein.

Fig. 8.2 and Plate 8.2 present an example of the cutting and the folding pathway of a complex folding protein, guanylate kinase (PDB code-1gky). In such a case, the building blocks that interact are not sequence-connected. Plate 8.3 shows the building block assignment of guanylate kinase at cutting level 2. Table 8.1 shows the coloring scheme. Inspection of the anatomy trees can immediately indicate if that is the case. In some complex-folding proteins, particularly in large ones, some building blocks are more critical for achieving the correct structure than others. These building blocks are dubbed "critical building blocks." Critical building blocks can be viewed as intramolecular chaperone-like building block fragments. Owing to the complexity of the fold, proteins containing critical building block(s) may be more prone to misfolding.

Table 8.1. The coloring scheme of the building block assignment for guanylate kinase shown in Plate 8.3. The unassigned fragments are colored in gray.

#	Color in figure	Size	C_α range
1	blue	25	6-30
2	red	48	34-81
3	orange	35	83-117
4	green	28	124-151
5	purple	20	162-181

There are three major considerations in selecting a candidate intramolecular chaperone-like critical building block fragment.[63,64] First, the candidate intramolecular critical building block mediates the interactions of other, sequentially-connected building blocks. Second, it should be in contact with most other building blocks in the structure; third, and most importantly, in its absence, non-native interactions between the building blocks themselves are likely to take place. However, the conformations of the assembled building blocks are likely to remain native-like. Under such circumstances, alternate less stable conformations may be selected in the combinatorial assembly, mutually stabilizing each other. The sequence of a critical building block is likely to be conserved in different organisms. This suggests that mutations occurring in critical building blocks are more likely to have deleterious effects on the protein conformation than those occurring elsewhere in the protein sequence.

Kumar *et al.*[64] have designed an algorithm to identify critical building blocks in the protein structure. Its input are the set of building blocks, obtained at different levels of cuttings of the native structure, in the generation of the anatomy tree.[18] The algorithm assigns a numerical value to each building block, based on its location in the protein, the identity and the number of other building blocks it contacts, and its surface area buried by such contacts. A building block whose critical building block index (CIndex) is greater by at least two standard deviations than the average building block Cindex value at a given hierarchical level is identified as critical for that level. If this building block is not cut further into component building blocks at lower levels, and if it has significantly high CIndex values at more than one level, it is considered as a critical building block (CBB) for the protein. We have applied this critical building block detection test to all the building blocks in the proteins in the PDB.

Clearly, if, when assigning building blocks to the target sequence, we are able to determine which fragment serves as a critical building block, the next bottle-neck task of combinatorial assembly will be greatly facilitated, as the number of potential arrangements of the building blocks in the three dimensional fold is reduced.

8.8. From the Building Block Folding Model to Structure Prediction: The Scheme

Our goal is to cut a given protein sequence into fragments, each of which constitutes a building block, i. e., a local minima. That is, our goal is to assign building blocks to an amino acid chain. The input is a protein sequence, and the collection of building blocks, created from the PDB. The tool is a graph theoretic sequence assignment algorithm, which finds the "best" building block assignments. The flow chart of the algorithm and an illustration of how it works are given in Fig. 8.3.

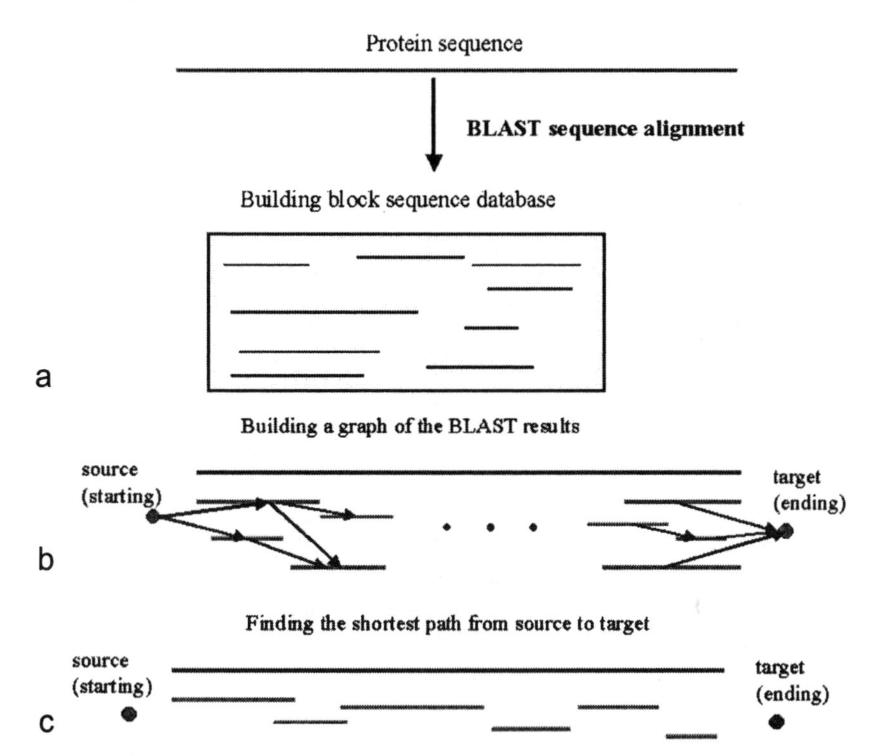

Figure 8.3. An illustration of the stages of the building blocks assignment algorithm.

The pool of building blocks undergoes a clustering procedure. To form the clusters, building blocks (i) similar in size (the size of the smaller protein is at least 70% of the size of the larger), (ii) similar in structure (for most under 1.8Å), (iii) from similar cutting levels, and (iv) belonging to the same fold (all-α; α+β; α/β; and all-β) are clustered. The clustering is done iteratively. Its details will be presented elsewhere.

Initially, the clusters contain building blocks from the same level of cutting. Since they are clustered by structural comparisons, all have similar structures. However, their sequences can be similar or dissimilar. Subsequently, the sequences within each cluster are compared, and if different ones are present, sub-clusters of these are formed. Additionally, all sequences from all clusters are compared, to make sure there are no highly similar sequences that belong to different clusters. Although unlikely, this may happen due to the (more rapid) clustering process which was selected. In this procedure, the probe building block is assigned into the first cluster it encounters if the structural similarity is under the RMSD threshold. Yet, similar sequences should have very similar structures. At the end of the multi-stage clustering process we are in possession of clusters of structurally similar building blocks. At the same time, we also have sub-clusters of similar sequences. That the structural similarity between members of the cluster is high can be seen in Figs. 8.4-8.7. In these figures a multiple structural comparison is carried out. However, in two of these clusters, the sequence alignments show that the sequences contributing to these similar structures are different. These are the more interesting cases.

```
                     *:****                  **.*** .*** **:*      :******::**:****:::*: ***:**
1gdd.196.294 --------FKMFDV----------------VTAIIFCVALSDYDLV------MNRMHESMKLFDSICNNKWFTDTSIILF   50
1git.189.293 -FTFKDLHFKMFDVGAQRSERKKWIHCFEGVTAIIFCVALSDYDLVLAEDEEMNRMHESMKLFDSICNNKWFTDTSIILF   79
1gfi.188.292 HFTFKDLHFKMFDVGGQRSERKKWIHCFEGVTAIIFCVALSDYDLVLAEDEEMNRMHESMKLFDSICNNKWFTDTSIILF   80
1tag.190.294 ------LNFRMFDVGGQRSERKKWIHCFEGVTCIIFIAALSAYDMVLVEDDEVNRMHESLHLFNSICNHRYFATTSIVLF   74
       ruler 1.......10........20........30........40........50........60........70........80

             *****:*.*****: *:**:*:* *
1gdd.196.294 INKKDLFEEKIKKSPLTICYPEYAGSN----    77
1git.189.293 INKKDLFEEKIKKSPLTICYPEYAGS-----   105
1gfi.188.292 INKKDLFEEKIKKSPLTICYPEYAG------   105
1tag.190.294 INKKDVFSEKIKKAHLSICFPDYNGPNTYED   105
       ruler ........90........100.......110.
```

Figure 8.4. An example of a cluster where the sequence identity among the members is very high. The sequence alignment is shown below.

Thus, we have in our possession clusters of building blocks, grouped by their structures and by their sequences. These form the input to our graph theoretic sequence assignment algorithm.

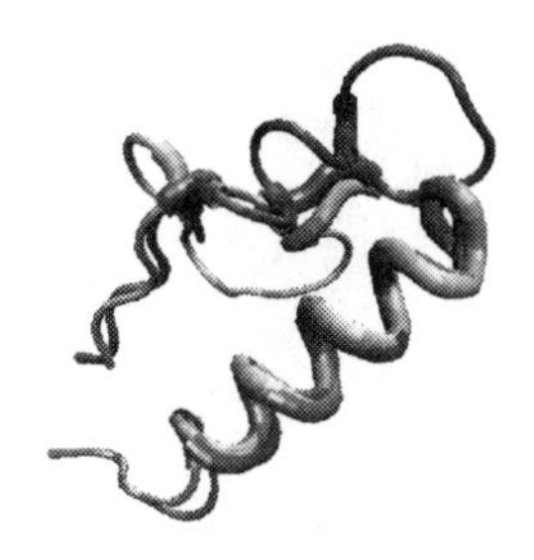

```
              .    : :   .    .    *:: :
1dbp.235.264 --QLPDQIG--AKGVETADKVLKGE--KVQAKYPVD    30
1drj.235.264 --QLPDQIG--AKGVETADKVLKGE--KVQAKYPVD    30
1grl.339.368 EAAIQGRVAQIRQQIEEATSDYDRE--KLQER----    30
1gca.260.293 -NQAKATFD-LAKNLAEGKGAADGTSWKIENKIVRV    34
       ruler 1.......10........20........30......
```

Figure 8.5. An example of a cluster where the sequence identity among the members is rather high. The sequence alignment is shown below.

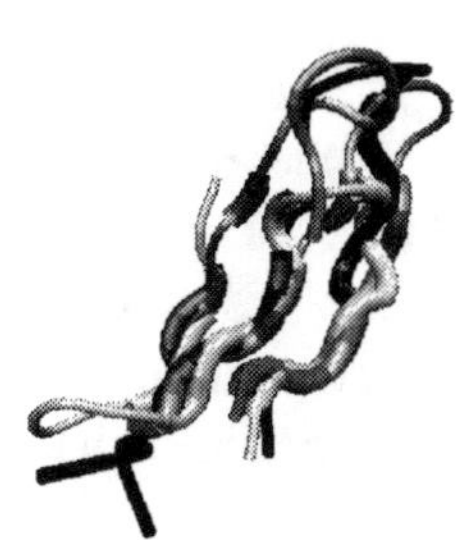

```
                .:   .:   .* : *    ::   .
1mmd.49.78 CGEIVSETSDSFTFKTVDGQDRQVKKDDAN    30
4ull.11.32 YTKYNDD--DTFTVK-V-G-DKELFTN---    22
 1sta.8.31 ----HKEPGGATLIKAIDGDTVKLMYKGQP    26
     ruler 1.......10........20........30
```

Figure 8.6. An example of a cluster where the sequence identity among the members is rather high. The sequence alignment is shown below.

For a given target sequence that we wish to fold, we carry out a sequence comparison with representatives of all building block clusters.

If a sequence similarity above a given threshold is found, this building block is represented as a weighted graph vertex. The weight can reflect several parameters—the sequence similarity, the similarity between the known secondary structure and the one predicted for the target sequence, the stability of the building block, a similarity in the location in the sequence, etc. The aligned building blocks are represented as a weighted directed acyclic graph (DAG). In the graph, two vertices can be connected if they are sequentially adjacent, and if they adhere to the rules followed in the generation of the building blocks from the native structures (no more than 7 residue-overlap, and not over 15 residues apart). The edge connecting the vertices is assigned the average weight of the two vertices. The next step involves finding the shortest paths of consecutive edges. To do that, we add two fictive vertices to the graph—the first one, called the "source" (starting) and the

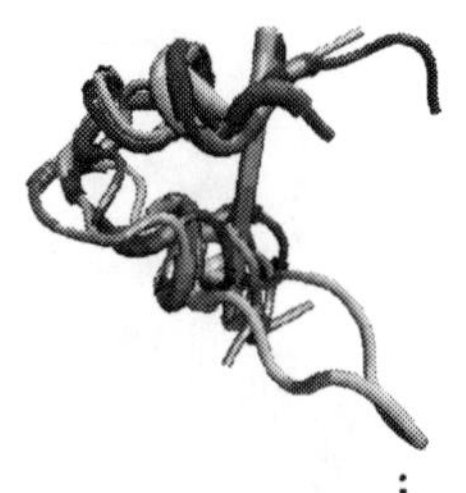

```
                                 :
1ddt.165.184 INFETRGKRGQD-AMYEYMAQ----     20
1sgk.165.184 INFETRGKRGQD-AMYEYMAQ----     20
  1bia.20.39 ---HSGEQLGETLGMSRAAINKH--     20
  1bib.19.38 --FHSGEQLGETLGMSRAAINK---     20
  1bib.39.60 -HIQTLRDWGVD--VFTVPGKGYSL     22
1aa6.401.420 ---MGEDPLQTDAELSAVRKAFE--     20
       ruler 1.......10.........20.....
```

Figure 8.7. An example of a cluster where the sequence identity among the members is not high. The sequence alignment is shown below.

last one, the "target" (ending) vertex (see Fig. 8.3b-c). A zero-weight edge connects each of these to a vertex which is up to a distance of 15 residues. These either follow (the starting "source" vertex) or are prior to (the ending "target" vertex). The shortest path algorithm[66] is used for this purpose. A path is actually a consecutive set of vertices that lead from the starting source vertex to the ending target vertex. Since each of these vertices stands for a building block, the path represents a building block cover of the protein sequence. That is a possible assignment of building blocks to the sequence. Among the obtained paths, the highest scoring ones are retained. The highest scoring paths are assumed to be better covers, with a higher probability of being a true possible building block assignment for the sequence. This stage of the work is still incomplete, and several weighting schemes will be tried. Figs. 8.8 and 8.9 give examples of a building block assignment to a trial protein sequence.

The reason for using a graph-based algorithm instead of a simple assignment scheme is the efficiency of the calculation. Using this scheme considerably shortens the time of the search and by reducing the number of assigned candidate building blocks. The graph-based algorithm is an efficient way to do an assembly process. Following the sequence assignment, the original conformation of the building block in the known native structure is assigned to the candidate building block. A simulation can then be applied to test its stability.

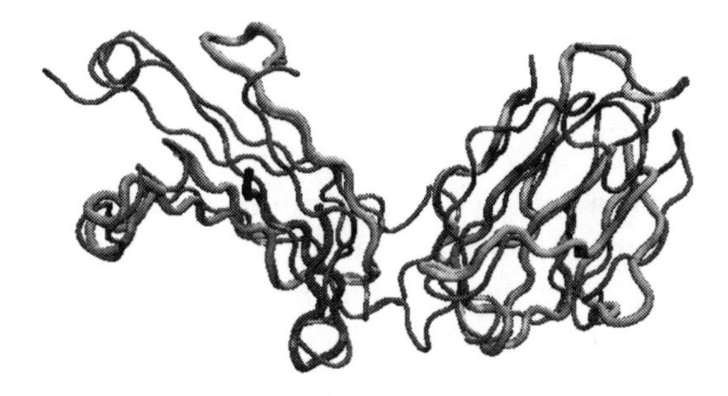

```
        1bac AVTQSPRNKVAVTGGKVTLSCQQTNNHNNMYWYRQDTGHGIRLIHYSYGAGSTEKGDIPDGYKASRPSQEQFSLIIELAT 80
   1mak.4-46 -MTQTPLSLPVSIGDQASISCRSSQSLVHSNGNTYINWYLQKAG------------------------------------
  1bra.33-70 -----------------------------LIWQQKLGKAPNLLIYDASTIETGVPSRFSGSGSGTE--------------
 1malA.81-98 -------------------------------------------------------------------------ILMNSLE

        1bac  PSQTSVYFCASGGGRGSYAEQFFGPGTRLTVIEDLRQVTPPKVSLFEPSKAEIANKQKATLVCLARGFFPDHVELSWWVN 160
 1malA.81-98  PEDTAIYYCAADY-------------------------------------------------------------------
4bj1B.117-133 ----------------------------------------PTVTLFPPSSEELQANK-----------------------
1bjm.135-159 ------------------------------------------------------------TLCLISDFYPGAVTVAWKAD

        1bac GKEVHSGVSTDPQAYKESNYSYCLSSRLRVSATFWHNPRNHFRCQVQFHGLSEEDKWPEGSPKPVTQNISAEAWGRAD   238
1bjm.135-159 GSPV--------------------------------------------------------------------------
1wba.145-160 ------GISTDPEGKKRLVVSY--------------------------------------------------------
```

Figure 8.8. An example of the building block assignment algorithm operated on the T cell antigen receptor from mouse (PDB code - 1bec). The fragments that were assigned by the algorithm are shown in thicker lines on the upper figure. The sequence alignment is shown below. The naming is <pdb code>. <start position>- <end position>. As shown, a good match was achieved even when the sequence similarity was not striking. Some of the matching fragments came from distantly related proteins.

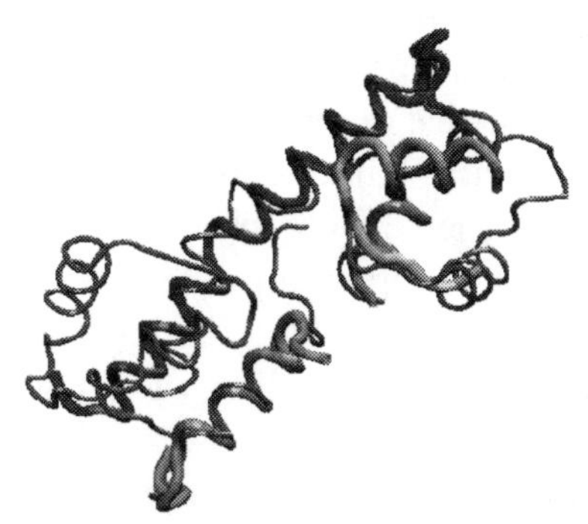

```
        1cfc ADQLTEEQIAEFKEAFSLFDKDGDGTITTKEIGTVMRSLGQNPTEAELQDMINEVDADGNGTIDFPEFLTMMARKMKDTD 80
 4tnc.94-111 -------SEEELADCFRIFDKNADG-------------------------------------------------------
1top.73-113 --------------------------------------------------------------IDFEEFLVKMVRQMKEDA

        1cfc ---SEEEIREAFRVFDKDGNGYISAAELRHVMTNLGEKLTDEEVDEMIREADIDGDGQVNYEEFVQMMTAK        148
1top.73-113 KGKSEEELANCFRIFDKNADGFI------------------------------------------------
  1gsq.14-32 ------------------------AELCRFVLAAHGEEFTDRV----------------------------
1cnpA.58-80 ------------------------------------------------DDLDRNKDQEVNFQEYITFIGAL
```

Figure 8.9. An example of the building block assignment algorithm operated on calmodulin from a frog (PDB code - 1cfc). (See more details in the text.)

8.9. Conclusions

Here we have outlined a potential way to reduce the computational time in protein folding. Our scheme mimics protein folding pathways. It proposes to initiate the computational folding by identifying "building blocks," i. e., fragments which form local minima along the protein sequence. These would be folded first. The next step is to combinatorially put them together in order to generate the entire protein fold. Clearly, there are two critical problems in such a scheme: First, how to identify such "building blocks" fragments along a protein sequence and second, how to put these together to generate a three-dimensional fold.

This scheme derives from the building block folding model. The model postulates that protein folding is a hierarchical process. It initiates from local conformationally fluctuating building blocks and proceeds via a combinatorial assembly to gradually reach the stable native fold. The building block folding model has been implemented in a computational scheme, to cut native protein structures into their building block fragments. The concept, and the results which have been obtained, correspond nicely with experiment.

In practice, our proposition is then to use the collection of building blocks which have already been generated from the structures in the PDB. These building blocks are clustered by their structures and by their sequences. The sequence of the target protein is compared against these clusters. Via an efficient building block assignment algorithm, and a reasonable weighting scheme, "optimal" building block assignments can be made. This graph theoretic algorithm finds the shortest paths connecting the candidate assigned building blocks. Candidate building block fragments either sequentially follow each other to cover the entire chain, or any specified (sub)domain.

The next steps in our scheme, not addressed here, are first assigning structures to these fragments from the original building block clusters, and simulations to probe their stability and conformations. The second step is to combinatorially assemble them, to yield the tertiary fold. This last step is the most time consuming in our scheme.

Acknowledgements

We thank Sandeep Kumar, Buyong Ma, and in particular J. V. Maizel for discussions and encouragement. The research of R. Nussinov and H. J. Wolfson in Israel has been supported in part by the Ministry of Science grant, and by the Center of Excellence in Geometric Computing and its Applications funded by the Israel Science Foundation (administered by the Israel Academy of Sciences). The research of H.J.W. is partially supported by the Hermann Minkowski-Minerva Center for Geometry at Tel Aviv University. This project has been funded in whole or in part with Federal funds from the National Cancer Institute, National Institutes of Health, under contract number NO1-CO-12400. The content of this publication does not necessarily reflect the view or policies of the Department of Health and Human Services, nor does mention of trade names, commercial products, or organization imply endorsement by the U.S. Government.

References

1. Bryngelson JD, Wolynes PG: **Intermediates and barrier crossing in a random energy model (with applications to protein folding).** *J Phys Chem* 1989, **93**:6902–6915.
2. Karplus M, Shakhnovitch EI: **Protein folding: theoretical studies of thermodynamics and dynamics.** In: *Protein Folding.* Creighton T, ed. New York: W. H. Freeman & Sons,. 1992:127–195.
3. Baldwin RL: **Matching speed and stability.** *Nature* 1994, **369**:183–184.
4. Baldwin RL: **The nature of protein folding pathways: The classical versus the new view.** *J Biomol NMR* 1995, **5**:103–109.
5. Dill KA, Bromberg S, Yue K, Fiebig KM, Yee DP, Thomas PD, Chan HS: **Principles of protein folding: A perspective from simple exact models.** *Protein Sci* 1995, **4**:561–602.
6. Karplus M, Sali A, Shakhnovitch E: **Comment: Kinetics of protein folding.** *Nature* 1995, **373**:664–665.
7. Wolynes PG, Onuchic JN, Thirumalai D: **Navigating the folding routes.** *Science* 1995, **267**:1619–1620.

8. Onuchic JN, Wolynes PG, Luthey-Schulten Z, Socci ND: **Towards an outline of the topography of a realistic protein folding funnel.** *Proc Natl Acad Sci USA* 1995, **92**:3626–3630.

9. Karplus M: **The Levinthal paradox: Yesterday and today.** *Folding & Design* 1997, **2**:S69–S75.

10. Dill KA, Chan HS: **From Levinthal to pathways to funnels.** *Nature Struct Biol* 1997, **4**:10–19.

11. Lazaridis T, Karplus M: **"New view" of protein folding reconciled with the old through multiple unfolding simulations.** *Science* 1997, **278**:1928–1931.

12. Frauenfelder H, Leeson DT: **The energy landscape in non-biological molecules.** *Nature Struct Biol* 1998, **5**:757–759.

13. Baldwin RL, Rose GD: **Is protein folding hierarchic? I. Local structure and peptide folding.** *Trends Biochem Sci* 1999, **24**:26–33.

14. Baldwin RL, Rose GD: **Is protein folding hierarchic? II. Folding intermediates and transition states.** *Trends Biochem Sci* 1999, **24**:77–84.

15. Tsai C-J, Xu D, Nussinov R: **Protein Folding via Binding, and vice versa.** *Folding & Design* 1998, **3**:R71–R80.

16. Tsai CJ, Kumar S, Ma B, Nussinov R: **Folding funnels, binding funnels and protein function.** *Protein Sci* 1999, **8**:1181–1190.

17. Tsai CJ, Maizel JV, Nussinov R: **Distinguishing between sequential and non-sequentially folded proteins: Implications for folding and misfolding.** *Protein Sci* 1999, **8**:1591–1604.

18. Tsai CJ, Maizel JV, Nussinov R: **Anatomy of protein structures: Visualizing how a 1-D protein chain folds into a 3-D shape. Proc Natl Acad Sci USA** 2000, **97**:12038–12043.

19. Tsai CJ, Ma B, Kumar S, Wolfson H, Nussinov R: **Protein folding: Binding of fluctuating building blocks via population selection.** *CRC Crit Rev Biochem Mol Biol* 2001, in press.

20. Tsai CJ, Ma B, Sham YY, Kumar S, Wolfson H, Nussinov R: **A hierarchical, building blocks-based computational scheme for protein structure prediction.** *IBM J Res Develop*, issue on *Life Sciences* 2001, **45**:513–523.

21. Bernstein FC, Koetzle TF, Williams GJB, Meyer EF Jr, Brice MD, Rodgers JR, Kennard O, Shimanouchi T, Tasumi M: **The protein databank: A computer-based archival file for macromolecular structures.** *J Mol Biol* 1977, **112**:535–542.

22. Tsai CJ, Nussinov R: **Transient, highly populated building blocks folding model.** *Cell Biochem Biophys* 2001, **34**:209–235.

23. Wu LC, Grandori R, Carey J: **Autonomous subdomains in protein folding.** *Protein Sci* 1994, **3**:359–371.

24. Tsai C-J, Nussinov R: **Hydrophobic folding units derived from dissimilar monomer structures and their interactions.** *Protein Sci* 1997, **6**:24–42.

25. Tsai CJ, Nussinov R: **Hydrophobic folding units at protein-protein interfaces: implications to protein folding and protein-protein association.** *Protein Sci* 1997, **6**:1426–1437.

26. Dill KA: **Dominant forces in protein folding.** *Biochemistry* 1990, **31**:7134–7155.

27. Xu D, Tsai CJ, Nussinov R: **Hydrogen bonds and salt bridges across protein-protein interfaces.** *Protein Eng* 1997, **10**:999–1012.

28. Wang Y, Shortle D: **A dynamic bundle of four adjacent hydrophobic segments in the denatured state of staphylococcal nuclease.** *Protein Sci* 1996, **5**:1898–1906.

29. Kuhlman B, Yang HY, Boice JA, Fairman R, Raleigh DP: **An exceptionally stable helix from the ribosomal protein L9: Implications for protein folding and stability.** *J Mol Biol* 1997, **270**:640–647.

30. Ramirez-Alvardo M, Daragam VA, Serrano L, Mayo KH: **Motional dynamics of residues in a beta-hairpin peptide measured by ^{13}C-NMR relaxation.** *Protein Sci* 1998, **7**:720–729.

31. Munioz V, Thompson PA, Hofrichter J, Eaton WA: **Folding dynamics and mechanism of beta-turn formation.** *Nature* 1997, **390**:196–199.

32. Alba ED, Jimenez MA, Rico M, Nieto JL: **Conformational investigation of designed short linear peptides able to fold into b-hairpin structures in aqueous solution.** *Folding & Design* 1995, **1**:133–144.

33. Briggs MS, Roder H: **Early hydrogen-bonding events in the folding reaction of ubiquitin.** *Proc Natl Acad Sci USA* 1992, **89**:2017–2021.

34. Lu J, Dahlquist FW: **Detection and characterization of an early folding intermediate of T4 lysozyme using pulsed hydrogenexchange and two-dimensional NMR.** *Biochemistry* 1992, **31**:4749–4756.

35. Dyson HJ, Merutka G, Waltho JP, Lerner RA, Wright PE: **Folding of peptide fragments comprising the complete sequence of proteins. Models for initiation of protein folding I. Myohemerythrin.** *J Mol Biol* 1992, **226**:795–817.

36. Waltho JP, Feher VA, Merutka G, Dyson HJ, Wright PE: **Peptide models of protein folding initiation sites. 1. Secondary structure formation by peptides corresponding to the G- and H-helices of myoglobin.** *Biochemistry* 1993, **32**:6337–6347.

37. Shin JC, Merutka G, Waltho JP, Tennant LL, Dyson HJ, Wright PE: **Peptide models of protein folding initiation sites. 3. The G-H helical hairpin of myoglobin.** *Biochemistry* 1993, **32**:6356–6364.

38. Richards FM, Kundrot CE: **Identification of structural motifs from protein coordinate data: secondary structure and first-level supersecondary structure.** *Proteins* 1988, **3**:71–84.

39. Sun Z, Jiang B: **Patterns and conformations of commonly occurring supersecondary structures (basic motifs) in protein data bank.** *J Protein Chem* 1996, **15**:675–690.

40. Boutonnet NS, Kajava AV, Rooman MJ: **Structural classification of α β β and β β α supersecondary structure units in proteins.** *Proteins* 1998, **30**:193–212.

41. Karplus M, Weaver DL: **Protein folding dynamics: The diffusion-collision model and experimental data.** *Protein Sci* 1994, **3**:650–668.

42. Panchenko AR, Luthey-Schulten Z, Wolynes PG: **Foldons, protein structural modules, and exons.** *Proc Natl Acad Sci USA* 1996, **93**:2008–2013.

43. Panchenko AR, Luthey-Schulten Z, Cole R, Wolynes PG: **The foldon universe: A survey of structural similarity and self-recognition of independently folding units.** *J Mol Biol* 1997, **272**:95–105.

44. Zehfus MH, Rose GD: **Compact units in proteins.** *Biochemistry* 1986, **25**:5759–5765.

45. Zehfus MH: **Improved calculations of compactness and a reevaluation of continuous compact units.** *Proteins* 1993, **16**:293–300.

46. Itzhaki LS, Otzen DE, Fersht AR: **The structure of the transition state for folding of chymotrypsin inhibitor 2 analyzed by protein engineering methods: Evidence for a nucleation condensation mechanism for protein folding.** *J Mol Biol* 1995, **254**:260–288.

47. Srinivasan R, Rose GD: **LINUS: A hierarchic procedure to predict the fold of a protein.** *Proteins* 1995, **22**:81–99.

48 Hamada D, Segawa S-I, Goto Y: **Non-native α-helical intermediate in the refolding of β-lactoglobulin, a predominantly β-sheet protein.** *Nature Struct Biol* 1996, **3**:869–873.

49. De Alba E, Jimenez MA, Rico M, Nieto JL: **Conformational investigation of designed short linear peptides able to fold into β-hairpin structures in aqueous solution.** *Folding & Design* 1996, **1**:133–144.

50. De Alba, Rico M, Jimenez MA: **Cross-strand side-chain interactions versus turn conformation in beta-hairpins.** *Protein Sci* 1997, **6**:2548–2560.

51. Ramirez-Alvarado M, Blanco FJ, Serrano L: **De novo design and structural analysis of a model b-hairpin peptide system.** *Nature Struct Biol* 1996, **3**:604–612.

52. Dill KA, Fiebig KM, Chan HS: **Cooperativity in protein-folding kinetics.** *Proc Natl Acad Sci USA* 1993, **90**:1942–1946.

53. Gutin AM, Abkevitch VI, Shakhnovitch EI: **Chain length scaling of protein folding time.** *Phys Rev Lett* 1996, **77**:5433–5436.

54. Wolynes PG, Luthey-Schulten ZA, Onuchic JN: **Fast-folding experiments and the topography of protein folding energy landscapes.** *Chem Biol* 1996, **3**:425–432.

55. Eaton WA, Munoz V, Thompson PA, Chan CK, Hofrichter J: **Submillisecond kinetics of protein folding.** *Curr Opin Struct Biol* 1997, **7**:10–14.

56. Sosnick Mayne L, & Englander SW: **Molecular collapse: The rate-limiting step in two-state cytochrome c folding proteins.** *Proteins* 1996, **24**:413–426.

57. Onuchic JN, Socci ND, Luthey-Schulten Z, Wolynes PG: **Protein folding funnels: The nature of the transition state ensemble.** *Folding & Design* 1996, **1**:441–450.

58. Martinez JC, Pisabarro MT, Serrano L: **Obligatory steps in protein folding and the conformational diversity of the transition state.** *Nature Struct Biol* 1998, **5**:721–729.

59. Grantcharova VP, Riddle DS, Santiago JV, Baker D: **Important role of hydrogen bonds in the structurally polarized transition state for folding of the scr SH3 domain.** *Nature Struct Biol* 1998, **5**:714–720.

60. Gruebele M, Wolynes P: **Satisfying turns in folding transitions.** *Nature Struct Biol* 1998, **5**:662–665.

61. Capaldi AP, Ferguson SJ, Radford SE: **The Greek key protein apo-pseudoazurin folds through an obligate on-pathway intermediate.** *J Mol Biol* 1999, **286**:1621–1632.

62. Bystroff C, Baker D: **Prediction of local structure in proteins using a library of sequence-structure motifs.** *J Mol Biol* 1998, **281**:565–577.

63. Ma B, Tsai CJ, Nussinov R: **Binding and folding: In search of intra-molecular chaperone-like building block fragments.** *Protein Eng* 2000, **13**:617–627.

64. Kumar S, Sham YY, Tsai CJ, Nussinov R: **Protein folding and function: The N-terminal fragment in adenylate kinase.** *Biophys J* 2001, **80**:2439–2454.

65. Cormen TH, Leiserso CE, Rivest RL: *Introduction to algorithms.* Boston: MIT Press, 1990.

9

Protein Threading Statistics: An Attempt to Assess the Significance of a Fold Assignment to a Sequence

Antoine Marin, Joël Pothier, Karel Zimmermann, and Jean-François Gibrat

Abstract

To assess the reliability of fold assignments to protein sequences we have developed a new fold recognition method called FROST (Fold Recognition Oriented Search Tool) and a database specifically designed as a benchmark for this new method under realistic conditions. This benchmark database consists of proteins for which there exists, at least, another protein with an extensively similar three-dimensional structure in a database of representative three-dimensional structures (i.e., more than 65% of the residues in both proteins can be structurally aligned). Since the testing of our method must be carried out under conditions similar to those of real fold recognition experiments, no protein pair with sequence similarity detectable using standard sequence comparison methods such as BLAST or FASTA is accepted in the benchmark database. While using FROST, we achieved a coverage of 42% for a rate of error of 1%.

9

Protein Threading Statistics: An Attempt to Assess the Significance of a Fold Assignment to a Sequence

Antoine Marin[†], Joël Pothier[‡], Karel Zimmermann[†], and Jean-François Gibrat[†]

[†]*INRA, Versailles, France*
[‡]*Atelier de Bioinformatique, Paris, France*

9.1 Introduction

Computational biology plays a key role in genomic projects. In particular, *in silico* functional analysis, whose purpose is to assign a function to the product of newly discovered genes, is essential. It is based on a search for homologous proteins in database. Proteins are said to be homologous when they are related by descent from a common ancestor. Homologous proteins have similar three-dimensional (3D) structures and, in general, also exhibit close functions.

Most molecular biologists are well acquainted with programs such as BLAST[1] or FASTA.[2] These methods make use of sequence comparisons to infer an evolutionary relationship between two proteins. They are fast, reliable, easy to use and, above all, their principal strength lies in the statistical analysis they provide. Sequence comparisons result in alignments that are characterized by a score. The authors of BLAST were the first to provide a rigorous answer to the question of the (statistical) significance of this score. They showed, for ungapped local alignment, that the best scores are distributed according to an extreme value law whose parameters can be calculated from the substitution matrix used to compute the score and the lengths of the sequences that are compared. It is thus possible, for an alignment with a given score, to calculate the probability that a score greater or equal occurs just by chance. In practice, in BLAST or FASTA outputs, the score expected-value (E-value) is used. It has been shown[3] that this E-value constitutes the best criterion to judge whether two proteins are homologous.

However, it has also been shown by the same authors[3] that these sequence comparison methods become very inefficient when one needs to compare sequences of proteins belonging to very distant organisms. In such a case, sequences may have greatly diverged and there is no longer enough information left in the sequences to reliably infer a homology relationship between the two proteins.

Fold recognition techniques (also called threading methods) have been developed to tackle this problem. It is well known that 3D structures are better conserved than sequences. 3D structures, therefore, constitute a better means to detect distant evolutionary relationship between proteins than sequences do.

Fold recognition techniques are based on alignments of sequences onto 3D structures. The fitness of the sequence to the structure is measured by an empirical score function. This approach would be hopeless if each sequence would adopt a different 3D structure. In fact, it is now widely accepted that only a limited number of folds exist,[4-7] although the published figures vary from 700 to 8000 different folds. In the databases there are currently 600–700 different folds. If we assume that the true figure is close to the lower boundary of the above estimations, we already know a sizeable fraction of all possible folds. Whatever the true answer to this question, the number of folds is, in any case, several order of magnitudes smaller than the number of known protein sequences. One goal of fold recognition methods is to use this corpus of data to improve the detection of homology between remote organism proteins.

Fold recognition techniques, by definition, permit the detection of the fold corresponding to the sequence under study. It may be difficult, for instance when several functions are associated to a particular fold, to infer the true function of the sequence. The question arises of knowing to what extent the function can be deduced from the knowledge of the 3D structure? An analysis of the database of protein folds shows that, in 66% of the cases, proteins having the same fold also share a similar function.[8] For the remaining cases the 3D structure constitutes, nevertheless, the best starting point for inferring the function. For instance, it is possible to check the conservation of active sites residues to try assigning a function to a sequence.

Fold recognition methods consist, usually, in four components[9]:

1. a library of "cores" representative of the known folds,
2. an empirical score function measuring the fitness of the sequences for the cores,
3. a combinatorial algorithm to optimally align sequences onto the cores,
4. a statistical analysis to assess the significance of the scores found.

Although all four components are important, the second and fourth are really critical to obtain an effective fold recognition method. The importance of the score function is reflected in the number of works that have been devoted to this problem, and in the fact that there are as many different score functions as fold recognition methods.[10-19] On the other hand, it seems to us, that the fourth element has not received all the attention it deserves. The reason for this might be that, until recently, fold recognition methods were principally used to find the 3D structures of a limited number of protein sequences for which, often, other biological information was known, helping in the fold assignment process. The work of analysis of the results, and integration of the known facts about the system, was thus left to human experts. This approach is very impractical when fold recognition techniques are applied to genomic data[20] since, on the one hand, there is no biological information available on the system other than the sequences and, on the other hand, this is a very burdensome task for a human expert to analyze in detail the results for all the proteins of a genome.

We believe that the success of sequence comparison methods, such as BLAST, is due, to a large extent, to the statistical analysis they provide. It is therefore crucial for a fold recognition method to have a similar reliable criterion to assess the significance of a given fold assignment to a sequence.

The desire to test such a criterion prompted us to develop yet another fold recognition method called FROST (for Fold Recognition Oriented Search Tool). In the following we will present the principles of this method, stressing more specifically our approach to the problem of significance of a fold assignment to a sequence.

9.2. Method

9.2.1. Library of "Cores"

The three-dimensional, or tertiary, structure of a protein consists in the position in space of all the (heavy) atoms. The fold of a protein is the conformation in space adopted by its backbone. The core of a fold is the parts of the fold whose relative orientations and positions are conserved among a family of similar folds. They constitute a structural invariant of the family. In general, one observes that secondary structure elements are conserved and that loops, especially surface loops, are more prone to variations. Of course, if loops are involved in the function, such as the canonical trypsin binding loop of trypsin inhibitor proteins,[21] their 3D structure is perfectly conserved.

When several members of a fold family are known it is straightforward to superimpose the 3D structures and to determine which parts are conserved.[22] An alternative is to extract locally similar segments in a set of related proteins.[23] These structurally conserved parts constitute the core. Note that the members of the family in question should be proteins with less than 40-50% residues conserved since, above this threshold, the 3D structures of the proteins are too similar to bring any useful information about the core of the fold.

When only one exemplar of a fold is known (or when all the known proteins are too similar) one has to resort to an educated guess concerning what might constitute the core. In absence of better evidence, the best choice for the core elements consists in selecting the secondary structure elements: α-helices and β-strands.

Our library of folds uses this latter definition which has the merit of simplicity, even though, for some folds, we might have used a consensus based on structural superpositions. We used PDB release 89 and grouped protein sequences into families having more than 35% sequence identity. The "best" representative of each family was then selected. Here "best" refers to a mixture of criteria ranked by increasing importance, such as the resolution (NMR structures were arbitrarily assigned a resolution of 3 Å), the percentage of missing residues, the percentage of residues with missing sidechains, the percentage of non-standard residues.

We ended up with 1175 representative structures. For each representative structure we performed a search for homologous proteins in the NCBI non-redundant sequence database. Sequences having a percentage of similar residues lying between 40 and 90% with the sequence of the representative structure and less than 80% identical residues with all other selected sequences were aligned using ClustalW.[24] This resulted in a profile characteristic of the representative structure.

For this release, our estimate of the number of unique folds was 600 approximately. Our database was a redundancy of 2 on average (of course, some folds exist only in one exemplar whereas others, for instance the immunoglobulin fold or the TIM-barrel fold, are present several times). We chose to keep this redundancy because, as will be described in the next section, our parameter set takes into account information about the sequence of the template. It appears therefore, that it is better to have several representatives of the same core with different sequences. We hope, in this way, to improve the sensitivity of the method for remote homologs.

9.2.2. Development of a Score Function

As mentioned earlier the development of the score function is a critical point of any fold recognition method development. In general, the score function is associated with a low resolution representation of the protein, i.e., a model where each residue is represented by a few sites, often only one: the position of the C_α, the C_β or the centroid of the side chain. Although many different score functions exist,[10-19] they can be divided into two categories.

The first category concerns score functions whose parameters take into account only one residue of the structure. Methods using this type of score functions do not really consider the 3D structure of the proteins but a degenerate version of it, where the structure has been "projected" on the sequence. These methods are essentially profile methods, except that in addition to considering amino acid types, they also make use of structural information. For instance a residue in the structure may be characterized by the type of secondary structure it belongs to, its solvent accessibility, the presence of disulfide bridges, the existence of a hydrogen bond with the backbone or with another sidechain, etc.

The second category concerns score functions that take into account more than one residue in the structure. In the vast majority of cases only two residues are considered simultaneously, although multi-body methods have been tested.[25] Methods using this type of score functions truly use the 3D structure of the protein, be it through the use of a distance between interacting sites or the fact that they only consider sites whose sidechains are in contact in the 3D structure. The score associated with an alignment is calculated as a sum over all the pairs of residues interacting in the structure. Several approaches have been considered in calculating the score function parameters of this latter category.

The first approach assumes that residues in the structure are distributed according to Boltzmann law.[10-11] Each residue feels a potential of mean-force due to its neighbors. This mean-force potential can be divided into a sum of interactions involving two residues separated by a given distance. This approach is based on a physical description of the problem, although the physical assumptions and ways of calculating the parameters have been questioned.[26]

The second approach is purely statistical. The problem of assigning a fold to a sequence is likened to a classification problem. One has a number of "objects" (the sequences) that must be classified into the proper "category" (the folds). To perform this classification one must learn from a database of examples what characterizes each category.

In practice, these two approaches are similar in that they both extract the required information from a database of known 3D structures. This information is intended to reproduce, in one case, physical energies and, in the other case, common structural properties of the 3D structures.

In the last approach, one considers that, since a classification problem has to be solved, one has better to specialize and optimize the score functions to fulfill this particular task even though the resulting score functions no longer reflect physical properties of the system.[27-29]

In FROST we use two sets of parameters that we call 1D parameters and 3D parameters. 1D parameters depend on one residue in the structure. 3D parameters involve two residues that are *in contact* in the 3D structure. Two residues are defined to be in contact if the following equation holds:

$$d_{AB} < r_A + r_B + 1 \text{ Å} \quad , \qquad (9.1)$$

where atom A belongs to the sidechain of a residue and atom B belongs to the sidechain of the other residue, d_{AB} is the distance between the centers of atoms A and B, r_A and r_B are the corresponding van der Walls radii. We use this definition for selecting interacting residues, rather than a distance cutoff, for two reasons. The first fundamental reason is that in using a cutoff distance, two residues are considered to be in interaction even though there is no clear physical contact between them. For instance, two sidechains can be within 8 Å distance, but point in the opposite direction. We think that such an interaction does not carry specific information and is just responsible for an increase of the background noise. The second, more prosaic, reason is that, with this definition, the number of pairs of interacting core elements is notably cut down. This facilitates considerably the task of the combinatorial optimization algorithm as will be discussed in the next section.

We use a low resolution model for the structure where each residue is represented by the position of the center of mass of its side chain.

Each position k in the structure is characterized by a conformational state S_k. A conformational state is defined by the type of secondary structure (H for α-helices, E for β-sheets, and C for coils) and the solvent accessibility of the residue occupying this position in the original structure. We consider only two levels for the accessibility, b for buried and e for exposed. Buried residues are those for which the relative accessible surface area is less than 25% of the corresponding surface area for the same residue in an extended Ala-R-Ala tripeptide (R representing the residue considered.) Surface areas are calculated using the program naccess.[30] There are thus six possible conformational states {Hb, He, Eb, Ee, Cb, Ce} for a given position. In addition, as mentioned in the previous section, since each structure comes with a related multiple sequence alignment, each position in the structure is characterized by a column of this multiple sequence alignment, in practice, a vector of 20 frequencies.

Parameters are calculated as information values.[31] Our method is thus developed in the frame of the second approach described above: we seek to solve a classification problem, not to mimic a physical situation. Information values are calculated with frequencies estimated from the database of known 3D structures. Altschul[32] has shown that scores obtained from any substitution matrix have a log-odds interpretation. The definition of information we use here, due to Fano,[33] can also be shown to be similar to log-odds.

Fano's definition of information is a measure of the "statistical constraint" between two events, x and y:

$$I(x;y) = \log (P(x|y)/P(x)) \quad . \tag{9.2}$$

Here $I(x;y)$ is the information event y carries about the occurrence of event x. $P(x|y)$ is the conditional probability of event x knowing that event y has occurred and $P(x)$ is the probability of event x.

This definition is very intuitive since, if event y has no effect whatsoever on the occurrence of event x, then $P(x|y)$ is equal to $P(x)$ and the information is naught. If the occurrence of event y favors the occurrence of event x then $P(x|y)$ is greater than $P(x)$ and the corresponding information is positive. Conversely if the occurrence of event y has an adverse effect on the occurrence of event x, the information is negative.

Using the definition of conditional probabilities:

$$P(x,y) = P(x|y) \times P(x) \quad , \tag{9.3}$$

it is straightforward to show that Fano's information value can be rewritten as a log-odds formulation:

$$I(x;y) = \log (P(x,y)/P(x)P(y)) \quad , \tag{9.4}$$

where $P(x,y)$ is the joint probability of events x and y. The second member of Eq. (9.4) corresponds to a classical log-odds formulation: The log of the ratio of the probability for the joint occurrence of events x and y divided by the corresponding probability of occurrence of these two events if they were independent.

In FROST, 1D parameters are thus calculated as:

$$score(r_i, R_k S_k) = -2 \log_2(\mathrm{P}(r_i, R_k S_k) \,/\, \mathrm{P}(r_i)\mathrm{P}(R_k S_k)) \quad , \tag{9.5}$$

where the base 2 log and the multiplying factor, 2, have been introduced in order that our scores be similar to the way BLOSUM substitution matrices are calculated.[34] The minus sign before the expression in Eq. (3.5) implies that the best alignment of a sequence onto a core is the one with the smallest score value. $score(r_i, R_k S_k)$ represents the score of aligning the ith residue in the sequence, say a glycine, with the kth site in the structure. This k^{th} site is characterized by state S_k, say Hb, and residue R_k, say a leucine.

Since positions in the structure are characterized by a column of an alignment, the score of aligning a residue with a position in the structure is given by:

$$score(r_i, position_k) = \sum score(r_i, R_k S_k) \times f(R_k) \quad , \tag{9.6}$$

where $f(R_k)$ is the observed frequency of residue R_k in the kth alignment column.

3D parameters are calculated similarly except that they involve two residues that are in contact in the 3D structure, as discussed above.

$$score(r_i r_j, R_k S_k R_l S_l) = -2 \log_2(\mathrm{P}(r_i r_j, R_k S_k R_l S_l) \,/\, \mathrm{P}(r_i r_j,)\mathrm{P}(R_k S_k R_l S_l)) \; . \tag{9.7}$$

1D parameters are completed with affine gap penalties composed of a gap opening (go) and a gap extension (ge):

$$Gp(l) = go + l \times ge \quad , \tag{9.8}$$

where Gp is the gap penalty cost and l the gap length.

Three gap penalty sets are used in FROST. One set for gap penalties in secondary structure elements, one set for gap penalties in loops and one set for gap penalties at both ends of the sequence. The values of these sets are chosen such that it is costly to make an insertion/deletion in a secondary structure element. Insertion/deletion in loops are more penalized than insertion/deletion at the extremities of the sequences. For extremity gaps, after some optimization, we use $go = 1.0$ and $ge = 0.5$, for loops we use $go = 12.0$. The value of gap extension penalties depends on the amino acid considered and is calculated from multiple alignments. We observe that, even when secondary structure elements are conserved in a family of folds, their length can be different. To take this fact into account, the gap opening value in secondary structure elements is modulated such that inserting or deleting residues at the extremities of an element is less costly than in the central part.

For 3D parameters, no gap penalty is used but we require that residues be aligned with all the positions located in secondary structure elements. No insertion/deletion is allowed in secondary structure elements. Positions corresponding to coils in the structure are ignored. These two sets of parameters (1D and 3D) are used independently of each other as will be explained in more detail below. We believe that no set of parameters, especially when one uses crude models of the polypeptide chain as those used in fold recognition methods, will perfectly describe all aspects of the problem. Therefore we designed our method to act as a series of filters, each filter capturing some aspect of the reality and each progressively enriching the list of cores in true positives as shown schematically in Fig. 9.1.

Hereafter parameters using only one residue will be referred as 1D parameters and parameters using two residues will be referred as 3D parameters.

Compared to 1D parameters, 3D parameters substantially increase the complexity of the method (as will be described in the next section). However, in theory, they provide a more detailed information on the system under study than 1D parameters can do. With 3D parameters, one can take into account, in the model, residues that are close in space even though they are located far apart in the amino acid sequence.

To illustrate this point let us consider the score for an aspartate and a lysine to be located in a buried helix position. The corresponding values: $score(D;Hb) = 0.782$ and $score(K;Hb) = 0.887$, are two the most unfavorable values, if we except proline $score(P;Hb) = 1.085$. These values are to be compared to the values for leucine, $score(L;Hb) = -0.603$, and alanine, $score(A;Hb) = -0.490$, that are the two most favorable

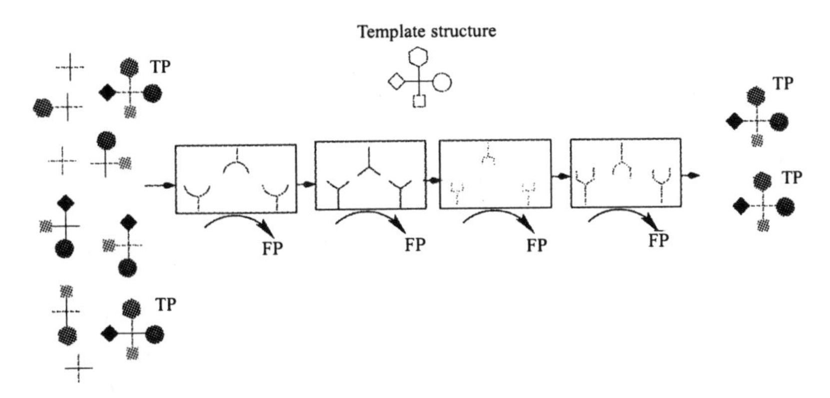

Figure 9.1. The threading algorithm seen as a series of filters. True positives are schematically represented by objects with a geometrical figure at each "arm" and are labeled with TP. Because the model for the polypeptide chain is crude and the score function is approximate, false positives may share, just by chance, some features with the true positives. For a given filter, it can happen, thus, that a false positive obtains a good score. However, it is unlikely that the same false positive will obtain good scores for all the filters. Applying several filter helps in discriminating true and false positives.

residues to occupy a buried position in an α-helix. In the present situation, this is the fact, for charged residues, to be buried that is penalized, not the fact of being in an α-helix since the score for a lysine to appear in an exposed helix site is *score*(*K*;*He*) = –0.694 (only glutamate is more favorable: *score*(*E*;*He*) = –0.758). When 1D parameters are used, it is clear that burying charged residues is unfavorable and should be avoided. However when one considers 3D parameters for the pair aspartate, lysine the score, *score*(*DK*;*HbHb*) = –0.670, turns out to be one of the most favorable values in the table of scores for pairs of *Hb* sites in contact. Of course, now, the situation has changed, the two residues are able to make a salt bridge which is energetically favorable. In such a case, 1D parameters are not only less accurate, they provide a completely wrong description of the phenomenon.

9.2.3. Combinatorial Optimization Algorithm

The role of the optimization algorithm is to find the alignment with the optimal score (in our case the alignment with the minimum score) for the parameter set, the query sequence and the particular core used.

When only 1D parameters are used a good algorithm exists, called dynamic programming algorithm. It provides an exact solution with a complexity of $O(NM)$. N and M are the sequence lengths. In molecular biology, the use of this algorithm goes back to the end of the sixties when it was first introduced by Needlemann and Wunsch[35] to perform sequence alignments.

Unfortunately, when 3D parameters are used, and variable length alignment gaps are allowed, there is no such a fast and convenient algorithm available. It has been shown[36] that, in the most general case, the problem is NP-complete. Recently, however, Xu *et al.*[37] showed that, when the number of interactions (links) between blocks is limited, as this is the case in most threading methods, the optimal score can be found in $O(Mn^{1.5c+1} + mn^{c+1})$ where n is the length of the sequence, m is the length of the core, M is the number of core elements, and c is a (hopefully small) integer constant depending on the maximum number of open links in the core structure. The complexity of the problem depends on the topology of the connections. The most difficult case is obtained when each block interacts with all other blocks, whereas the easiest case is obtained when there is no interaction (or link) between the blocks. For the first case the complexity of the algorithm is exponential, for the second case it is quadratic. In general the problem complexity is intermediate between these two cases. In the divide and conquer algorithm proposed by Xu *et al.*[37] the complexity is measured by the parameter c, that, roughly defined, indicates the maximum number of links that are severed when the template is partitioned (we refer the reader to the original paper for the technical definition of c). When interaction between blocks are mediated through the existence of contacts between residues, as in FROST, a block interacts only with a few blocks that are close neighbors and, therefore, the topology of the problem is simplified.

Two methods providing an exact solution to the combinatorial optimization problem have been proposed, the first one based on a divide and conquer algorithm[37] already mentioned and the second based on a branch and bound algorithm.[38]

We have programmed both algorithm. For the most difficult cases, these algorithms are extremely slow. The alternative to these exact algorithms is the use of heuristics. Several types of heuristic methods exist, one of the most used is based on stochastic algorithms.[16,39]

In our method, in order to test the significance of the scores, as will be described in the next section, the amount of work that has to be spent for the alignments is multiplied by a factor 100. We cannot use exact algorithms, but for the simplest cases. We have thus developed a very simple, but quite effective heuristic method. We will give elsewhere a comparison of the exact algorithms and various heuristic methods.

The heuristic method we use in FROST is analogous to optimization methods for continuous multi-dimensional functions. In these latter methods, one calculates the gradient, i.e., locally, the direction of steepest descent. A linear search for a minimum is performed along the direction in space defined by the gradient. At this minimum the gradient is once again calculated and the process is iterated until a local minimum is found. In our case, the variables are the blocks (in what follows a block refers to a core element) and the space is discrete. Starting from a valid alignment, an analog of the gradient is calculated, i.e., for each block, in the field of all other blocks, one searches for a feasible position having the minimum score. The system is then moved to the "minimum" point along the direction so defined. Unlike the continuous case, in a discrete space, the "minimum" point can belong to a domain of space that is not feasible (for instance if two blocks overlap, or if the block order is not conserved along the sequence). The point we select finally corresponds to the closest feasible point from the "minimum" point originally defined. This procedure is iterated until the score of the alignment generated cannot be improved. Since we are interested in the global minimum of the score function, we perform a number of such "runs" starting each time from a valid, randomly generated, alignment.

9.2.4. Empirical Distribution of Scores

Alignments produced by threading methods are characterized by scores that are a measure of the fitness of sequences to cores. The value of these scores, in addition to the particular amino acid sequence under study, depends also on the sequence length, the core length, and the 3D structure of the core. Scores obtained with 1D parameters are principally sensitive to the difference between the core and the sequence lengths since gap penalties are used for insertions/deletions. Scores obtained with 3D parameters are sensitive to the size of the core (the number of pairs of positions in contact) but, above all, they are extremely dependent on the particular features of the core 3D structure. In our experience, raw 3D scores are totally useless in ranking cores for a given sequence. One needs a means of interpreting these raw scores.

As mentioned in the introduction, the score interpretation, based on statistical considerations in sequence comparison programs such as BLAST, constitutes the principal strength of these methods. Regarding threading methods, it seems rather difficult for the time being, to obtain analytical results for the distribution of scores, especially when 3D parameters are used. The main reason for this fact is that the random model for the sequence, that is at the basis of sequence comparison statistical analysis for example, constitutes here, a very poor model. From a biochemical standpoint it is clear that any random amino acid permutation will not necessarily code for a functional protein. Functional proteins need to reach a unique 3D structure in a relatively short time. The implication of this observation is the existence of specific constraints in the amino acid sequence. It has been calculated[40] that only approximately 10% of all potential amino acid sequences can adopt a unique fold (although this calculation has been obtained with a very crude lattice model and, thus, has to be taken with a grain of salt, it is indicative of the order of magnitude). Therefore, to obtain accurate statistical data on the score distribution one has to resort to an empirical approach.

With respect to a given core, for instance a cytochrome *c'* as shown in Fig. 9.2, all existing protein sequences can be divided into two populations: population 1 containing proteins that adopt this particular fold and population 2 gathering all other proteins that adopt another fold.

Ideally, for an optimal set of parameters, if we align members of these two populations with this core we should get a bimodal distribution as shown on Fig. 9.2. Unfortunately, it is not possible to determine the score distribution of population 1 empirically. This would require the knowledge of enough proteins whose sequence shows no relationship with the cytochrome *c'* sequence but whose 3D structure is known to be similar to the cytochrome *c'* fold. With the existing database (PDB), only the score distribution of the immunoglobulin fold, could, perhaps, be calculated in this way.

On the other hand it is relatively straightforward to calculate the score distribution for members of population 2. For this purpose it is sufficient to extract randomly, from a database of protein sequences, for instance the NCBI non-redundant database, a number of sequences

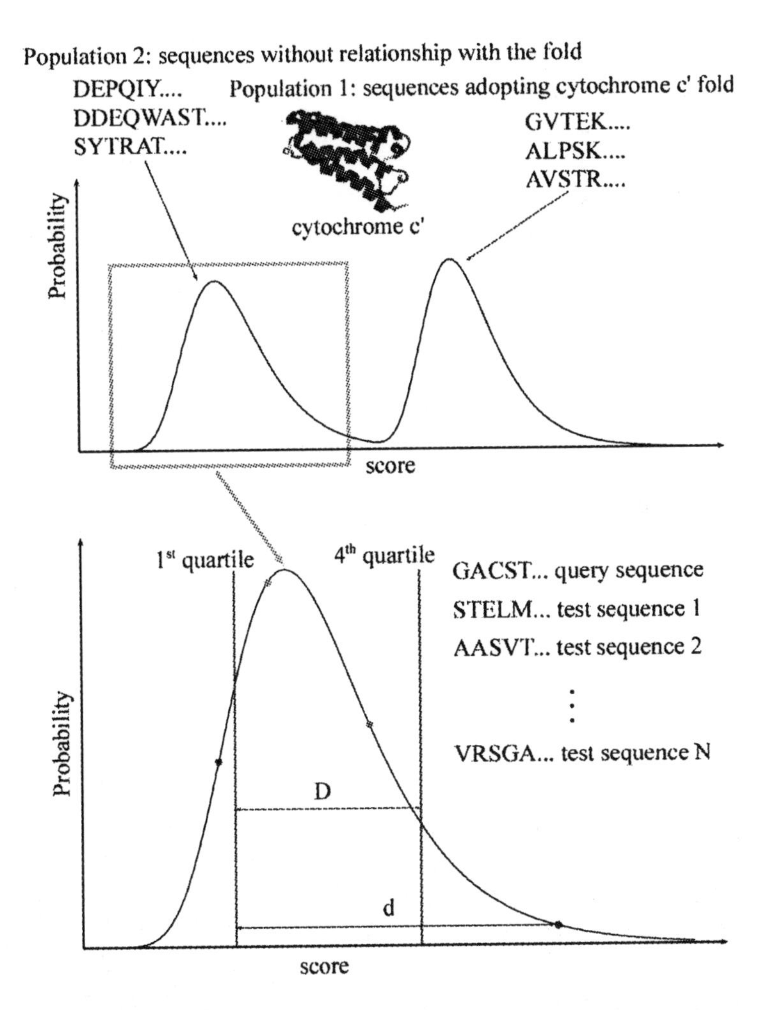

Figure 9.2. For a given fold, here the cytochrome *c'* fold, the sequences can be divided into two populations. Population 1 gathers sequences that adopt this particular fold and population 2 contains sequences that adopt any other existing fold. For a given set of parameters, the distribution of scores of these two populations is schematically represented in the upper part of the figure. Only the score distribution of population 2 can be empirically determined. For this purpose, a number *N* of test sequences (see text) are aligned onto the fold. The scores obtained in this way provide an approximation of the score distribution. Using this distribution, the score of the query sequence is normalized using the distance D between the first and last quartile and the distance d between the first quartile and the score obtained by the query sequence. This normalized distance: $d_N = d / D$ is used to rank the fitness of a particular query sequence for all the cores of the library.

without any obvious relationship to the core and the query sequence (a rule of thumb is to keep sequences having less than 25% identical residues when aligned with the core or the query sequence). It is also required that any pair of these sequences has to show less than 25% identical residues after alignment. Finally, in order to avoid a bias due to the length of the sequences, only sequences with a length identical to the query sequences are kept.

Occasionally, one might select sequences related to the core even though this is not detectable at the sequence level. However, since sequences are randomly selected, we expect this kind of problem to be rare enough so as not to systematically bias the score distribution.

In any case, we take care of this potential problem in the following way (see Fig. 9.2): We disregard the first and last quartiles of the distribution (i.e., the first 25% and last 25% scores of the distribution). These quartiles contain scores that we do not "trust"; in the first quartile because we can suspect sequence possibly related to the core to be present.

As shown in Fig. 9.2 the distance, D, between the first and the last quartile is used as a reference distance. The distance, d, between the first quartile and the score of the query sequence is normalized using this reference distance as:

$$d_N = d \,/\, D\,. \tag{9.9}$$

This normalized distance, d_N, allows us to compare the alignment results of a given query sequence onto different cores. With this approach, we do not need to make unjustified hypotheses regarding the shape of the underlying distribution.

For a particular fold, the greater the normalized distance, the less chance the query sequence has to belong to population 2 and, therefore, the more likely it is to belong to population 1 (protein sequences that adopt this fold). Of course, since we do not know the analytical form of the distribution, we cannot calculate directly the corresponding probability. However, an upper bound for this probability can be empirically estimated as will be shown in the next section.

9.2.5. Development of a Benchmark Database

In order to test our approach under realistic conditions we need a set of protein pairs with similar 3D structures that shows no detectable relationship according to the usual sequence comparison methods (BLAST or FASTA). In real cases, fold recognition techniques are only used when sequence comparison methods fail to give a meaningful answer. Our only requirement for pairs was to have similar 3D structures, not necessarily to be homologous.

Using VAST[16,22] we performed all the pairs of 3D structure comparisons with the 1175 representatives of the groups we described in the section "library of cores." VAST provides a statistical criterion to judge whether a structural comparison between two proteins is significant. We chose a relatively loose threshold of 95% confidence. With such a threshold we obtain a number of marginal matches that often correspond to the presence of similar supersecondary structure elements in both structures. We do not consider such matches as true positives. A loose threshold was chosen in order to be able to spot these marginal matches that might constitute difficult cases for the threading algorithm. We define for each 3D structure comparison the minimum percentage of aligned residues, pmin. It is is defined as:

$$pmin = \mathbf{min}(p1,p2) \quad , \tag{9.10}$$

where $p1 = N_{aln}/N_{res1}$, $p2 = N_{aln}/N_{res2}$, N_{aln} is the number of residues in the structural alignment, N_{res1} the number of residues in the first protein and N_{res2} the number of residue in the second protein.

When pmin is greater or equal to 50%, two proteins have similar 3D structures. If *pmin* is less than 50% we have to consider the case where the second protein is a domain of the first protein: $p1 < 50\%$ and $p2 > 50\%$. If we exclude this case and when pmin is in the range 40% to 50%, the 3D structure superposition needs to be examined. In this range we carried out a visual inspection of the structural superposition and we checked the SCOP classification[41] for confirmation. Below 40% the match corresponds most likely to a super secondary structure element.

We focused on pairs of proteins for which *pmin* was at least 65%. This definition excludes proteins that are domains of a larger protein. It excludes as well, difficult cases where only a small nucleus of the structure is conserved whereas numerous insertions/deletions of segments are observed (Plate 9.1 illustrates such a case: structural alignment of mamba intestinal toxin—PDB code 1IMT, in blue, and porcine pancreatic procolipase B—PDB code 1PCN, in gold. 46 residues are aligned with a root-mean-square deviation (RMSD) of 1.54 Å, representing 49% of 1PCN residues and 57% of 1IMT residues. The upper part of the plate shows the structural alignment of the complete proteins, the lower part presents only the aligned region for the sake of clarity. This region consists in a pair of triple-strand β-sheets, each sheet being cross-linked by two disulfide bridges. Outside this conserved region, other parts of the structures are different. This pair of proteins is not included into our benchmark database since the smallest percentage of aligned residues, 49%, is less than the 65% threshold we chose.) We, nonetheless, obtained in this way, a rather large list of protein pairs with similar 3D structures.

We then carried out a homology search with these pairs using FASTA. For this purpose, the sequence of one element of each pair was given to FASTA, leaving to the algorithm the task of finding the second member of the pair. All pairs, for which the FASTA E-value was less than 1, were discarded, i.e., only pairs for which the score was really non significant were kept. The number of pairs getting through this stage was 316. A unique exemplar of all proteins appearing in these pairs was selected, leaving us with 280 proteins. It is important to notice that the above E-value was obtained with the NCBI non redundant sequence database that contains about $5 \cdot 10^5$ sequences. The E-value is approximated for small p-values by E-val = Dp, where D is the size of the database, used and p is the probability of the observed score.[42] We have repeated the experiment with BLAST, testing, in turn, each protein against the 279 others. None of the above pairs had an E-value less than $1 \cdot 10^{-6}$. The thresholds used are thus consistent for both programs.

For each of these 280 proteins we know that there exists at least one good structural match (*pmin* = 65%). Of course other matches can exist including what we called above marginal matches. These 280 proteins constitute our benchmark database.

Fig. 9.3 presents a "phase diagram" showing the target difficulty.[43] This diagram plots the percentage of structurally aligned residues according to VAST, against the percentage of identical residues according to Align. Align is a global sequence alignment program of the FASTA suite.[42] Parameters used here are the BLOSUM50 substitution matrix and gap penalties (−12, −2). Let us note that the choice of Align, that penalizes insertion/deletion at both extremities of the sequences, rather than Align0 that does not, has a significant impact on the figures for the percentages of identity. Values obtained with Align represent an upper bound of the corresponding values obtained with Align0, or, as in the article,[43] the percentage of identical residues in the structural alignment. Each pair is labeled according to the SCOP classification: fold, superfamily, and family.[41]

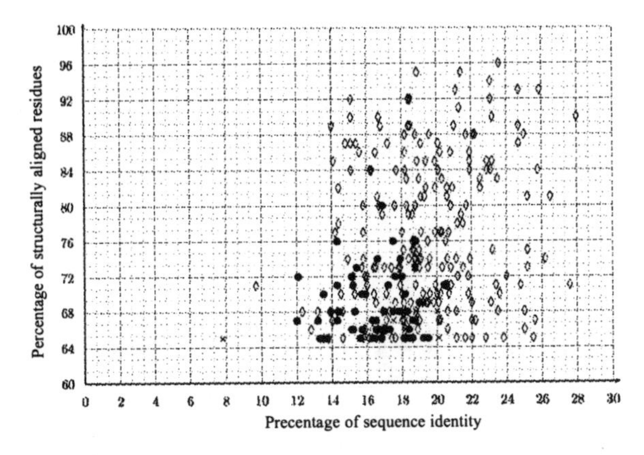

Figure 9.3. "Phase diagram" of the benchmark database. The 316 structurally similar pairs (see text) are plotted according to the percentage of structurally aligned residues and the percentage of identical residues after sequence alignment. Pairs are labeled according to the SCOP classification: cross for folds, filled circle for superfamilies, diamond for families.

9.3. Results

We gave to the fold recognition program the sequence of each of the 280 proteins in turn. Each query sequence came with 100 sequences randomly extracted from the non redundant database and used to compute the normalized score as explained in the section 9.2.4 *Empirical distribution of scores*. These 101 sequences were aligned against the database of 1175 cores. In practice, since we require the alignment to be global, only cores for which the length was ±30% of the length of the query sequence were really taken into consideration.

An example of the program output for the query sequence 1ISUA0 (PDB code 1ISU chain A: High potential iron-sulfur protein from the bacterium *rhodocyclus tenuis*) is shown in Fig. 9.4.

****** 1D alignment ******

V%al	Core	Nqry	Sqry	Smin	Sc25	Sc50	Sc75	Smax	Id	Definition	Rank
64.8	1HPI_0	5.6	-83.4	-46.9	-20.7	-14.7	-7.1	11.1	22.1%	High-Potential Iron-Sulfur Protein From	1
0.0	1VIB_0	3.2	-38.6	-34.3	-8.8	-1.8	4.6	21.0	18.3%	Neurotoxin B-Iv, Toxin, Hydroxylation	2
25.8	1FCA_0	3.1	-18.4	-10.2	5.1	11.1	16.3	26.4	19.6%	Ferredoxin	3
0.0	1ATA_0	2.7	-25.7	-19.8	-3.0	3.5	10.1	17.9	20.6%	Trypsin Inhibitor	4
0.0	2TGF_0	2.7	-26.8	-36.4	-5.3	0.8	7.3	25.7	15.3%	Transforming Growth Factor-Alpha	5
0.0	1HOE_0	2.7	-19.4	-19.3	1.8	7.6	14.4	25.7	11.0%	Alpha-Amylase Inhibitor Hoe-467A	6
0.0	1GKS_0	2.6	-22.3	-21.8	-5.0	0.4	6.1	20.6	7.7%	Ectothiorhodospira Halophila Cytochrome	7
0.0	5PTI_0	2.5	-13.9	-175.2	4.3	10.7	16.1	32.5	20.0%	Trypsin Inhibitor	8
0.0	1AHO_0	2.5	-4.7	-8.5	9.5	14.5	18.9	31.0	16.2%	Scorpion Protein Toxin Toxin Ii	9
0.0	1IML_0	2.4	-34.0	-48.7	-15.6	-9.0	-2.3	11.3	11.7%	Cysteine Rich Intestinal Protein	10
19.4	1IGD_0	2.1	-32.5	-32.6	-18.7	-11.9	-6.1	14.0	21.0%	Protein G	15
35.5	1NXB_0	1.9	-13.7	-133.8	-2.4	3.1	9.9	23.4	12.7%	Neurotoxin b	20
16.7	1XXBB0	1.8	-14.1	-27.6	-0.1	8.7	18.0	43.4	16.2%	C-Terminal Domain Of Escherichia Coli Ar	26
17.9	2PTL_0	1.5	-23.4	-35.7	-16.6	-10.3	-3.0	13.7	10.8%	Protein L	38
25.8	1FXD_0	1.3	0.1	-16.0	3.4	8.1	13.0	30.5	10.2%	Ferredoxin Ii	44
27.1	1ERG_0	1.2	-16.9	-66.8	-13.3	-5.3	2.8	43.2	13.7%	Human Complement Regulatory Protein Cd5	51
17.1	1TNS_0	0.9	-26.1	-49.0	-27.3	-20.6	-13.5	0.8	14.1%	Mu Transposase	67
20.6	1AA3_0	0.7	-7.3	-27.7	-10.6	-4.3	1.5	13.4	11.1%	C-Terminal Domain Of The E. Coli Reca	77
22.6	1APQ_0	0.3	-2.0	-31.4	-11.8	-5.7	1.4	18.3	13.3%	The Egf-Like Module Of Human C1r	116

****** 3D alignment ******

V%al	Core	Nqry	Sqry	Smin	Sc25	Sc50	Sc75	Smax	Id	Definition	Rank
64.8	1HPI_0	2.4	-54.1	-68.3	-25.7	-14.5	-5.0	13.0	6.0%	High-Potential Iron-Sulfur Protein From	1
0.0	1AHO_0	2.3	-37.2	-52.9	-6.9	4.0	16.3	54.9	5.3%	Scorpion Protein Toxin Toxin Ii	2
0.0	1ATA_0	2.1	-34.3	-45.4	-23.0	-17.4	-13.2	-1.9	5.6%	Trypsin Inhibitor	3
0.0	1HOE_0	1.4	55.8	14.6	74.5	99.0	121.8	184.9	6.7%	Alpha-Amylase Inhibitor Hoe-467A	4
0.0	2TGF_0	1.4	-26.4	-47.2	-20.3	-12.1	-4.3	13.1	2.7%	Transforming Growth Factor-Alpha	5
0.0	5PTI_0	1.0	19.9	-248.4	18.5	35.4	49.0	81.4	2.8%	Trypsin Inhibitor	6
0.0	1IML_0	0.7	-4.5	-75.4	-17.5	2.1	19.7	50.2	2.4%	Cysteine Rich Intestinal Protein	7
0.0	1VIB_0	0.6	-0.8	-43.4	-15.4	3.8	19.7	67.3	12.7%	Neurotoxin B-Iv, Toxin, Hydroxylation	8
25.8	1FCA_0	0.3	-17.4	-87.8	-38.1	-20.0	-7.9	45.5	10.1%	Ferredoxin	9
0.0	1GKS_0	0.1	85.4	-15.3	42.1	66.0	90.6	158.2	6.4%	Ectothiorhodospira Halophila Cytochrome	10

Figure 9.4. *See description on next page.*

Figure 9.4. Example of output for the query sequence 1ISUA0 (PDB code 1ISU, chain A, domain 0) Note that since only whole protein are considered the domain is always 0. When there is only one chain in the PDB file we use the symbol "_" to name the chain instead of the PDB blank character. Column V%al corresponds to the percentage of aligned residues with the core according to VAST. Column Core indicates the core name. Column Ndist, corresponds to the normalized distance d_N, (cores are sorted according to this normalized distance), column Sqry corresponds to the score of the query sequence when aligned with the core, column Smin corresponds to the minimum score obtained for the alignment of one of the 100 test sequences with the core, column Sc25 corresponds to the value of the score for the first quartile, column Sc50 corresponds to the value of the score for the median, column Sc75 corresponds to the value of the score for the last quartile, column S_{max} corresponds to the maximum score obtained for the alignment of one of the 100 test sequences with the core, column Id gives to the percentage of identical residues in the threading alignment, column Definition provide a short definition of the core and column rank indicates the rank of the core in the list. Note: for stage 1 (1D parameters) all cores for which the structural alignment with the query sequence is significant at a threshold of 95% according to VAST are listed. Only the first 10 cores of stage 1 are processed in stage2 when 3D parameters are used.

Corresponding alignments are omitted. Cores are ranked by decreasing normalized distance, d_N. Here only core 1HPI_0 is a true positive, all other cores with non-zero value in the first column correspond to marginal matches.

For instance, for core 1FCA_0 that appears in third position in the list in Fig. 9.4, 16 residues are structurally aligned (1ISUA0 has a length of 62 residues and 1FCA_0 has a length of 55 residues). These residues belong to two beta-sheets that happen to have about the same orientation in both structures but the rest of the structures is very different. Not surprisingly, the probability according to VAST to observe the occurrence of such a structural similarity by chance is marginal: 0.05. Cores with a value of zero in the first column have probability greater than this 0.05 threshold.

Note that for the second stage (3D parameters), only the first 10 cores ranked after the first stage (1D parameters) are considered. Fig. 9.5 shows, for the 280 proteins, the cumulative probability for a true positive to appear before or at a given rank. After the first stage, the probability of observing at least one true positive among the first 10 cores is 0.8. Note also that with this test database, for 60% of the query proteins, a true positive appears in first position in the list. Of course, here, for

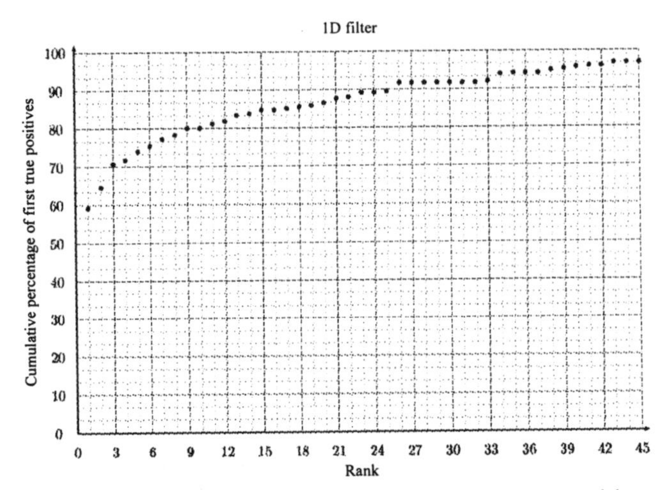

Figure 9.5. Plot of the cumulative probability for a true positive to appear up to, or at, a given rank after the first stage (1D parameters). In 60% of the cases a first positive appears in first position. The probability of meeting at least one true positive amongst the first 10 cores after the first stage is 0.8. Only the first 10 cores are used in the second stage (3D parameters).

each query protein there is at least one true positive. In general, we do not know whether the fold of a query protein is similar to a known fold and therefore the rank does not constitute a good measure of success. Fig. 9.6a shows the cumulative distributions of percentages of true and false positives as a function of the normalized distance for 1D parameters (Fig. 9.6b for 3D parameters). For each query protein the normalized distance for the first true positive and first false positive appearing in the core list are collected. For instance, for the query protein 1ISUA0 (see Fig. 9.4), the normalized distance for the first true positive is 5.6 (1D) and 2.4 (3D) and the normalized distance for the first false positive is 3.2 (1D) and 2.3 (3D). For 1D parameters we see that if the normalized distance is greater than 3.9 the number of false positives is less than 1% whereas the coverage (the number of true positives that are beyond this score) is 38%. For 3D parameters, for the same threshold of 1% false positives, we only cover 12% of the true positives.

Fig. 9.7 shows the same "phase diagram" as above but this time, pairs are labeled according to the normalized distance obtained after stages 1 (Fig. 9.7a) and 2 (Fig. 9.7b) of the threading algorithm. For stage 1D (resp. 3D) , 26% (resp. 10%) of the pairs have a score above 3.9. These figures are different from the figures (38% and 12%) obtained from

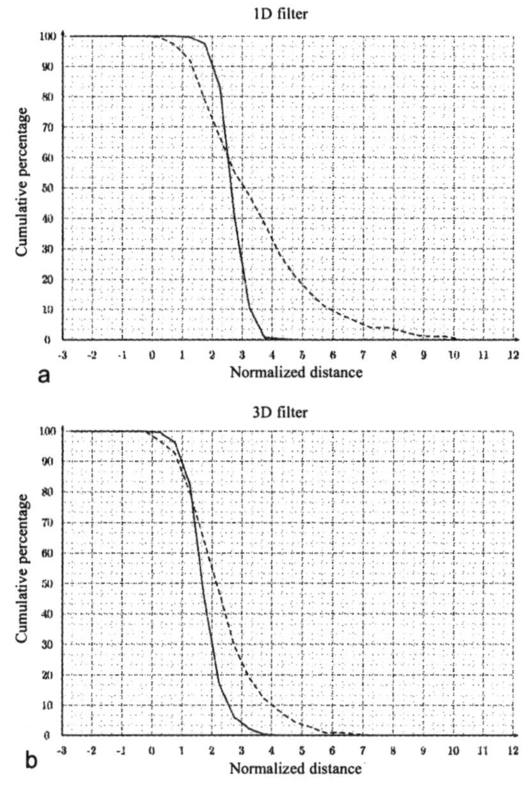

Figure 9.6. Cumulative distribution of false positives (solid line) and true positives (dashed line) as a function of the normalized distance, d_N. These curves are obtained as follows: For each of the 280 target proteins of the benchmark database the score of the first true positive and the score of the first false positive met in the core list are compiled. Fig. 9.6a corresponds to the first stage, Fig. 9.6b to the second stage.

Fig. 9.6. For a number of query proteins, quite surprisingly, it turns out that this is not always the best structural match (*pmin* > 65%) that appears in first position in the core list. In fact, after stage 1 the first true positive appearing in the core list corresponds to the pair partner only for 53.4% of the pairs (57.0% for stage 3D). Fig. 9.7a shows that pairs with a score above 3.9 are preferentially those for which the percentage of sequence identity is greater than 20% or the percentage of structurally aligned residues is greater than 80%. The same pattern, less pronounced, is visible for stage 3D (Fig. 9.7b).

FROST, has been designed as a series of filters, the role of each filter being to capture some (hopefully different) aspect of the reality. For the moment the series of filters is minimal since we only have two filters (1D and 3D parameters). We need, nonetheless, a general method of combining the results of the various filters. Our approach is similar to the technique known as support vector machine.[44] If there are M filters, cores are represented by points in an M-dimensional space. A simple

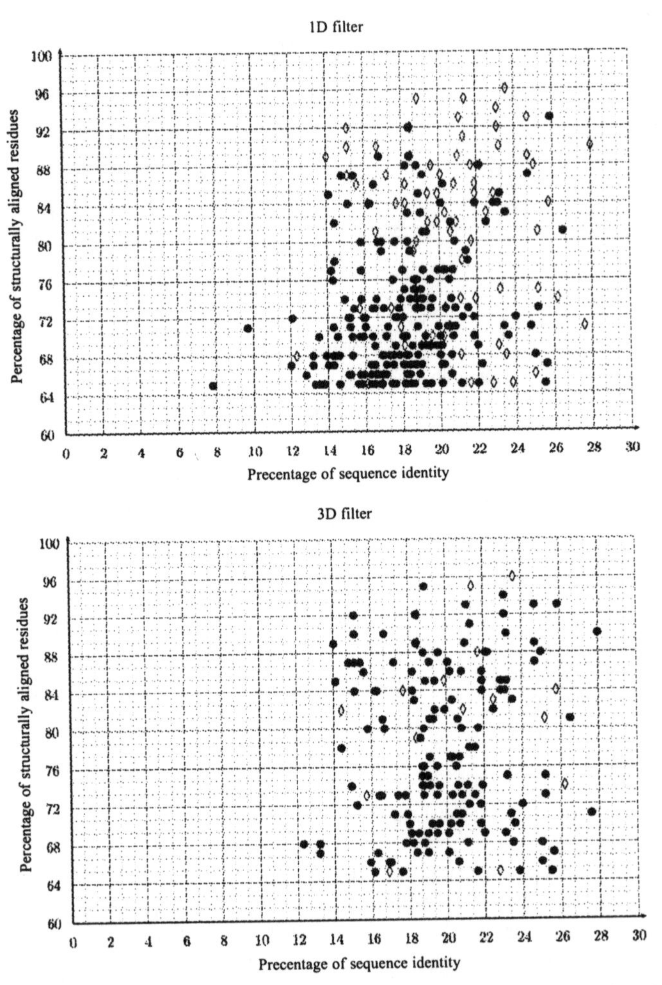

Figure 9.7. "Phase diagram" as in Fig. 9.4. Pairs are labeled according to the value of the score obtained after stage 1 and stage 2. Diamonds indicate a score greater than 3.9 and filled circles a score less than 3.9. For 1D stage 26% of the pairs have a score above 3.9, for 3D parameters 10% have a score above 3.9.

way to build a binary classifier is to construct hyperplanes separating class members from non-members in this space (i.e. true positives from false positives). In practice our problem involves non-separable data, for which there does not exist hyperplanes, that successfully separates the class members from non-class members in the training set. We therefore use a weaker condition which is to find hyperplanes beyond which the number of false positives is less than a specified value, typically 1%. In two dimensions hyperplanes are just lines, the problem to solve is very simple and one does not really need to use of support vector machine techniques. We manually adjusted the lines in order to define a region of the plane where the number of false positives is less than 1%. Fig. 9.8 shows the plot of the first 10 cores for 278 query proteins as a function of their 1D and 3D scores. On this plot there are 403 true positives and 2372 false positives.

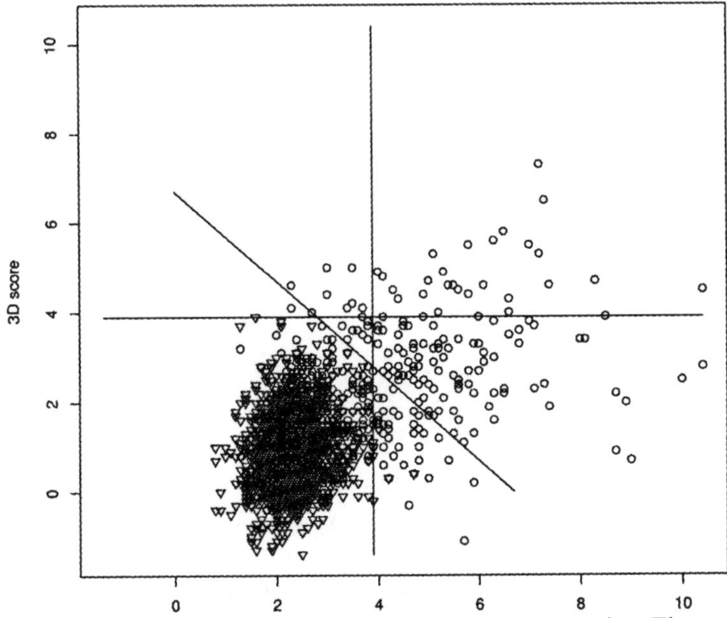

Figure 9.8. Plot of the first 10 cores for 278 query proteins. There are 2372 false positives (triangles) and 403 true positives (circles) plotted on this diagram. The vertical line $x = 3.9$, the horizontal line $y = 3.9$, and the oblique line (slope -1, going through point: 3.9, 2.8) divide the plane into two regions. Beyond these lines, in region II, there are 184 true positives and 3 false positives.

The three lines shown in Fig. 9.8 divide the plane into two regions. Region II, beyond these lines, contains 184 true positives and 3 false positives. A few query proteins, such as the globins, can have several true positives in region II. Such cases make the results look better than they really are. Therefore, we calculated the percentage of true and false positives on the basis of the number of query proteins, i.e., if a query protein has several true positives in region II it is counted only once. The percentage of query proteins for which at least one true positive (respectively false positive) lies in region II is 42.1% (respectively 0.7%). Two of the three false positives in region II correspond to the same query protein 1FCA_0 (a ferredoxin from *Clostridium acidurici*). These false positives are 1TGSI0 (porcine trypsin inhibitor) and 1AHO_0 a scorpion toxin. This latter core correspond to a marginal match (19% of the residues are structurally aligned with 1FCA_0). The scores for these two cores are (4.7, 0.4) for 1TGSI0 and (4.2, 0.3) for 1AHO_0. Note that the core the most similar to 1FCA_0 (a ferredoxin, 1FXD_0, with *pmin* = 0.71) appears in first position both after the 1D and the 3D stages with scores (10.4, 2.8).

9.4. Discussion

9.4.1. Use of Filters

Two reasons, one practical and the other fundamental, have led us to design the method as a series of filters.

The alternative to the filter approach is to combine the scores for 1D and 3D parameters. The combination of scores requires finding adequate weights for each component involved. With two components the task is relatively easy but, the greater the number of components, the more involved the optimization task. Moreover this optimization of the weights must be performed all over again when one of the components is modified, or a new component is added. With a series of filters, modifications brought to a particular component do not perturb others. This latter point is very important if, as we will discuss later, we wish to calculate in advance score distributions used to obtain statistical criteria.

The second reason is the fear that a combination of scores might lead to detrimental averaging. For instance, if a core gets a good score with 1D parameters, but for some reason, a bad score with 3D parameters, the resulting combined score might be some intermediate value that is likely to be non significant. As shown on Fig. 9.8, if the score for one of the filters crosses the threshold along one of the axes, it is successfully detected even though it might utterly fail other filters. This averaging problem is likely to get even worse when more than two filters are considered.

As illustrated in Fig. 9.8 one can (literally) "corner" cores (e.g., cores beyond the oblique line). Unlike the case of combination of scores, the more filters the better since more filters implies more "corners," where one is potentially able to trap cores.

9.4.2. Difficulty of the Benchmark

Following Panchenko *et al.*,[43] the benchmark database can be characterized by the proportion of hard, medium, and easy targets. Hard targets are those for which the percentage of structurally aligned residues is less than 60%, there is no detectable sequence motif according to PSI-BLAST,[45] and the percentage of sequence identity in the structural alignment is less than 12%. Medium targets are those for which the percentage of structurally aligned residues is more than 60%, with no detectable sequence motif, and the percentage of identical residues in the structural alignment is less than 12%. Easy targets are those for which the percentage of structurally aligned residues is more than 60%, and the percentage of identical residues in the structural alignment is more than 12%, that may or may not have a detectable sequence motif with PSI-BLAST.

Note that our definition of sequence identity is different to the one used by Panchenko *et al.*[43] We estimate that their 12% threshold on sequence identity corresponds roughly to our 20% threshold.

By these criteria, our test database does not contain hard targets (since we require that there exists for each query at least one protein for which there is more than 65% structurally aligned residues). Our rationale for choosing a threshold of 65% is that fold recognition methods, as indicated by their very name, need some minimal resemblance between 3D structures in order to work properly. Folds with less than 50% structurally aligned residues are more dissimilar than similar and we believe that fold recognition techniques, in this regime, are not the right tool to use (we are rather in the realm of *ab initio* or, as they are called since CASP4 meeting, new fold methods). We therefore concentrate on targets for which there is, a priori, a chance to succeed. Although, this is not always the core with the best structural match that is picked out by the threading algorithm as is apparent in Fig. 9.7.

Our database contains 70% medium targets and 30% easy targets. We emphasize that these targets are "easy" by fold recognition standard only; they are not detected by conventional sequence comparison methods.

Our test procedure is as close as possible to a real fold recognition experiment, in that each query sequence is aligned with the whole database of existing cores that are of comparable size. The criterion is that the length of the core is ±30% the length of the query sequence. Beyond these limits it is very unlikely that the query sequence adopts the fold of the core as a *whole*. We observe that, in our benchmark database, only 1.5% of the 316 pairs have a length difference greater than 30%.

The use of whole protein structures has the drawback, however, of excluding cases where a core corresponds to a structural domain of a long query protein, or, conversely, the query sequence corresponds to a domain of a multi-domain core.

Unlike the work of Panchenko *et al.*,[43] this procedure allows us to check not only the sensitivity (the coverage) of the method but also the specificity (the number of false positives that are beyond a specified threshold) since each query may recruit cores not related to its fold. The knowledge of a method specificity is important if this method is intended to be used for an automatic annotation of genomic sequences.

9.4.3. Statistical Criterion

We have already abundantly stressed the importance of a statistical criterion to accurately distinguish true positives from false positives. Threading scores depend on the query sequence amino acid composition, the length of the query sequence, the size of the core and, when 3D parameters are used, the particular 3D structure of the core. All these parameters affect the score distribution to various extents, but in our experience, the influence of the core 3D structure is paramount.

Bryant and Altschul[46] and Mirny *et al.*[47] have proposed schemes to measure the statistical significance of protein structure prediction by threading. Bryant and Altschul compare the score of the query protein with the distribution of scores of shuffled versions of this sequence realigned on the cores. Implicitly, since they use Z-scores, they assume that the distribution of these scores is Gaussian. Mirny *et al.*[47] use a Random Energy Model that makes the assumptions that energies of "decoys" (i.e., wrong alignments of a sequence onto a 3D structure) are statistically independent random values and that the probability density of energies of decoys is Gaussian. They calculate a parameter, ε, that takes into account the sequence composition, the query length and the number of decoys. This parameter is a measure of the probability that the best alignment found is native-like. Under these conditions they show that the probability of successful prediction in threading follows an extreme value distribution.

We believe that random sequences do not constitute a good model of a protein sequence when one has explicitly to take 3D structure into account. Random sequences are likely to lack the subtle patterns common to all true protein sequences that must fold to a unique 3D structure in a rather short time. When comparing alignments of a true protein sequence with shuffled versions of it onto a 3D structure, the alignment of the true protein sequence is likely to appear more significant than it really is. This is the reason we use real protein sequences for calculating the statistics, in order to avoid potential bias.

Empirically it is difficult to determine the form of the distribution of scores for aligning a sequence onto a 3D structure. By construction, most of the points are close from the mean value; very few belong to the tail of the distribution which is the region of interest. We tried to fit the empirical distributions obtained on known distributions: the Gaussian and the extreme value distributions. With only 100 sequences to calculate the empirical distribution, the fit of the empirical distribution with these two distributions is, in both cases, not significant. We therefore decided to bypass this problem with the scheme described in Section 9.2.4 *Empirical Distribution of Scores*. The drawback of this approach is that now, probabilities can only be estimated empirically using the benchmark database. Since this database contains 280 proteins, the smallest p-value that can be estimated is of the order of 1/280, about 0.004.

Mirny *et al.*[47] noted that the shuffling and realigning approach of Bryant and co-workers is computationally demanding (the same remark is, of course, valid for our technique). Indeed the amount of work required by our approach is multiplied by 100 just to obtain the statistics.

However parameters needed for the statistics do not need to be calculated on the fly, they can be calculated beforehand. Since, in our method, sequences can only be aligned onto cores having approximately the same size (within ±30% of the core length), it is possible to classify sequences according to their length in bins. All the sequences in the bins are aligned onto the corresponding core. Parameters of the statistics for each bin (i.e., for a given core, the values of the score for the first and fourth quartile) are then calculated from the empirical distribution of scores. For instance, if we consider a core of length 100, then the allowed test protein lengths may vary between 70 and 130. Test proteins can be gathered in bins of length 10, say, from 70 to 79, from 80 to 89, etc. Since the calculation is done once and for all for each bin, more than 100 test proteins can be used to calculate the parameters allowing, we hope, a more accurate estimation of the empirical distribution.

9.4.4. Present Limits of the Method

It is well known that many proteins are modular, each module along the sequence having a particular 3D structure and function. For the moment our method expects the query sequence to correspond roughly to a core of the database. If, for instance, a query protein corresponds to two 3D domains that exist independently in the database of cores, the method in its present state will miss the corresponding matches. This is a major problem since most sequences of length greater than 250–300 residues are likely to contain more than one structural domain. If one is interested in a limited number of query sequences, it is possible to try various partition of these sequences randomly. This becomes very cumbersome when one has to analyze many genomic sequences. The ability of splitting sequences into sub-sequences corresponding to structural domains is therefore becoming crucial.

The most common approach to tackle this problem consists in using sequence alignments to define modules along the sequence of a protein in a way similar to what is done in ProDom,[48] for instance. This approach, being based on sequence alignments, suffers from the same drawbacks, i.e., the homology might not be apparent from the sequence alone. Another possibility consists in trying to identify linker sequences that connect structural domains.[49] The last approach would be to use directly the threading algorithm to detect subdomains. To do this, several modifications of the method presented here are required. First, the database of cores must include, not only whole proteins, but also structural domains. Second, instead of a global alignment, one must perform a local alignment that allows part of a sequence to be aligned onto a structural domain.

9.5. Conclusion

Our goal in this chapter was to provide an assessment of the significance of fold assignments to protein sequences using threading methods. This assessment depends strongly on the data used to carry out the test. Using release 89 of the PDB, we developed a benchmark database that is characterized by the percentage of hard, medium and easy target pairs that it contains. Our database contains 69% medium and 31% easy target pairs. We did not consider hard target pairs (with less than 60% structurally aligned residues) since we believe that, in this regime, the use of *ab initio* methods is more appropriate than the use of fold recognition methods.

Our threading method is based on the concept of filters. Threading methods use a crude model for the polypeptide chain. It is unlikely that a single score function will capture all aspects of the relationship sequence-3D structure. Several score functions must be used to best describe this relationship. In such a case, we think that filters are not only more convenient to use than a weighted combination of scores, they also allow a better discrimination between false and true positives. Using two types of parameters, 1D parameters that correspond to a profile, and 3D parameters that make use of the 3D structure of proteins we obtained for our benchmark database, 42% coverage for a number of false positives less than 1%.

In our hands, 1D parameters appear more effective than 3D parameters in terms of balance between sensibility and specificity. In theory 3D parameters should provide a more accurate model than 1D parameters for fold recognition. However 3D parameters, for reasons we have discussed above, are much more difficult to use in practice than 1D parameters. For the moment the drawbacks of using 3D parameters outweigh the advantages.

We think there is still room for improvement since, after stage 1, in 60% of the cases a true positive appears in first position in the core list. Unfortunately for some of them, their score lies in regions where the number of false positives is high. We hope that improvement in the 3D parameters and the use of new filters will help to push these cores beyond the border lines, in a region where false positives are scarce.

References

1. Altschul SF, Gish W, Miller W, Myers EW, Lipman DJ: **Basic local alignment search tool.** *J Mol Biol* 1990, **215**:403–410.
2. Pearson WR: **Effective protein sequence comparison.** *Methods Enzymol* 1996; **266**:227–258.
3. Brenner SE, Chothia C, Hubbard TJP: **Assessing sequence comparison methods with reliable structurally identified distant evolutionary relationships.** *Proc Natl Acad Sci USA* 1998, **95**:6073–6078.
4. Orengo CA, Jones DT, Thornton JM: **Protein superfamilies and domain superfolds.** *Nature* 1994, **372**:631–634.
5. Chothia C: **One thousand families for the molecular biologist.** *Nature* 1992, **357**:543–544.
6. Wang Z-X: **A re-estimation for the total numbers of protein folds and superfamilies.** *Protein Eng* 1998, **11**:621–626.
7. Govindarajan S, Recabarren R, Goldstein RA: **Estimating the total number of protein folds.** *Proteins* 1999, **35**:408–414.
8. Koppensteiner WA, Lackner P, Wiederstein M, Sippl MJ: **Characterization of novel proteins based on known protein structures.** *J Mol Biol* 2000, **296**:1139–1152.
9. Smith TF, Lo Conte L, Bienkowska J, Gaitatzes C, Rogers RG, Lathrop RH: **Current limitations to protein threading approaches.** *J Comput Biol* 1997, **4**:217–225.
10. Jones DT, Taylor WR, Thornton JM: **A new approach to protein fold recognition.** *Nature* 1992, **358**:86–89.
11. Sippl MJ: **Calculation of conformational ensembles from potentials of mean force. An approach to the knowledge-based prediction of local structures in globular proteins.** *J Mol Biol* 1990, **216**:859–883.
12. Bryant SH, Lawrence CE: **An empirical energy function for threading protein sequence through the folding motif.** *Proteins* 1993, **16**:92–112.
13. Nishikawa K, Matsuo Y: **Development of pseudoenergy potentials for assessing protein 3D-1D compatibility and detecting weak homologies.** *Protein Eng* 1993, **6**:811–820.
14. Di Francesco V, Garnier J, Munson PJ: **Protein topology recognition from secondary structure sequences.** *J Mol Biol* 1997, **267**:446–463.

15. Fischer D, Eisenberg D: **Protein fold recognition using sequence derived predictions.** *Protein Sci* 1996, **5**:947–955.

16. Madej T, Gibrat J-F, Bryant SH: **Threading a database of protein cores.** *Proteins* 1995, **23**:356–369.

17. Godzik A, Kolinski A: **Sequence-structure matching in globular proteins: Application to supersecondary and tertiary structure predictions.** *Proc Natl Acad Sci USA* 1992, **89**:12098–12102.

18. Burke DF, Deane CM, Nagarajaram HA, Campillo N, Martin-Martinez M, Mendes J, Molina F, Perry J, Reddy BVB, Soares CM, Steward RE, Williams M, Carrondo MA, Blundell TL, Mizuguchi K: **An iterative structure-assisted approach to sequence alignment and comparative modeling.** *Proteins* 1999. **Suppl 3**:55–60.

19. Kelley LA, MacCallum RM, Sternberg MJ: **Enhanced genome annotation using structural profiles in the program 3D-PSSM.** *J Mol Biol* 2000, **299**:499-520.

20. Fischer D, Eisenberg D: **Predicting structures for genome proteins.** *Curr Opin Struct Biol* 1999, **9**:208–211.

21. Bode W, Huber R: **Natural protein proteinase inhibitors and their interaction with proteinases.** *Eur J Biochem* 1992, **204**:433–451.

22. Gibrat J-F, Madej T, Bryant SH: **Surprising similarities in structure comparison.** *Curr Opin Struct Biol* 1996, **6**:377–385.

23 Jean P, Pothier J, Dansette P, Mansuy D, Viari A: **Automated multiple analysis of protein structures: Application to homology modeling of cytochromes P450.** *Proteins* 1997, **28**:1–16.

24. Jeanmougin F, Thompson JD, Gouy M, Higgins DG, Gibson TJ: **Multiple sequence alignment with CLUSTALX.** *Trends Biochem Sci* 1998, **23**:403–405.

25. Munson PJ, Singh RK: **Multi-body interactions within the graph of protein structure.** *Proc Int Conf Intell Syst Mol Biol* 1997, **5**:198–201.

26. Thomas PD, Dill KA: **Statistical potentials extracted from protein structures: How accurate are they?** *J Mol Biol* 1996, **257**:457–469.

27. Torda AE: **Perspectives in protein-fold recognition.** *Curr Opin Struct Biol* 1997, **7**:200–205.

28. Koretke KK, Luthey-Schulten Z, Wolynes PG: **Self-consistently optimized statistical mechanical energy functions for sequence structure alignment.** *Protein Sci* 1996, **5**:1043–1059.

29. Crippen GM: **Easily searched protein folding potentials.** *J Mol Biol* 1996, **260**:467–475.

30. Hubbard SJ, Campbell SF, Thornton JM: **Molecular recognition. Conformational analysis of limited proteolytic sites and serine proteinase protein inhibitors.** *J Mol Biol* 1991, **220**:507–530.

31. Gibrat J-F, Robson B, Garnier: **Further developments of protein secondary structure prediction using information theory.** *J Mol Biol* 1987, **198**:425–443.

32. Altschul SF: **Amino acid substitution matrices from an information theoretic perspective.** *J Mol Biol* 1991, **219**:555–565.

33. Fano R: *Transmission of Information*. New-York: Wiley, 1961.

34. Henikoff S, Henikoff JG: **Amino acid substitution matrices from protein blocks.** *Proc Natl Acad Sci USA* 1992, **89**:10915–10919.

35. Needlemann SB and Wunsch CD: **A general method applicable to the search for similarities in the amino acid sequence of two proteins.** *J Mol Biol* 1970, **48**:443–453.

36. Lathrop RH: **The protein threading problem with sequence amino acid interaction preference is NP-complete.** *Protein Eng* 1994, **7**:1059–1068.

37. Xu Y, Xu D, Uberbacher C: **An efficient computational method for globally optimal threading.** *J Comput Biol* 1998, **5**:597–614.

38. Lathrop RH, Smith TF: **Global optimum protein threading with gapped alignment and empirical pair score functions.** *J Mol Biol* 1996, **255**:641–665.

39. Yadgari J, Amir A, Unger R: **Genetic algorithms for protein threading.** *ISMB* 1998, **6**:193–202.

40. Shakhnovich EI, Gutin AM: **Implications of thermodynamics of protein folding for evolution of primary sequences.** *Nature* 1990, **346**:773–775.

41. Hubbard TJ, Ailey B, Brenner SE, Murzin AG, Chothia C: **SCOP: A structural classification of proteins database.** *Nucleic Acids Res* 1999, **27**:254–256.

42. Pearson WR: **Effective protein sequence comparison.** *Methods Enzymol* 1996, **266**:227–258.

43. Panchenko AR, Marchler-Bauer A, Bryant SH: **Combination of threading potentials and sequence profiles improves fold recognition.** *J Mol Biol* 2000, **296:**1319–1331.

44. Vapnick V: *Statistical Learning Theory.* New York: Wiley, 1998.

45. Altschul SF, Madden TL, Schaffer AA, Zhang J, Zhang Z, Miller W, Lipman DJ: **Gapped BLAST and PSI-BLAST: A new generation of protein database search programs.** *Nucleic Acids Res* 1997, **25**:3389–3402.

46. Bryant SH, Altschul SF: **Statistics of sequence-structure threading.** *Curr Opin Struct Biol* 1995, **5**:236–244.

47. Mirny LA, Finkelstein AV, Shakhnovich EI: **Statistical significance of protein structure prediction by threading.** *Proc Natl Acad Sci USA* 2000, **97**:9978–9983.

48. Corpet F, Gouzy J, Kahn D: **Recent improvements of the ProDom database of protein domain families.** *Nucleic Acids Res* 1999, **27**:263–267.

49. Argos P: **An investigation of oligopeptides linking domains in protein tertiary structures and possible candidates for general gene fusion.** *J Mol Biol* 1990, **211**:943–958.

10

Protein Structure Prediction by Threading: Force Field Philosophy, Approaches to Alignment

Thomas Huber and Andrew E. Torda

Abstract

Protein threading can be regarded as a search for the most appropriate known structure for a sequence of interest. This chapter discusses some of the fundamental issues that arise from score functions or force fields which are based on through-space interactions between particles. We consider different levels of resolution and representation of a protein and the rationalizations of score functions. These range from the statistical mechanical to purely empirical and even ad hoc.

Protein threading also involves calculation of a sequence to structure alignment. Although this is conceptually separate from the score function, we consider ways in which the two aspects interact.

Given the possibility of non-physical score functions, we discuss how these may be parameterized and even combined with functions such as sequence comparison terms which do not act through space. Continuing in this vein, we show how the library of known structures, can be manipulated so as to optimize fold recognition.

10

Protein Structure Prediction by Threading: Force Field Philosophy, Approaches to Alignment

Thomas Huber and Andrew E. Torda

The Australian National University,
Canberra ACT, Australia

10.1 Introduction

If you are given a protein's sequence, you might have all the information you need to predict its structure. You have the composition and (bond) topology of the system, so you only have to rearrange its atoms so they are somewhere in the major free energy basin and the problem is solved. There might be some problems with this approach. The search space grows exponentially with the number of particles. If you are able to search the conformational space for a five-residue peptide this year, it might be another year or two until you can tackle six residues when your computer is several fold faster. Then, you have to have an energy or score function that really can discriminate between correct and incorrect conformations. Any score function that is fast enough to apply to more than a few hundred atoms will be full of approximations and no longer close to the best level of theory. It is also worth remembering that we are only assuming that the native protein conformation really is a free energy minimum, and that some native conformations may only be of low energy because of prosthetic groups or unusual interactions (with ligands, ions or other proteins). Finally, if we believe that proteins find a free energy minimum, then we should remember that while potential energy is a property of a conformation, free energy is a function of many conformations. Considering all these problems, it would be fair to say that the history of protein structure prediction is one of approximation, optimism and cunning heuristics.

Our particular interest is in the set of approximations and heuristics that underpin the methodology or set of methods known as protein threading. Specifically, we are most interested in a class of score functions built for this application, how they may be constructed to ease the conformational search problem and any other devices which we can employ within a threading framework.

Protein threading grew out of the observation that often when a protein structure is solved, it is remarkably similar to one already in the protein data bank—even when it would not be expected from sequence similarity. Perhaps, it was reasoned, it would be a major advance if one could take a sequence of interest and just find the most compatible structure from the protein data bank. Looking back, it seems that the threading is the child of two camps:

1. The biologist's approach to structure prediction:

   by comparison and induction—if sequence1 is similar to sequence2 then structure1 is similar to structure2 and there is probably an evolutionary explanation.

2. The physicist's approach to structure prediction:

   proteins form structures according to fundamental rules that we call energies or free energies.

It is clear why the schools merged. Biologists learned that protein structures are much more conserved than protein sequences, so comparing sequences is not sufficient. Physicists, on the other hand, saw that the set of known conformations is much smaller than the set of all possible conformations. Threading as we know it now contains elements from both camps. From the physicists' side come elements such as low-resolution or coarse-grained force fields. Classical biology / bioinformatics has provided techniques such as sequence to structure alignment methods based on dynamic programming.

10.1.1. Common Methodology

For the rest of this chapter, we repeatedly refer to threading and its properties. We assume several elements:

- One has a sequence whose structure is to be predicted and this sequence does not have significant homology to anything of known structure, otherwise the predictor would probably be better off with some other method.
- There is a library of known structures / templates. The sequence will be aligned to and tested against each in turn. The original sequence of the template is known, but may not be used.
- There is some kind of score function or force field that is capable of returning the happeness of a sequence residue at any position on a template.
- One has a method for calculating sequence to structure alignments with gaps. This method must be able to handle gaps and insertions. The most common methods are dynamic programming and Monte Carlo.

Now, some points need to be expanded. There are several ways to align protein sequences to structures (templates). We describe one diagrammatically. Fig. 10.1 shows a sequence whose structure we do not know at the start (left of diagram) and some template structure from the library (right of diagram). This template is not exactly the same shape or size as the unknown answer, but it is the best library member. The library has hundreds or thousands such templates, most of which are completely wrong for the sequence. The computational problem is summarized in Fig. 10.2. On the one side we have the sequence (unknown structure) and on the right, the template. We then construct a matrix which gives the score that every sequence residue would have if it were placed at any numbered position from the template (Fig. 10.3). A dynamic programming method is used to find the path through the matrix which preserves sequence ordering, while allowing gaps and insertions. In this example, there is a gap, omitting residues 7, 8 and 9 from the template. Having calculated the sequence to structure alignment, we now have a prediction for the sequence shown in Fig. 10.3. It is not a perfect guess (left hand side), but it is as good as one can do using the available template.

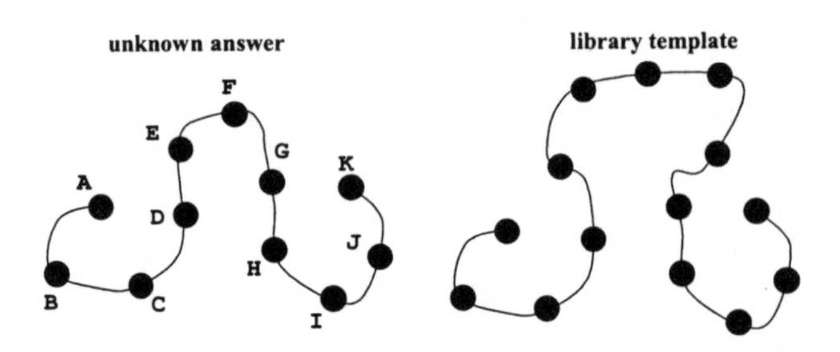

Figure 10.1. Sequence of unknown structure and candidate library template.

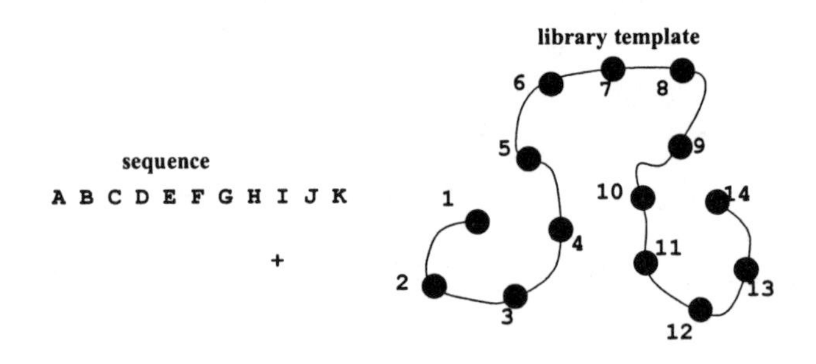

score matrix

	A	B	C	D	E	F	G	H	I	J	K
1	23	8	22	7	...	...	...	...	...	...	...
2	7										
3	8										
4	9										
5	12										
6	7										
7	...										
8	...										
9	...										
10	...										
11	...										
12	...										
13	...										
14	...										

Figure 10.2. Aligning a sequence to a template by constructing a score matrix. Some example matrix elements are filled in and a path is marked for the best sequence to structure alignment.

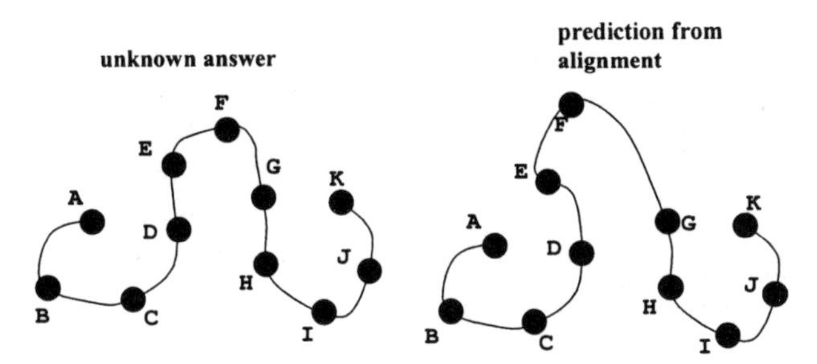

Figure 10.3. Correct answer and predicted structure. The prediction corresponds to the template and alignment path from Fig. 10.2.

With this background, we can be more specific about the contents of this chapter. We are interested in methods that use score functions that look (mathematically) like force fields, although they may not model real physics very closely. We are also interested in the approximations and methods one uses to find a sequence to structure alignment. Lastly, we discuss some of the methods that could be called elegant heuristics or computational tricks, but seem particularly tied to protein threading.

10.2. Force Field Based Scoring

As described above, one is going to need a function that can score a sequence residue at a position on a template. For the moment, we concentrate on what could be called pairwise, through-space interaction functions. By this we mean the situation shown in Fig. 10.4. We want the score associated with a residue "A" at the position on the template, and it will be calculated by considering interactions with its neighbours through space as shown by the arrows. This kind of problem is not unique to threading, but has been at the heart of every method to model or simulate molecules. Before considering threading specific score functions, they can be placed in context by considering the history of through-space force fields in general.

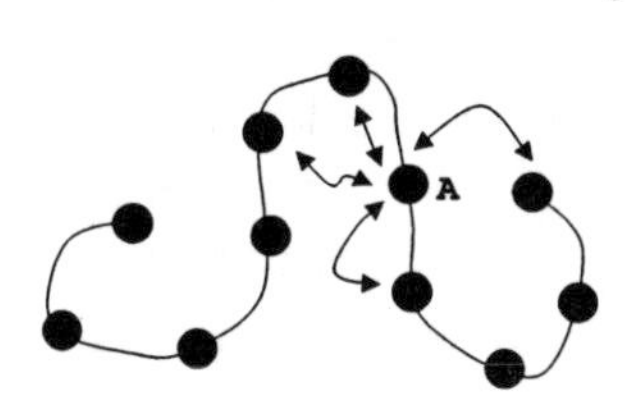

Figure 10.4. Calculation of pairwise, through-space scores.

The difference between methods lies in the choice of elementary unit (i.e., particle with no explicit internal degrees of freedom), how an environment is incorporated, what empirical functional forms are used and how they are parameterized. In purely *ab initio* methods, for example, elementary units are wave functions of electrons from the isolated system in vacuo, and the only "empirical parameters" are fundamental constants, such as Plancks constant. Semi-empirical *ab initio* methods still work at an electronic resolution, but use parameterized integrals to reduce the computations. A further step of approximations in the representation is made in molecular mechanics force fields. The smallest entity is an atom and electronic degrees of freedom are (in a time-average sense) implicitly considered in empirically parameterized atomic interaction functions. After omitting electronic detail, there is a significant gain in computational simplicity and it often becomes feasible to include the local environment explicitly in calculations.

An atomic representation is intuitively appealing for modelling molecular properties, but a much lower resolution description is required to simulate macroscopic behaviour of complex systems.[1] This leads to the concept of smooth particle simulation, in which the properties of a mesoscopic volume element of a system are captured by a single particle-like object. Just as in a molecular mechanics force field, inter-particle interactions are then based on physical concepts, and their parameters are fine tuned empirically.

Regardless of these details, whether a method uses a sub-atomistic or a mesoscopic description, it will contain approximations of some kind. The best method for a given application domain is the result of a trade-off between computational expense and performance.

In the particular application domain of modeling the overall structure of proteins, the representation of an amino acid residue by a single (or few) interaction sites has become popular for several reasons. First, there is a rationalization based on beliefs of protein folding. It is often believed that a protein chain collapses into its correct fold before an annealing of side chain conformations forms the exact native structure. If this is true, then it may not be necessary to model the details of side chain interactions and it could be adequate to treat them in a mean field manner. Secondly, there is a purely pragmatic justification. Omitting side chain degrees of freedom greatly simplifies calculations. With this simplification, however, one is faced with the problem of finding a score or potential energy function which summarizes mean-field, many-body interactions from amino acids into a few pairwise interactions.

Given an idea of what level of force field resolution one wants, we need to know what methods are available to build and parameterise force fields?

10.3. Parameterizing Force Fields

It is useful to consider three kinds of force field:

1. Physically-based potential energies;
2. Potentials of mean force;
3. Optimized force fields.

These are discussed in turn.

10.3.1. Physically-Based Potential Energies

By physically-based, we mean force fields which try to mimic the true physics of a system. If we follow physics, the answers will be correct and the methods will be transferable from system to system. At the risk of being too simple, consider a score function with atom-atom bonds. We can find bond lengths from X-ray crystallography, model the bond as a harmonic spring and say the energy for a single bond between particles at positions $\vec{r}_i$ and $\vec{r}_j$ is given by

$$V^{bond}\left(\vec{r}_i, \vec{r}_j\right) = k_{ij}^{bond}\left(\left|\vec{r}_i - \vec{r}_j\right| - r_0\right)^2, \qquad (10.1)$$

where k_{ij}^{bond} is the spring constant appropriately chosen for the bond between particles i and j, and r_0 is the ideal bond length. To find the energy of the system as a whole, we sum over all V_{bond} terms for each bonded ij pair. Similarly, if we know the partial charges on atoms, we could just use Coulombs law to calculate the electrostatic energy. Continuing in this vein, one could add in Lennard-Jones terms, bond angles, maybe dihedral angles and come up with a full force field, suitable for an application such as molecular dynamics calculations or energy minimization.[2-4]

Because we believe in physics, one must ask why this kind of force field is not more often used in protein threading. First, there has been a widespread belief that these empirical, atomistic force fields are not very good at discriminating between correct and incorrect models for a protein whenever the incorrect model is basically well folded and reasonably packed.[5] This has led some groups to incorporate extra pairwise terms to at least account for solvation.[6-9] Next, there is an issue of computational expense. Calculating the energy of a protein conformation is fast, but sequence to structure alignment calculations can involve huge numbers of these calculations. There is also the issue of potential versus free energy. Classical force fields are designed to yield potential energies, but in the introduction, we stated that proteins are more interested in free energy minima. One could search for approximations to entropic effects, but there will be fundamental limitations as to how well a single conformation can approximate an ensemble property such as free energy.[10] Lastly, there is a very good reason these force fields will always be problematic. To work at atomic resolution, one must know where all the atoms are. In protein sequence to structure threading and alignment, one usually does not know the location of the side-chain of any residue, let alone the neighbours it interacts with. This is discussed at length under sequence to structure alignments.

10.3.2. Potentials of Mean Force

Potential of mean force are certainly the most popular kind of interaction function for protein threading and come, literally, from textbook statistical mechanics.[11] Basically, the philosophy is that we can look at some property of a system, like the typical distance between two particles. If the particles are always close to each other, we might think there is some attractive force between them. If some other pairs of particles are never near each other, we might think that they repel each other. This can be formalized by considering the radial distribution function, $g(r)$, which really tells you the ratio of the probability seeing two particles at distance r and what one would expect from random placement. Then, one simply follows the Boltzmann relation to say

$$A(r) = -k_B T \ln(g(r)) \quad , \tag{10.2}$$

where $A(r)$ is a potential of mean force, k_B is Boltzmanns constant and T is the temperature. The result of this calculation is usually a tabulated interaction function. Most implementations collect observations in distance (r) bins and have corresponding values for $A(r)$ at each discrete distance. Philosophically, the results of the method are more interesting. Obviously, there are entropic contributions, but as the name implies, there are mean contributions from every possible source. For example, the interaction between two particles may be physically very much influenced by solvent. In this formulation, the solvent (along with other contributions) is present in an average sense.

Potentials of mean force do not rely on any particular representation of the system. In the statistical mechanical literature, they are usually atomistic, but there is nothing to stop someone collecting data at the level of whole amino acids. This is exactly what Tanaka and Scheraga did more than 25 years ago.[12] They treated a set of protein structures as if it were a statistical mechanical ensemble and extracted interaction functions between whole residues in proteins. The philosophy was applied to a larger set of proteins nearly a decade later[13] and by the 1990s there were many more implementations of the principle.[14-16]

There are many variations on the formulations for the collection of potentials of mean force and they are covered in this volume by authors that use them. While they are the most popular, they are not totally without critics.[17,18] Fundamentally, protein structures from the protein data bank are not strictly a real statistical mechanical ensemble; although many would argue that they are a good enough approximation. Furthermore, free energies are definitely not additive quantities.[19] Perhaps one can avoid this debate by considering the philosophy of the next section.

10.3.3. Optimized Force Fields

Force fields based on physics have useful properties. Because they are in real units, they can be compared against properties that depend on time or temperature. Sometimes, however, one does not care about kinetic or energetic effects. For protein structure prediction, you only need a function that takes a sequence and structure and returns the best score when the coordinates are correct (native). There is absolutely no need for this function to work in conventional energy units or for its values to scale to energy in any way. This has led many groups to pursue a direction that is not bound by conventional physics. Can one build a score function that is simply able to tell good from bad structures or reliably pick the best candidate from a set of trial configurations? You could call this a fold recognition, sequence-structure compatibility or discrimination function. Under duress, it could even be called a quasi-energy function or force field. To build this kind of score function, you do need: (a) a set of native proteins (sequences with their correct structures) and (b) for each native protein structure, some set of non-native, decoy conformations; (c) some set of energy / score functions with some adjustable parameters.

Now, we consider that class of score functions that are built by optimizing their score parameters, rather than by following any rules from physics or statistical mechanics. As an example, consider the approach suggested more than a decade ago in various forms by Crippen and co-workers.[20-22] First, they picked some general form for interactions. This may have been of an almost Lennard-Jones form like

$$V(r_{ij}, a_{ij}, b_{ij}) = a_{ij} r_{ij}^{-12} + b_{ij} r_{ij}^{-10} \ , \qquad (10.3)$$

where r_{ij} refers to the distance between some point on each residue, i and j and a_{ij} and b_{ij} are some adjustable parameters depending on the types of residues i and j. The force field may typically contain tens or hundreds of these a and b parameters. Assuming additive interactions, one can make the obvious summation over all pairs ij to get a total energy:

$$V^{tot}\left(\vec{r}, \vec{a}, \vec{b}\right) = \sum_{j>i} V\left(r_{ij}, a_{ij}, b_{ij}\right) , \qquad (10.4)$$

where we note that the final energy is a function of all parameter values a and b as well as coordinates. Crippen and co-workers then asked whether they could adjust the various a and b values of Eq. (10.3) so that the energy of a native protein is always lower than that of a decoy conformation. If we think of a and b as variables instead of parameters, we could say that we want to solve an inequality:

$$V_{nat}^{tot}(\vec{r}, \vec{a}, \vec{b}) < V_{dec}^{tot}(\vec{r}, \vec{a}, \vec{b}) \qquad (10.5)$$

with respect to a and b. Of course, there would be a very large set of inequalities to be solved, considering one protein in the parameterization set and its decoy structures, then every protein in the parameterization set and its corresponding decoys.

In this description, there has been no discussion of what the exact form of Eq. (10.3) should be, nor details such as what atoms (interaction centres) are used for calculating the r_{ij} values. These are just examples to make the point that this kind of score function can be a very artificial creation. It may be possible to explain the final a and b values in Eq. (10.3) in physical terms, but this would be a rationalisation after the fact. It is not implicit in the procedure.

Continuing in this vein, there is now a volume of work along these lines that we can try to view in some unified way. First, assume there is a set of native structures, each with decoys. These will not be changed, so we do not explicitly write the vectors r of coordinates. Similarly, we will not bother to label total energies *tot* since energies are assumed to be totals over a protein. Instead, we will consider energies as functions of parameters and use terms like $V_{nat}(\vec{p})$ and $V_{dec}(\vec{p})$ for native and decoy energies respectively. If a protein has N_{dec} decoy structures, then perhaps

$$c(\vec{p}) = \sum_{i=1}^{N_{dec}} V_{nat}(\vec{p}) - V_{dec}^{i}(\vec{p}) \tag{10.6}$$

would be a useful cost function to minimize. To work on many proteins, we should, of course, have

$$C(\vec{p}) = \sum_{j=1}^{N_{nat}} c_j(\vec{p}) \quad , \tag{10.7}$$

where we sum the cost function over N_{nat} native proteins. Now, it appears that one has the ingredients for building a force field. For some set of interaction functions, one wants to minimize Eq. (10.7) by adjusting the parameter vector, $\vec{p}$. One way to approach this is to see that for many types of interaction function, including those of Eq. (10.3), one can take the partial derivatives with respect to parameters. This could lead to using a minimizer such as steepest descents or conjugate gradients, but perhaps there are multiple minima on the cost function surface. In that case, it would be better to use an even more powerful minimizer such as simulated annealing.[23] This approach was taken to extremes in one piece of work that borrowed a method from density functional theory.[24] The parameters were given fictitious masses and velocities in parameter space. Then, by analogy with molecular dynamics, a function like Eq. (10.7) was treated as if it were a potential energy and parameter dynamics was used in parameter space.[25]

This approach is entertaining, but perhaps somewhat intractable. The quasi-forces experienced by a parameter depend on a potentially huge number of native protein conformations, decoy conformations and all the interactions within each structure. A better method would be to try to gather the relevant properties of native and decoy structures into a pre-calculated set of properties. Consider the energy of misfolded decoy structures for some sequence. Under certain conditions, their energy will follow a normal distribution as shown in Fig. 10.5. This is characterized by the mean, $\langle E_{dec} \rangle$ and standard deviation, σ. We want a score function that gives an energy for the native structure that is much lower than the mean energy of the decoy structures. One could maximize the difference by simply scaling the energies, but this would not be very helpful. Instead, one wants to adjust parameters so as to move the energy of the native structure to the left of the picture, while simultaneously minimising the standard deviation. This can be seen as optimizing a general statistical property, the z-score which just measures how many standard deviations an observation is from the mean. In the case of building score functions, we are not interested in optimizing the z-score for one protein, but in optimizing the z-score for many proteins simultaneously. Typical numbers would be 300–400 native structures and 10^7 total misfolded decoys.[26]

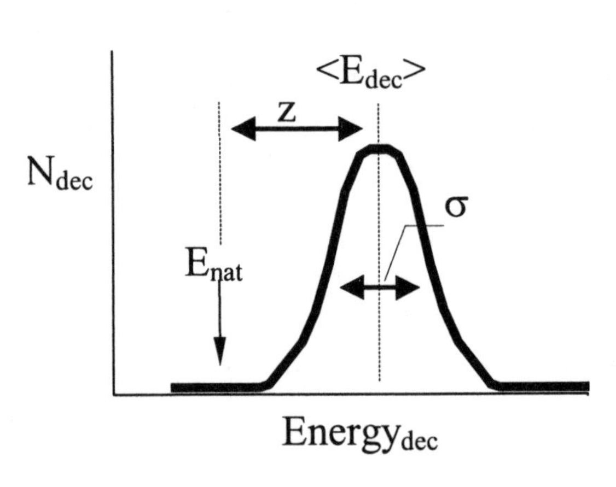

Figure 10.5. Distribution of energies of misfolded proteins.

Z-score optimized force fields appeared from several groups almost simultaneously, with differences in functional forms and implementation.[26-29] Compared to other methods for optimizing force fields, the methodology has several attractive features. First, it can be quite fast. Many properties, depending on the coordinates, can be calculated before any minimization is applied. Next, the approach is, in principle, capable of finding something which could be called the best possible score function. The parameters should be sent to the best possible values without the restriction of following some assumed distribution.

There are now, in some ways, many score functions for protein fold recognition, although they differ widely in what type of interaction function they use and the quantity they optimize.[30-35] Some are based on optimizing a penalty function and some on constraint satisfaction. Interaction functions include Lennard-Jones like interactions,[20,22,25] tabulated distance-dependent terms,[32,36] sigmoidal contact functions[21,26] and combinations of very general, almost polynomial basis functions.[37] The score functions also differ in their level of detail (one or more than one site per residue), where they place the interactions (backbone C_α, sidechain centre of mass, C_β) and how they treat distance along the protein backbone. For example, should one parameterize interactions between residues very near in the sequence separately from those with many intervening residues?

It is certain that not all of these score functions are equally good, but it is not possible to say which are best. There is no agreed upon measure for testing, nor consensus as to what they should be able to do. Most workers would like to be able to recognize a structural homologue for a protein given only its sequence, but there is no accepted definition of structural homologue. It could be defined in terms of structural difference and the amount of structural overlap. Furthermore, score functions are rarely compared in isolation. The results that a group observes will depend on the alignment or other testing method they use.

Rather than say what the best score function is, one might evade the question by saying there is probably no ideal force field. Different formulations will probably perform best on different problem domains.[37] For example, a force field may be trained on a set of native structures and decoys where the decoys can be quite close to the native structure. This could be too difficult, so one may ignore decoys if they are similar to the native[25] or be more sophisticated and ask that the energy function be sensitive to just how similar a decoy is to the native.[21] One could simply ignore the issue and hope that if a few decoys are similar to native structures, then they disappear in the statistical noise.[26]

Ultimately, nobody knows the limits of this kind of approach to parameterizing force fields. If people cannot reliably predict protein structure, it could be because they have used poor interaction functions, they have optimized the wrong penalty function or satisfied the wrong constraints. Most ominously, it has even been shown that some forms of interaction function will never be able to distinguish native structures from certain decoys![38]

10.4. Alignment Philosophy

10.4.1. Common Alignment and Score Methods

In the introduction, we stated that one needs a means of finding sequence to structure alignments. In its most general form, this problem is surprisingly difficult and is actually NP-complete.[39] This means that probably no one can say there is no deterministic method guaranteed to find the best alignment in reasonable time. Instead, there is a selection of approximations in the literature.

Although we described the problem in terms of a score matrix in Fig. 10.2, there are other heuristics one could try. One could place sequence residues on the template, take random steps and apply a conventional Monte Carlo / simulated annealing method to find a good alignment.[40] One could just as well use a genetic algorithm.[41,42] There are two reasons we do not use these methods. First, in such a complex search space, one cannot guarantee finding the optimal solution. Second, making Monte Carlo tractable requires putting restrictions on the placement of gaps and insertions. These methods may work well enough in practice, but in the pursuit of elegance we would prefer a method which can put a gap or insertion of any length at any position.

The second major class of alignment methods relies on dynamic programming. It is deterministic and guaranteed to find the best possible alignment for the problem as posed. The price, however, is that restrictions are placed on the scoring. The most popular methods have their roots in sequence comparison and can only use what one would call a single body term. Through-space scoring functions require a scheme as in Fig. 10.4 but this cannot be used. In the figure, we know where the residue of type "A" is, but not it's neighbours. They have not yet been aligned. One approximation is a two-level dynamic programming method, but it is computationally very expensive.[16] Alternatively, one could remember that we do know the residues that were present on the original template structure. In the diagram, we could label the neighbours of "A" with their residue types from the original template and then calculate a score directly.[15,43,44] The quality of this so-called frozen approximation is obviously good in the case of homologous proteins with high sequence similarity. How good this approximation is for proteins with very distant homology or for orthologous proteins is, however, debatable. In the next section, we describe a different approximation which avoids the problem of sequence memory of the template.

A new class of methods has been proposed which are closer in spirit to branch and bound algorithms.[45-47] Despite their elegance and potential power, they have not become popular, probably due to difficulty of implementation.

10.4.2. Sausage Alignments

Our code, travelling under the name of sausage, attempts to operate at the best possible trade-off between scoring and searching methods, while simultaneously remaining as simple as possible.[48,49] The approach is to split the prediction into two very separate steps of sequence to structure alignments and ranking the models that are produced. These are not just conceptually separate, but usually done with different force fields and different gap penalties. There are distinct advantages to the approach. The tasks of aligning sequences and ranking models are fundamentally different and the best score function for one may not be the best for the other. Significantly, it is easy to fine tune the details for each step and each task can be optimized independent of the other.

The most unusual feature of sausage is a force field approximation in the first step which allows an optimal alignment to be calculated in polynomial time. The approximation is only used for alignments, and final ranking scores are calculated with a force field with full pairwise interactions. The aim is to find a function that allows scoring a single residue at any position on a template as shown in Fig. 10.4. Looking at the picture, we know the location of the residue's neighbours, but not their identity. This leads naturally to building a special score function that uses the identity and coordinates of one particle ("A" in the diagram), but only the coordinates of its neighbours. In other words, the particle interacts with neighbours with some kind of average particle type.

For example, a conventional neighbour specific score function for three amino acid types would have parameters for pairs AA, AB, AC, BB, BC, CC. In a neighbour non-specific (alignment) score function there are only 3 parameters for pairs AX, BX, CX, where X is a generic amino acid type. The X residue is conceptually an average amino acid, but its parameters result from numerical optimization and not averaging over an existing score function. Despite the rather drastic approximation of treating all neighbours uniformly, the method works remarkably well. It is particularly attractive in the case when one thinks of proteins which adopt similar structures due to convergent evolution and may have no similarity at the amino acid level.

10.5. Beyond Pairwise Terms

Most threading force fields are dominated by pairwise terms and approximations to solvation terms which one could call single-body terms. Maybe these score functions get incrementally better every year, but maybe they will never be sufficient to recognize protein folds.[38] Given the startling array of functional forms listed above, there is no reason to think that any group has found the correct way to encode certain kinds of information. For example, we know that there are statistical propensities for small stretches of amino acids to prefer certain secondary structures. Looking at typical interaction functions formulated in terms of distances, is there any reason to believe that they will (statistically) drive backbone angles to the correct conformations? Perhaps it would be better to find some other way of encoding secondary structure preferences into a protein prediction scheme. Similarly, some of the most adept fold recognition methods do not use any pairwise, distance dependent interaction terms at all. Hidden Markov[50-58] models are well described in this volume and they, along with PSI-BLAST[59] work entirely on proteins which have been reduced to one-dimensional strings. Perhaps a computational chef will find a delicate blend of terms from different fields that most economically capture available information. For the moment, we consider a bucket-chemist's approach of throwing terms together and stirring with optimism. Specifically, what issues arise when one tries to add different kinds of information together?

First, consider the case of sequence similarity and the toy example of Fig. 10.6. We know our sequence of interest, but we also know the original sequence from the template. Looking at the example, it is easy to calculate a sequence similarity score by looking up the AH element in an amino acid substitution matrix, then the DK element and so on. We now suggest that this can be added into an existing, pairwise, through-space score function. This is never going to be an elegant process. If we use the language of force fields, then our score function is returning some kind of energy and any term we add is also an energy. To do this, is as outrageous as saying you can measure sequence similarity in the same units as steric overlap.

```
sequence A  D  E  G  A  -  F  . . .
template H  K  L  -  P  Q  R  . . .
```

Figure 10.6. Simple sequence alignment for scoring.

Aside from this offence against scientific aesthetics, there are technical reasons to be careful. When adding terms together, we would like them to have the property we might call orthogonality. That is, a through-space score function term should provide different and independent information to something like a sequence comparison term. We can say, in advance, that this is certainly not the case and make the point with a simple example. Most through-space, pairwise score functions encode something like surface exposure / burial preferences. Amino acid substitution matrices encode the hydrophobicity by saying that similar residues can be swapped for each other. Clearly, both the bare score function and a substitution matrix are going to contain some similar information and the point could have been made with other properties such as size, aromaticity, or polarity. Now, we do not know in advance the extent of the overlap or independence, so there is little guide as to how the different components should be combined. One idea is simple empiricism or trial and error.[60,61]

An example of this trial and error was proposed by Panchenko *et al.* who added a sequence profile-based similarity term to a Boltzmann-based, pairwise set of interaction functions.[61] In order to scale the relative contributions, they simply titrated the sequence term against the rest of the score function, measuring success across a set of test proteins. When mixing terms, one can do better than this. If one has a rapid way of scoring force field quality, we can use straightforward numerical optimization to find the weight given to the sequence similarity term. This can be shown by example from the article.[62] We can define a sequence similarity score for a site simply by the score from a BLOSUM62 amino acid substitution matrix.[63] Next, we construct a measure of alignment quality based on 572 pairs of proteins with structural similarity, but no significant sequence identity.[64] Basically, we align the sequence of one member of each pair to the structure of the other and use a measure of structure quality to judge the alignment. This many alignments can be calculated quickly and a simplex optimization method can be used to adjust the weight on the sequence similarity term. Fig. 10.7 shows the result of this kind of calculation. One should define the parameterization set and quality function formally, but the figure is sufficient to make several points.

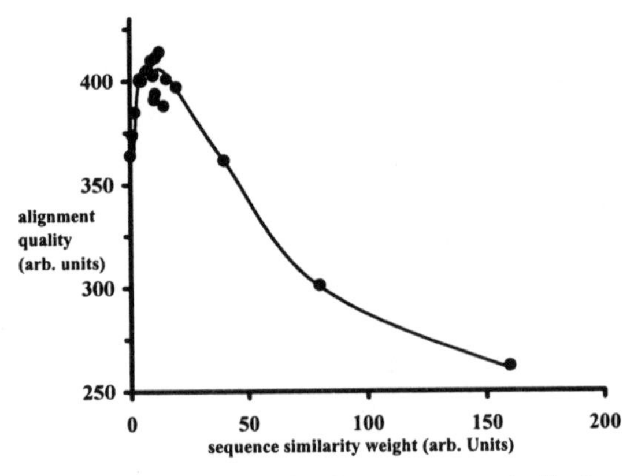

Figure 10.7. Optimization of weight for sequence similarity within overall force field.

Most clearly, the performance of the bare force field (left hand side of plot) is improved by adding a sequence similarity term. This suggests that we are adding information which was not encoded in the original score function. At the same time the term cannot be too high, otherwise performance decreases again. We deliberately do not give any more details on this calculation, because the exact result will be very force field specific. From the point of view of force field construction, there are other lessons to be learned. If we had a proper, self-consistent approach, the force field would have been optimized with the sequence similarity term built in and not added as an accessory at the end. At the same time, it could be that there is no single ideal weight for the extra term. If a sequence has significant sequence similarity to a template, then this term should be weighted very highly (this is the domain where sequence alignments are reliable). As similarity between sequence and template decreases, the term will add noise and its weight should be gradually decreased. To this end we note that one popular threading code already incorporates a sequence term which is switched on or off, depending on the degree of sequence/template similarity.[65]

Another source of extra information is the secondary structure known or predicted for a sequence. This should not be necessary since a perfect force field would naturally prefer the correct secondary structure for a residue. In practice it appears that no such perfect score function exists, so it is common to bias a score function with the output from some secondary structure prediction method.[66] The reason is probably a matter of how the encoding is performed. Popular and successful methods for predicting secondary structure are usually based on neural networks that consider a window of residues centered on the site to be predicted.[67-72] Through-space score functions have not, generally, been parameterized in these terms.

If one is going to include secondary structure information, there are several forms one might try. We think of secondary structure in terms of backbone ψ, and ϕ angles, but these could be used to put restraints on some interatomic distances. One could try simply matching the type (α-helix, β-sheet) predicted for the sequence to the type observed in the structure and constructing a score term which rewards a correct match. In our experience, neither of these approaches works well with real structures.[62] α-helices are easy to encode in terms of specific inter-residue distances, but β-sheets are not (the neighbour is not specified). Simple switching functions have been similarly unsuccessful since they require some threshold for recognising a type of secondary structure, but backbone angles from real structures too often sit near the borders of classic secondary structure ranges. A far better approach is to use a continuous function that gently rewards a residue for agreeing with a prediction. In our code, we have used[73]

$$V^{\mathrm{sec}}(\psi) = -\cos(\psi_0 - \psi) - 1 \,, \qquad (10.8)$$

where ψ is the backbone angle at the site on the template and ψ_0 is the literature value for the predicted secondary structure for the residue ($\psi_0 = -47°$ for α-helix and $\psi_0 = 124°$ for β-sheet). One probably should use an extra weighting depending on the confidence in the secondary structure assignment, but in our implementation, we have used only high confidence predictions and added a single weighting coefficient to balance this term against the rest of the scoring function. As with the sequence similarity term described above, we have used numerical optimization to balance this term against the rest of the scoring function and find a similar pattern. First, we do find a significant improvement in alignment quality after incorporating predictions from a popular server.[74] Second, this term should not be weighted too high, otherwise performance drops. While predictions are sometimes wrong, they do provide information which was not present in the original score function.

Given the utility of other forms of information, it would seem that protein fold recognition tools should be able to incorporate more terms from predictions or experiment. Correlated mutations may provide information about residues near in space.[75,76] Perhaps predictions of solvent accessibility are better than the hydrophobic preferences in current score functions.[77] It is probably more exciting to consider what can be done to take advantage of experimental data. If an experiment is going to produce a reliable structure, it might be best to avoid theoretical speculation. If, however, an experiment provides sparse and perhaps even noisy information, then it may be very valuable to combine it with protein fold recognition / threading. Nuclear magnetic resonance (NMR) spectroscopy is a prime example. Although NMR is known for producing final structures, there is a wealth of data which is never used to produce a final structure. Even if a protein is too large for a structure determination, the early resonance assignments may be enough to predict secondary structure at more than 90% reliability, far in excess of the best prediction methods.[78-81] It was shown some years ago, with early fold recognition codes, that even a relatively small amount of reliable secondary structure information from NMR can make an enormous difference to the reliability of fold recognition.[73]

Certain kinds of experiments can also yield sparse distance information. Conventional protein chemistry may locate disulfide bonds, and cross-linking combined with advances in protein mass-spectrometry, will put limits on the distance between sites in proteins.[82] Currently, the challenge is to computationally take advantage of this data. It is easy to filter proteins as to whether they agree with distances, but it may be harder to use the information directly. Young et al checked threading predictions for agreement with mass-spectrometry data, but there was no attempt to incorporate the information into sequence to structure alignments.[83]

10.6. Template Libraries

Throughout this chapter we have spoken of aligning a sequence to a template structure from a library. The library itself will have a distinct impact on the success of a method. It must include all possible candidate structures and maybe it should avoid duplication of very similar proteins. This requires some sampling or clustering method that can be based on either sequences or structures. This is not a simple process since there is tremendous redundancy (there are now more than 400 variations on T4 lysozyme) and the selection should be biased toward high quality structures. Furthermore, a particular threading implementation may work best if the library is biased to prefer larger or smaller structures. At the more detailed level, there are possible improvements to be made to the actual coordinates or representations used in the library. Here we consider two possibilities. First, can one average over structures and second, can one use numerical optimization of the structures so as to make them more suitable for recognition by a sequence?

Despite the use of coarse-grained models and simple interaction functions, a major problem with threading scores is their sensitivity to small changes in structure and/or sequence. It is very easy to recognize the correct structure for a sequence. It is usually difficult to recognize a structure which looks similar, by eye, to the correct one, but which differs in many details. In the case of sequence changes, the problem has been well studied. Basically, one should perform calculations on sets of aligned sequences whenever possible. Taking advantage of structural similarities is much more challenging and has proved less popular.

There are good technical reasons for this. First, it is not clear what constitutes a similar structure. Should one require that two proteins are superimposable to within a coordinate RMSD of 3 Å for 90 % of their residues or should one say that a helix-strand-helix motif provides a similar environment to the residues within? Next, there is the problem of using aligned three-dimensional structures. Averaging Cartesian coordinates results in structures with disturbed bonds, while averaging in internal angles or even coordinates in transformed space (such as Fourier space routinely used in X-ray crystallography) results in similar problems. One way around the problem is to largely ignore gapped alignments. Finkelstein et al have shown several times that one can average scores from different structures, but have not managed to put this in the context of sequence to structure alignments using dynamic programming.[84-86] In order to tackle this problem, we have taken a somewhat indirect approach and asked if we can average environments within a protein. First, we have taken protein structural alignments from the literature[87-90] and from each set of aligned structures, declared one to be the parent or representative protein. For each site on this protein, one can calculate the score that a residue of each of the 20 types would experience. One can then look at corresponding sites on other proteins from the structural alignments and calculate the analogous scores. These can now be directly averaged for each type of amino acid. The approach requires a score function that can be calculated for a sequence residue at an arbitrary position on a template without knowing the final alignment, but this is straightforward. One can either use a neighbour non-specific score function[91] described above, or even the frozen approximation favoured by other groups.[15,43,44]

As an informal example, we can show some results for alignment quality, using the same quality function and test set as in Fig. 10.7. Fig. 10.8 shows the results using the bare, z-score optimized force field, the same force field with structure averaging and the same force field, but using a simple averaging over sequences aligned by BLAST[92] searches. Structure averaging provides a clear improvement in the quality of models produced (published elsewhere).[62] In this specific implementation, the benefit is not as much as with multiple sequences. This might be a general finding or it may simply reflect more experience setting thresholds for multiple sequence alignments.

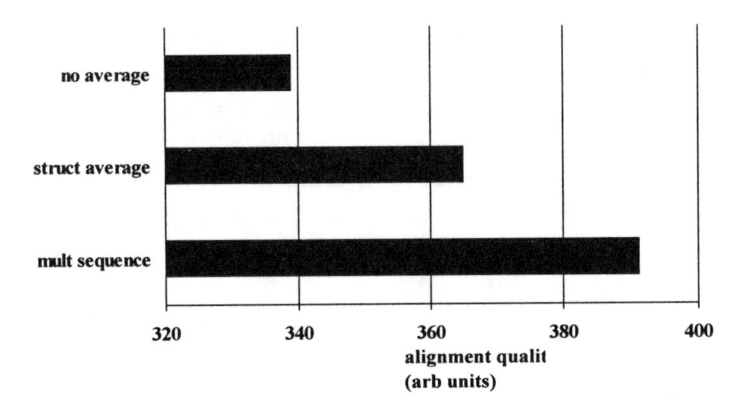

Figure 10.8. Effect of averaging over structures and sequences. *no average* refers to the bare score function, *struct average* to averaging over structures and *mult sequence* to averaging over sequences.

The second idea for manipulating templates in a library is to change the protein structures so as to support the scoring function in discriminating a family of native sequences from these same sequences aligned on any other structures. Conceptually, the idea is very simple. One could, for example, have a library of two different proteins with (families of) sequences A1, A2 and their corresponding structures S1, S2. What one would like to do is to change the coordinates of the first structure S1 in order to maximize the score of sequence(s) A1 on structure S1 and minimize the score of sequence(s) A1 aligned on the other structure S2, while simultaneously performing a similar optimization for the second sequence(s) A2. One then could think of optimizing the two structures independent from each other, trying to increase the gap between sequences A2 and A1 on structure S1, and the gap between A1 and A2 on structure S2.

We have performed such structure optimizations for a complete fold library of 893 non-redundant structures. For each protein, a PSI-BLAST search was performed to find a family of up to 20 homologous sequences with less than 95% sequence identity to each other. To speed up otherwise intractable calculations, not all 893 aligned sequences, but only the 20 best scoring alignments from structurally unrelated proteins were included as the unwanted negative cases in the calculation. During the optimization it was necessary to include other additional scoring terms. The sausage score function is based on a simple smoothed contact term and was parameterized for fold recognition only. This means that it lacks conventional force field terms which prevent residues overlapping or adopting conformations not found in proteins. To repair this deficiency, a molecular mechanics (GROMOS[2]) force field term for a generic poly-Ala chain was used to keep structures protein-like. Furthermore, a harmonic restraint was used to keep proteins within 1 Å of their native coordinates. These terms were brought together into an energy-like score which was then minimized with respect to structure coordinates using 100 steps of quasi-Newton minimization followed by 500 steps of momentum biased (molecular dynamics like) optimization. During this procedure, the gap between homologous sequences and alternative sequences generally widens, as one would have hoped from the design of the calculation.

The effect of the optimization is demonstrated in Plate 10.1 for the example of lyzozyme (PDB code 1531). The left side of the figure shows the energy spectrum of sequences placed on the native X-ray structure before the optimization. The green bars correspond to 20 homologous sequences, whereas the red bars indicate the 20 best scoring alignments of structurally unrelated proteins. The right side of Plate 10.1 shows the recalculated energy spectrum of the same sequences on the optimized structure. Clearly, our aim was achieved and now homologous sequences are well separated from alternative sequences. In our experience, fold recognition with the optimized library generally performs better than using a library of X-ray structures. This could be due to regularizing structures from the protein databank, since they do contain errors,[93] and the molecular mechanics term will help remove these. Much more significantly, the optimization intensifies features particular to each fold, at least in terms of this score function, while simultaneously deprecating common features. Furthermore, the coordinates can even adjust to compensate for weaknesses in the score function such as cut-off effects. In this scheme there is a mutual dependence of the scoring function (which is derived from structures) and the optimized library (optimized with respect to some quality of the scoring function). It would therefore be desirable to bring both parts into harmony by iterating the procedure to self-consistency. Until now we have limited the optimization to only one cycle due to the high computational costs of force field generation and library optimization. In future we will build a new fold library and force field that will be iteratively refined to self-consistency.

So far, this description has overlooked a flaw in the structure optimization. There is no term which enforces absolute values for structures. Imagine that, during optimization, the gap between homologous and alternate scores increases as it should. Within the framework, there is nothing to stop the absolute set of scores on a structure shifting together. This could mean that some structure ends up as very preferred by its native sequence, but also preferred by every other sequence. So far, this phenomenon has never been seen and we mainly see changes in the energy/score gaps between correct / incorrect sequences. Ultimately, it should be possible to use a scheme where structures are optimized simultaneously and connected to each other. This will be even more computationally expensive, but will guarantee that, for all homologous sequences, the native structure is higher in score than the same sequence aligned to any other structure.

10.7. Outlook and Speculation

The most remarkable feature of protein threading is its popularity and immaturity. By this, we mean the number of publications, often proposing new methods. This is quite different from other fields. Consider protein sequence analysis where there are a number of accepted methods for fast or accurate alignment and the statistics are relatively well understood. Score matrices do not change much from year to year. Consider molecular dynamics (MD) simulations. They are longer every year and there are regular proposals for improvements to terms such as electrostatics or treatment of solvent. Nobody, however, expects that MD simulators will abandon their classic integrators or basic pairwise force fields. In threading, there is no such consensus. There are alignments calculated by dynamic programming and some by Monte Carlo. There are force fields based on Boltzmann statistics and others on numerical optimization. Despite public comparisons of results, it is very difficult to say what the best approach is. Rarely, does one group's sequence to structure alignment method get used with another group's score function and a different group's method for assessing reliability.

Can one at least guess where the biggest weaknesses in current methods are? It is not clear if the best combination of multiple sequences, structures, and score functions has been found. Most of the work in this direction has come from groups with through-space score functions adding terms for sequence similarity. Given the success of sequence analysis, without any use of coordinates, it is quite possible that they will improve further as they make better use of through space relations.

Boltzmann-based or pure physically-based force fields are changing relatively slowly, but score functions which come from optimization or constraint satisfaction still appear in different forms. One obvious change would be a more holistic approach where terms such as secondary structure prediction and sequence similarity are cast as part of the original score function construction problem. Currently, these kinds of terms are treated as decoration on the main score function.

Finally, threading may not be the best approach at all. There may be a simple force field waiting to be built which will allow proteins to swiftly find their native states with a method such as dynamics simulation, genetic algorithm or other search method. Quite possibly, a fragment-based approach will end up as the most successful.[66]

More modestly, some improvement may come without much change to the score functions or search methods, but simply from better estimates of reliability. Sequence analysis has a good statistical basis and the statistics of structure prediction are the basis of another chapter in this volume.

The biggest threat to protein predictors comes from experimentalists. Advances in automation, robotics and chemistry mean that structures are solved at an ever increasing rate. The question is whether protein structure predictors will be able to make a useful contribution before they are outdated by the flood of experimental data.

References

1. Kremer K, Müller-Plathe F: **Multiscale problems in polymer science: Simulation approaches.** *MRS Bull* 2001, **26**:205–210.

2. van Gunsteren WF, Billeter SR, Eising AA, Huenenberger PH, Krueger P, Mark A, Scott WRP, Tironi IG: *Biomolecular Simulation: The GROMOS96 manual and user guide.* Zurich and Groningen: vdf Hochschulverlag AG an der ETH Zurich and BIOMOS b.v., 1996.

3. MacKerell AD, Bashford D, Bellott M, Dunbrack RL, Evanseck JD, Field MJ, Fischer S, Gao J, Guo H, Ha S, Joseph-McCarthy D, Kuchnir L, Kuczera K, Lau FTK, Mattos C, Michnick S, Ngo T, Nguyen DT, Prodhom B, Reiher WE, Roux B, Schlenkrich M, Smith JC, Stote R, Straub J, Watanabe M, Wiorkiewicz-Kuczera J, Yin D, Karplus M: **All-atom empirical potential for molecular modeling and dynamics studies of proteins.** *J Phys Chem B* 1998, **102**:3586–3616.

4. Cornell WD, Cieplak P, Bayly CI, Gould IR, Merz KM, Ferguson DM, Spellmeyer DC, Fox T, Caldwell JW, Kollman PA: **A second generation force field for the simulation of proteins, nucleic acids, and organic molecules.** *J Am Chem Soc* 1995, **117**:5179–5197.

5. Novotny J, Bruccoleri R, Karplus M: **An analysis of incorrectly folded protein models—implications for structure predictions.** *J Mol Biol* 1984, **177**:787–818.

6. Lazaridis T, Karplus M: **Discrimination of the native from misfolded protein models with an energy function including implicit solvation.** *J Mol Biol* 1999, **288**:477–487.

7. Janardhan A, Vajda S: **Selecting near-native conformations in homology modeling: The role of molecular mechanics and solvation terms.** *Protein Sci* 1998, **7**:1772–1780.

8. Wang YH, Zhang H, Scott RA: **A new computational model for protein-folding based on atomic solvation.** *Protein Sci* 1995, **4**:1402–1411.

9. Wang YH, Zhang H, Li W, Scott RA: **Discriminating compact nonnative structures from the native structure of globular-proteins.** *Proc Natl Acad Sci USA* 1995, **92**:709–713.

10. Reith D, Huber T, Müller-Plathe F, Torda AE: **Free energy approximations in simple lattice proteins.** *J Chem Phys* 2001, **114**:4998–5005.

11. Chandler D: *Introduction to modern statistical mechanics.* New York: Oxford University Press, 1987.

12. Tanaka S, Scheraga HA: **Statistical mechanical treatment of protein conformation. 1. Conformational properties of amino-acids in proteins.** *Macromolecules* 1976, **9**:142–159.

13. Miyazawa S, Jernigan RL: **Estimation of effective interresidue contact energies from protein crystal structures: Quasi-chemical approximation.** *Macromolecules* 1985, **18**:534–552.

14. Sippl MJ: **Calculation of conformational ensembles from potentials of mean force. An approach to the knowledge-based prediction of local structures in globular proteins.** *J Mol Biol* 1990, **213**:859–883.

15. Sippl MJ: **Boltzmann's principle, knowledge-based mean fields and protein folding. An approach to the computational determination of protein structures.** *J Comput-Aided Mol Des* 1993, **7**:473–501.

16. Jones DT, Taylor WR, Thornton JM: **A new approach to protein fold recognition.** *Nature* 1992, **358**:86–89.

17. Ben-Naim A: **Statistical potentials extracted from protein structures: Are these meaningful potentials?** *J Chem Phys* 1997, **107**:3698–3706.

18. Thomas PD, Dill K: **Statistical potentials extracted from protein structures: How accurate are they?** *J Mol Biol* 1996, **257**:457-469.

19. Dill KA: **Additivity principles in biochemistry.** *J Biol Chem* 1997, **272**:701–704.

20. Crippen GM, Snow ME: **A 1.8 Angstrom resolution potential function for protein folding.** *Biopolymers* 1990, **29**:1479–1489.

21. Maiorov VN, Crippen GM: **Contact potential that recognizes the correct folding of globular-proteins.** *J Mol Biol* 1992, **227**:876–888.

22. Seetharamulu P, Crippen GM: **A potential function for protein folding.** *J Math Chem* 1991, **6**:91–110.

23. Kirkpatrick S, Gelatt Jr. CD, Vecchi MP: **Optimization by simulated annealing.** *Science* 1983, **220**:671–680.

24. Car R, Parrinello M: **Unified approach for molecular dynamics and density-functional theory.** *Phys Rev Lett* 1985, **55**:2471–2474.

25. Ulrich P, Scott W, van Gunsteren WF, Torda AE: **Protein structure prediction force fields—Parametrization with quasi-Newtonian dynamics.** *Proteins* 1997, **27**:367–384.

26. Huber T, Torda AE: **Protein fold recognition without Boltzmann statistics or explicit physical basis.** *Protein Sci* 1998, **7**:142–149.

27. Hao MH, Scheraga HA: **How optimization of potential functions affects protein folding.** *Proc Natl Acad Sci USA* 1996, **93**:4984–4989.

28. Koretke KK, Luthey-Schulten Z, Wolynes PG: **Self-consistently optimized statistical mechanical energy functions for sequence structure alignment.** *Protein Sci* 1996, **5**:1043–1059.

29. Mirny LA, Shakhnovich EI: **How to derive a protein folding potential—A new approach to an old problem.** *J Mol Biol* 1996, **264**:1164–1179.

30. Chiu TL, Goldstein RA: **Optimizing energy potentials for success in protein tertiary structure prediction.** *Folding & Design* 1998, **3**:223–228.

31. Xia Y, Levitt M: **Extracting knowledge-based energy functions from protein structures by error rate minimization: Comparison of methods using lattice model.** *J Chem Phys* 2000, **113**:9318–9330.

32. Tobi D, Elber R: **Distance-dependent, pair potential for protein folding: Results from linear optimization.** *Proteins* 2000, **41**:40–46.

33. Tobi D, Shafran G, Linial N, Elber R: **On the design and analysis of protein folding potentials.** *Proteins* 2000, **40**:71–85.

34. Micheletti C, Seno F, Banavar JR, Maritan A: **Learning effective amino acid interactions through iterative stochastic techniques.** *Proteins* 2001, **42**:422–431.

35. Vendruscolo M, Najmanovich R, Domany E: **Can a pairwise contact potential stabilize native protein folds against decoys obtained by threading?** *Proteins* 2000, **38**:134–148.

36. Ayers DJ, Huber T, Torda AE: **Protein fold recognition score functions: Unusual construction strategies.** *Proteins* 1999, **36**:454–461.

37. Ohkubo YZ, Crippen GM: **Potential energy function for continuous state models of globular proteins.** *J Comput Biol* 2000, **7**:363–379.

38. Vendruscolo M, Domany E: **Pairwise contact potentials are unsuitable for protein folding.** *J Chem Phys* 1998, **109**:11101–11108.

39. Lathrop RH: **The protein threading problem with sequence amino acid interaction preferences is NP-complete.** *Protein Eng* 1994, **7**:1059–1068.

40. Madej T, Gibrat JF, Bryant SH: **Threading a database of protein cores.** *Proteins* 1995, **23**:356–369.

41. Pedersen JT, Moult J: ***Ab initio* protein folding simulations with genetic algorithms: Simulations on the complete sequence of small proteins.** *Proteins* 1997, **Suppl 1**:179–184.

42. Pedersen JT, Moult J: **Protein folding simulations with genetic algorithms and a detailed molecular description.** *J Mol Biol* 1997, **269**:240–259.

43. Godzik A, Kolinski A, Skolnick J: **Topology fingerprint approach to the inverse protein folding problem.** *J Mol Biol* 1992, **227**:227–238.

44. Wilmanns M, Eisenberg D: **Inverse protein folding by the residue pair preference profile method: Estimating the correctness of alignments of structurally compatible sequences.** *Protein Eng* 1995, **8**:627–639.

45. Lathrop RH, Smith TF: **Global optimum protein threading with gapped alignment and empirical pair score functions.** *J Mol Biol* 1996, **255**:641–665.

46. Lathrop RH: **An anytime local-to-global optimization algorithm for protein threading in $\mathcal{O}(m^2\tilde{n}^2)$ space.** *J Comput Biol* 1999, **6**:405–418.

47. Xu Y, Uberbacher EC: **A polynomial-time algorithm for a class of protein threading problems.** *Comput Appl Biosci* 1996, **12**:511–517.

48. Huber T, Torda AE: *Sausage program.* 2001, http://www.rsc.anu.edu.au/~torda/sausage.

49. Huber T, Russell AJ, Ayers D, Torda AE: **SAUSAGE: Protein threading with flexible force fields.** *Bioinformatics* 1999, **15**:1064–1065.

50. Eddy SR: **Hidden Markov models.** *Curr Opin Struct Biol* 1996, **6**:361–365.

51. Eddy SR: **Profile hidden Markov models.** *Bioinformatics* 1998, **14**:755–763.

52. Hughey R, Krogh A: **Hidden Markov models for sequence analysis: Extension and analysis of the basic method.** *Comput Appl Biosci* 1996, **12**:95–107.

53. Karplus K, Sjolander K, Barrett C, Cline M, Haussler D, Hughey R, Holm L, Sander C: **Predicting protein structure using hidden Markov models.** *Proteins* 1997, **Suppl 1**:134–139.

54. Karplus K, Barrett C, Hughey R: **Hidden Markov models for detecting remote protein homologies.** *Bioinformatics* 1998, **14**:846–856.

55. Karplus K, Barrett C, Cline M, Diekhans M, Grate L, Hughey R: **Predicting protein structure using only sequence information.** *Proteins* 1999, **Suppl 3**:121–125.

56. Park J, Karplus K, Barrett C, Hughey R, Haussler D, Hubbard T, Chothia C: **Sequence comparisons using multiple sequences detect three times as many remote homologues as pairwise methods.** *J Mol Biol* 1998, 284:1201–1210.

57. Sjolander K, Karplus K, Brown M, Hughey R, Krogh A, Mian IS, Haussler D: **Dirichlet mixtures: A method for improved detection of weak but significant protein sequence homology.** *Comput Appl Biosci* 1996, **12**:327–345.

58. Sonnhammer ELL, Eddy SR, Birney E, Bateman A, Durbin R: **Pfam: Multiple sequence alignments and HMM-profiles of protein domains.** *Nucleic Acids Res* 1998, **26**:320–322.

59. Altschul SF, Madden TL, Schaffer AA, Zhang J, Zhang Z, Miller W, Lipman DJ: **Gapped BLAST and PSI-BLAST: A new generation of protein database search programs.** *Nucleic Acids Res* 1997, **25**:3389–3402.

60. Rost B, Schneider R, Sander C: **Protein fold recognition by prediction-based threading.** *J Mol Biol* 1997, **270**:471–480.

61. Panchenko AR, Marchler-Bauer A, Bryant SH: **Combination of threading potentials and sequence profiles improves fold recognition.** *J Mol Biol* 2000, **296**:1319–1331.

62. Russell AJ, Torda AE: **Protein sequence threading—averaging over structures.** *Proteins* 2002, in press.

63. Henikoff S, Henikoff JG: **Amino acid substitution matrices from protein blocks.** *Proc Natl Acad Sci USA* 1992, **89**:10915–10919.

64. Torda AE: *List of proteins for alignment testing.* 2001, http://www.rsc.anu.edu.au/~torda/mult_strct/alignment_pairs.html.

65. Jones DT: **GenTHREADER: An efficient and reliable protein fold recognition method for genomic sequences.** *J Mol Biol* 1999, **287**:797–815.

66. Bonneau R, Baker D: ***Ab initio* protein structure prediction: Progress and reports.** *Annu Rev Biophys Biomol Struct* 2001, **30**:173–189.

67. Cuff JA, Barton GJ: **Evaluation and improvement of multiple sequence methods for protein secondary structure prediction.** *Proteins* 1999, **34**:508–519.

68. Cuff JA, Barton GJ: **Application of multiple sequence alignment profiles to improve protein secondary structure prediction.** *Proteins* 2000, **40**:502–511.

69. Chandonia JM, Karplus M: **New methods for accurate prediction of protein secondary structure.** *Proteins* 1999, **35**:293–306.

70. Jones DT: **Protein secondary structure prediction based on position- specific scoring matrices.** *J Mol Biol* 1999, **292**:195–202.

71. Rost B, Sander C: **Prediction of protein secondary structure at better than 70-percent accuracy.** *J Mol Biol* 1993, **232**:584–599.

72. Petersen TN, Lundegaard C, Nielsen M, Bohr H, Bohr J, Brunak S, Gippert GP, Lund O: **Prediction of protein secondary structure at 80% accuracy.** *Proteins* 2000, **41**:17–20.

73. Ayers DJ, Gooley PR, Widmer-Cooper A, Torda AE: **Enhanced protein fold recognition using secondary structure information from NMR.** *Protein Sci* 1999, **8**:1127–1133.

74. Rost B, Sander C, Schneider R: **PHD—An automatic mail server for protein secondary structure prediction.** *Comput Appl Biosci* 1994, **10**:53–60.

75. Göbel U, Sander C, Schneider R, Valencia A: **Correlated mutations and residue contacts in proteins.** *Proteins* 1994, **18**:309–317.

76. Shindyalov IN, Kolchanov NA, Sander C: **Can three-dimensional contacts in protein structures be predicted by analysis of correlated mutations?** *Protein Eng* 1994, **7**:349–358.

77. Rost B, Sander C: **Conservation and prediction of solvent accessibility in protein families.** *Proteins* 1994, **20**:216–226.

78. Wishart DS, Sykes BD: **The 13C chemical-shift index: A simple method for the identification of protein secondary structure using 13C chemical-shift data.** *J Biomol NMR* 1994, **4**:171–180.

79. Wishart DS, Sykes BD: **Chemical shifts as a tool for structure determination.** *Method Enzymol* 1994, **239**:363–392.

80. Wishart DS, Sykes BD, Richards FM: **The chemical shift index: A fast and simple method for the assignment of protein secondary structure through NMR spectroscopy.** *Biochemistry* 1992, **31**:1647–1651.

81. Wishart DS, Sykes BD, Richards FM: **Relationship between nuclear magnetic resonance chemical shift and protein secondary structure.** *J Mol Biol* 1991, **222**:311–333.

82. Aebersold R, Goodlett DR: **Mass spectrometry in proteomics.** *Chem Rev* 2001, **101**:269–295.

83. Young MM, Tang N, Hempel JC, Oshiro CM, Taylor EW, Kuntz ID, Gibson BW, Dollinger G: **High throughput protein fold identification by using experimental constraints derived from intramolecular cross-links and mass spectrometry.** *Proc Natl Acad Sci USA* 2000, **97**:5802–5806.

84. Finkelstein AV: **3D protein folds: homologs against errors—A simple estimate based on the random energy model.** *Phys Rev Lett* 1998, **80**:4823–4825.

85. Reva BA, Skolnick J, Finkelstein AV: **Averaging interaction energies over homologs improves protein fold recognition in gapless threading.** *Proteins* 1999, **35**:353–359.

86. Dykunov D, Lobanov MY, Finkelstein AV: **Search for the most stable folds of protein chains: III. Improvement in fold recognition by averaging over homologous sequences and 3D structures.** *Proteins* 2000, **40**:494–501.

87. Holm L, Sander C: **The FSSP database of structurally aligned protein fold families.** *Nucleic Acids Res* 1994, **22**:3600–3609.

88. Holm L, Sander C: **The FSSP database: Fold classification based on structure-structure alignment of proteins.** *Nucleic Acids Res* 1996, **24**:206–209.

89. Holm L, Sander C: **Dali/FSSP classification of three-dimensional protein folds.** *Nucleic Acids Res* 1997, **25**:231–234.

90. Holm L, Sander C: **Touring protein fold space with Dali/FSSP.** *Nucleic Acids Res* 1998, **26**:316–319.

91. Huber T, Torda AE: **Protein sequence threading, the alignment problem and a two step strategy.** *J Comput Chem* 1999, **20**:1455–1467.

92. Altschul SF, Gish W, Miller W, Myers EW, Lipman DJ: **Basic local alignment search tool.** *J Mol Biol* 1990, **215**:403–410.

93. Hooft RWW, Vriend G, Sander C, Abola EE: **Errors in proteins structures.** *Nature* 1996, **381**:272.

11

Predicting Protein Structure Using SAM, UCSC's Hidden Markov Model Tools

Kevin Karplus

Abstract

The protein-folding problem, in its purest form, is too difficult for us to solve in the next several years, but we need structure predictions now. One solution is to try to recognize the similarity between a target protein and one of the thousands of proteins whose structure has been determined experimentally. For very similar proteins, the relationships are easy to find and good models can be built by copying the backbone (and even some sidechains) from the homologous protein of known structure. For less similar proteins (in the "twilight zone"), the fold-recognition problem is more challenging, but it is often possible to find useful similarities.

Using evolutionary information helps enormously in recognizing remote relationships, and one convenient way to summarize a family of homologs is with a hidden Markov model (HMM). Homologs can be found and an HMM built by an iterated search, starting from a single target sequence. The resulting target HMM can be used to score the sequences of all proteins of known structure.

Similarly, homologs can be found and HMMs built for template proteins of known structure and used to score the target sequence. Combining both target-model and template-library results reduces the false positive rate. Some further improvements can be made by predicting local structural properties of the target sequence (such as secondary structure or solvent accessibility) and adding these predictions to the HMM used to score the template sequences.

Fold-recognition techniques based on these HMMs have performed quite well in blind prediction experiments (CASP2, CASP3, and CASP4) and are doing better than threading techniques based on pairwise potentials.

11

Predicting Protein Structure Using SAM, UCSC's Hidden Markov Model Tools

Kevin Karplus

University of California, Santa Cruz

11.1. A Naive View of Protein Structure Prediction

A complete solution of the "protein folding problem" would take the amino acid sequence of a protein and produce its native fold(s) from physical first principles, getting not only the correct structure, but the entire energy landscape for the protein, and consequently all other low-energy conformations and the dynamic behavior of the protein.

There are only five atom types in the standard amino acids (hydrogen, carbon, oxygen, nitrogen, and sulfur), so if these atoms and the interactions between them could be adequately modeled, it should not require too many physical parameters to get a complete solution of the folding problem.

This problem is intellectually important, but is much too difficult for our current knowledge and computational capacity. In practice, no one has made this approach work for anything more than toy problems. (More precisely, there have been occasional successes on small proteins, but not any consistent success.)

There are several problems with this naive view:

1. We cannot currently model the various atomic interactions with sufficient accuracy. Because real proteins have evolved to be only

barely stable at their operating temperatures, the energy computations involve subtracting terms that nearly cancel. Relatively small errors in any part of the computation completely change where in the conformation space the minima are located, thus changing which conformations of the protein appear to have the minimum free energy.

2. The energy function must model phenomena such as Van der Waals attraction and repulsion of atoms. The very narrow range of distances over which interactions like this are strongly favorable means that small changes in conformation result in large changes in free energy. That is to say, the energy function is very rough, with many local minima.
3. A typical protein domain has about 200 residues and 600 degrees of freedom (two degrees of freedom for each residue for the ϕ and ψ torsion angles and about one per sidechain). The solvent around the protein also needs accurate modeling, raising the number of degrees of freedom by six for each of hundreds of water molecules. Searching such a high-dimensional space with a fractal energy function takes enormous amounts of computation—far beyond current computer capability.

Because this pure *ab initio* approach to protein folding seems out of reach for a long time to come, but protein structure prediction is needed now, we must adopt a less purist approach to make progress.

Although one approach that has been tried is to approximate the energy function with smoother, more optimizable functions, the most pragmatic approach is to use all available information about a protein—not just its amino-acid sequence—in order to predict its structure. The most voluminous information is sequence information—the high-throughput DNA sequencing projects continue to produce an exponentially expanding supply of sequence information from a variety of organisms. Also increasing rapidly (though not as rapidly as the sequence information) is the database of solved protein structures. Other, more specific, information (such as disulfide bridges, active-site residues, proximity from cross-linking experiments, chemical protections, etc.) may be known for specific target proteins.

Our goal is to make the best use we can of this wealth of information.

11.2. Fold Recognition

The most successful structure-prediction techniques rely on fold recognition—finding a known structure (the *template*) that is similar to the protein whose structure is to be predicted (the *target*) and aligning the target sequence to the template sequence. The expectation is that aligned residues will play similar roles in the protein, producing similar structures. If finding the template is trivially done with simple sequence-based search (such as BLAST[1]or FASTA[2]), then the problem is referred to as *homology modeling*, and the emphasis is mainly on the quality of the alignment. Here we will concentrate on those cases where the template is not trivially found.

Fold recognition reduces the dimensionality of our search enormously. Instead of 100s or 1000s of degrees of freedom, we have only a few thousand templates to examine. There may be many possible alignments of a target to a template, but if we use efficient alignment procedures, we need only examine a few alignments for each template.

This chapter will focus on one fairly successful approach to fold recognition that relies on hidden Markov models (HMMs) for both selecting the template and for aligning the target to the template. One of the advantages of HMMs is that they allow the same efficient algorithms to be used for both tasks, while providing a rigorous probabilistic framework. The technique has been used successfully in three of the Critical Assessment of Structure Prediction (CASP) experiments.[3-5]

In its most basic form, a hidden Markov model, M, is just a *stochastic model* for a family of protein sequences, that is, it is an efficient way to compute a probability density function over all possible amino-acid sequences, $P_M(\text{seq})$.

Stochastic modeling of proteins is useful, because natural proteins are not designed, but are the results of a stochastic process of mutation and selection. If we can combine the information from several different proteins that result from the same ancestral sequence and essentially the same selection pressures, then we can have a much clearer picture of what is being selected for and what other proteins might result from the same (or sufficiently similar) process.

The stochastic modeling technique can be applied in two ways: We can model the family of the target sequence and search for a template belonging to that family, or we can build a library of template models and see if the target fits any of them. Both methods can be combined, to reinforce the weak signal from remotely related proteins.

An HMM is not a fully general stochastic model of a protein. Many important effects, such as long-range interactions and disulfide bridges, cannot be captured in an HMM, which treats each position of the protein sequence as essentially independent of the other positions. We accept this weakness of the modeling power in return for very efficient algorithms for doing computation with the HMMs. Many attempts have been made to create models that capture the tertiary interactions, most notably "threading" models that use pairwise terms,[6-10] but these have met with limited success[11] and relatively high computational cost.[12] Other than the disulfide bridges, the pairwise terms appear not to contain as much additional information as is commonly believed.[13]

To build a good HMM of the family of a protein, we need several representatives of the family. Ideally, we would like to have sequences that represent the full diversity possible for the family. Note that we have a bit of a circular problem here—our goal is to find a template sequence similar to the target sequence, but we need to know what sequences are similar to the target to build the HMM that we will search with.

As with other such circular problems, we can avoid the problem by using iteration. We first build an HMM based on a single sequence, then use that model to find similar sequences in a large database of proteins. These sequences can be used to build a better stochastic model of the protein family, which is used in turn for a new search and so on. This iterated search technique for building multiple alignments is the basis for the popular PSI-BLAST tool at NCBI[14] and for the SAM-T98,[15] SAM-T99, SAM-T2K,[5] etc. techniques developed at the University of California, Santa Cruz.

11.3. Hidden Markov Models

Let us look now at how a hidden Markov model (HMM) represents a protein family, how it is used for searching a database, how an alignment is made with an HMM, and how an HMM can be created from a family of protein sequences. The method described here is used with only minor variations in SAM-T98, SAM-T99, and SAM-T2K. The differences between the methods are mainly in the parameter settings and in minor improvements in the readability and maintainability of the scripts.

There are two approaches for explaining HMMs: for computer scientists and computer engineers, we start with finite-state machines and add probabilities, and for biologists and chemists, we start with multiple alignments. Although the first approach is more general, we are primarily interested in profile HMMs for this chapter, and so will take the second approach. A *profile* HMM is an HMM with a particularly simple structure explained in the next few paragraphs.

We can view a profile HMM as a condensed summary of the information in a multiple alignment of a family of proteins. It does not contain all the information of the multiple alignment—any information about individual sequences is lost, so procedures that rely on individual sequences (such as phylogenetic analysis) cannot be done with just the HMM. Fortunately, for the tasks we want to do (identifying whether a protein is a member of the family and aligning the protein to a member of the family), the information in the HMM is usually sufficient.

In summarizing a multiple alignment we distinguish between two classes of residues—those in *alignment columns*, which are assumed to have corresponding structure or function in the different proteins of the family, and *insertions*, which occur between the alignment columns and about which the multiple alignment makes no claims of similarity. Some proteins in the multiple alignment may not have any residue in a particular alignment column—the protein is said to have a *deletion* in that position. Note that insertions and deletions are defined here relative to the multiple alignment, not to a known or hypothesized ancestral sequence—there is no evolutionary relationship implied by these terms here.

Our profile HMMs will have two *states* for each alignment column: the *match state* and the *delete state*. Associated with each match state we have a table giving the probability of seeing each amino acid in that position of the multiple alignment. These probabilities are referred to as the *emission* probabilities for the state.

Simple dynamic programming algorithms can be used to compute the probability of a specific sequence given an HMM and to provide the most probable alignment of the sequence to the states of the HMM, hence to the alignment columns of the multiple alignment. These algorithms are beyond the scope of this chapter, but there is an excellent text available.[16]

One detail that is not well presented in that book is how to compute the probability vector for each state from the multiple alignment. We have found it best to weight the sequences, using any of several strategies,[15,17-20] then compute the mean posterior probability of each amino acid given the (weighted) observed counts and a prior that is a mixture of Dirichlet distributions.[21] A detailed discussion of Dirichlet mixtures is beyond the scope of this chapter, and many of the papers (including the recommended text)[16] get the formula for Dirichlet mixtures wrong—the tutorial paper by Sjölander *et al.*[21] is probably the most accessible presentation and has no major errors.*

A suite of tools for building and using profile HMMs, the Sequence Alignment and Modeling (SAM) suite, is available from the University of California, Santa Cruz (http://www.soe.ucsc.edu/research/compbio/sam.html). Another tool suite for profile HMMs, HMMer,[22,23] is available as open source, but provides somewhat less functionality, especially for building HMMs.

An unusual feature of the SAM methods is the use of *reverse-sequence null models*, explained in the next few paragraphs.

In any stochastic modeling approach, one wants to compare the probability of a model, M, with the probability of some null model, N, given some sequence, s. This is usually expressed as the logarithm of the *posterior odds ratio* (that is, the ratio of the probabilities we assign to the model and the null model after we have seen the sequence):

$$\log \frac{P(M|s)}{P(N|s)} = \log \frac{P(s|M)\,P(M)}{P(s|N)\,P(N)} = \log \frac{P(s|M)}{P(s|M)} + \log \frac{P(M)}{P(M)} \; .$$

Note that this formulation reduces to calculating the probability of the sequence given the model and given the null model and choosing a constant, the log prior odds ratio. A sequence is well fit by a model if the posterior probability of the model is much higher than the posterior probability of the null model.

---

* The only "error" I am aware of in the tutorial is that Sjölander chose to give the likelihood formulas for multisets rather than for strings, introducing a multinomial correction that is rarely needed. This choice does not affect the formulas for the mean posterior estimate, nor for the optimization of Dirichlet mixtures, so is rarely of any importance.

When simple null models are chosen, there are some anomalous effects. For example, a cysteine-rich protein (such as metallothionein) will have a high probability with any HMM for a protein family that has several conserved cysteines, even if the protein is unrelated.* As another example, amphipathic helices have a fairly standard 7-residue repeating pattern that extends for many residues, giving high probabilities for any sequence with a long amphipathic helix if the HMM is for a family with a long amphipathic helix. These anomalies cause some false positives with very strong scores in almost any homology-detection method, because the sequences are indeed much more similar than chance, even though they are not homologous.

Sequence-based search techniques, such as BLAST,[24] often rely on low-complexity filters and compositional correction to try to remove these anomalies, but neither the cysteine-richness nor the amphipathic-helix anomalies are effectively removed by these standard techniques.

We can greatly reduce the number of such false positives by using a more sophisticated null model that uses the reversal of the sequence:

$$P(s|N) = P(s^r|M) \,.$$

Since the reversed sequence s^r has the same composition and length as s, and since the reversal of the periodic pattern of an amphipathic helix also looks like an amphipathic helix, most of the misleading signals are canceled by this null model, and the log-odds ratio is centered at 0 for any reasonable definition of random sequences of similar size (that is, a definition that makes the probability of sequences and their reversals the same).

This reverse-sequence null model can be used with almost any scoring system, including BLAST[1] and Smith-Waterman,[25] but we have applied it mainly to HMM scoring.

11.3.1. Multitrack Hidden Markov Models

We can get improved fold-recognition results by incorporating more information about our templates than just the amino-acid sequence. For

* Not all sequences with biased composition cause such problems. For example, hisactophilin, with histidine for over a quarter of its residues, rarely turns up as a false positive, since few protein families are characterized primarily by patterns of conserved histidines.

example, we could use the known secondary structure of the template and match it against the predicted secondary structure of our target. One mechanism for including such information in an HMM-based fold-recognition scheme is the *multitrack* HMM.

A multitrack HMM has the same basic structure as an ordinary HMM, but instead of one table of emission probabilities for each match or insertion state, there are multiple tables, one for each *track* of the HMM. A two-track HMM might have one table for amino-acid probabilities and another for probabilities of secondary-structure codes (see Fig. 11.1). When the HMM is used for scoring, we need to have two input strings—the amino-acid sequence and the string of secondary-structure codes for the protein.

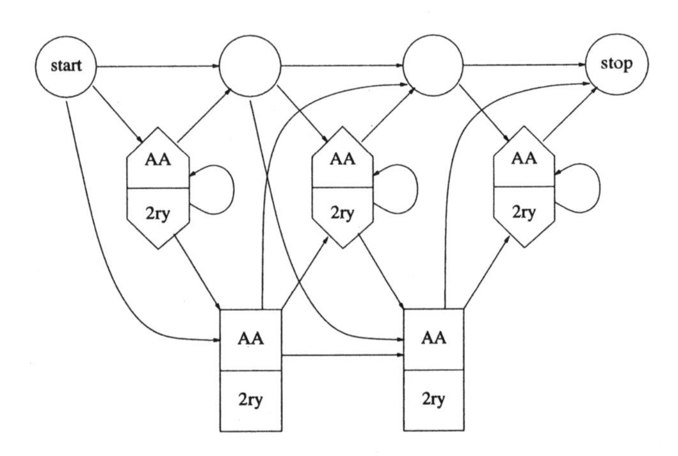

Figure 11.1. A profile HMM with two tracks (AA for amino acids and 2ry for secondary structure). Each match state (rectangle) corresponds to one column of a multiple alignment, each insert state (hexagon) corresponds to an insertion position, and each delete state (circle) provides a way to skip a particular alignment column.

For the match and insert states, both an amino acid and a secondary-structure code is matched each time the state is entered. This figure has only two match states, corresponding to a two-column multiple alignment, but typical profile HMMs have hundreds of states (one for each column of multiple alignment.)

To build a two-track HMM for a target sequence, we can create a multiple alignment using SAM-T2K, then build an amino-acid-only HMM in the usual way. The multiple alignment can also be used to predict the secondary structure of the target (say using a neural net that outputs a probability vector for each alignment column), and the probabilities can be put into the tables for the second track.

If we have a library of template sequences with associated known secondary structure codes, we can use the two-track HMM to score each of the templates. In Section 11.5, Fig. 11.4, we will see how much improvement one such two-track scheme provides.

11.3.2. Statistical Significance for Hidden Markov Models

When searching for remote homologs, one has to examine the statistical significance of the results carefully, as it is difficult to distinguish a weak signal from noise. Although many methods use some arbitrary internal scoring scheme, most search methods provide assistance in interpreting the results by converting the internal scoring system into *E-values*, the expected number of sequences that would score this well by chance when scoring the database. Of course, "by chance" requires a definition of a null model, and a poorly chosen null model could make the E-values uninformative.

In SAM version 3, the E-value is computed using the reverse-sequence null model, using the following formula:

$$E(a) = N\,P(\text{cost} < a) = N(1 + e^{-\lambda a})^{-1},$$

where N is the size of the database being searched, and λ is a scaling parameter. The derivation of the formula (still unpublished) implies that the scaling parameter λ should be 1, and this is the default. Forthcoming versions of the SAM package will allow calibration of λ (and fitting more complicated distribution functions) to improve the E-value computations. This calibration is essential for multitrack HMMs, as the default parameter values are not close to being correct for the multitrack HMMs.

One other important warning: the E-value reported is an estimate of how many sequences would score that well with the final model in a random database. A very low E-value means that the sequence or a very similar one was included in the training set from which the model was built—it does not necessarily mean that the sequence is very similar to the seed sequence. With any iterated search method (SAM-T98, SAM-T99, SAM-T2K, PSI-BLAST,[14] ISS,[26] etc.) including a false positive on any iteration results in much too strong a score for that sequence and similar sequences on subsequent iterations.

Since the default thresholds for SAM-T2K include sequences with E-values as large as 0.005 (measured in the non-redundant database) for the last iteration, all E-values less than

$$0.005\,\frac{\text{size(searched database)}}{\text{size(nonredundant protein database)}}$$

are roughly equivalent to each other.

11.4. Using SAM-T2K for Superfamily Modeling

The SAM-T2K method has been encapsulated in a single perl script: `target2k`. This script accepts a single protein sequence as input (the *target* sequence), does a search of a non-redundant protein database, and returns a multiple alignment of sequences similar to the target. The default parameters have been adjusted to give good performance at recognizing sequences related at the superfamily level of the SCOP database.[27]

For example, let us start with a single hemoglobin (say SWISSPROT sequence HBA_HUMAN), and try to find as many other globins as we can.

The command

```
target2k -seed hba_human.seq -out hba_human-t2k
```

will search the non-redundant protein database[28] and return a multiple alignment of protein sequences similar to hba_human. The seed sequence is provided first, followed by the similar sequences sorted in order of how well they fit the HMM constructed from the multiple alignment of the penultimate iteration. The best-scoring sequence is quite frequently not the sequence that was used as the seed. Note that the multiple alignment hba_human-t2k.a2m, with almost 1300 sequences, contains not only hemoglobins, but also myoglobins and leghemoglobins.

Sometimes this multiple alignment of putative homologs is the primary goal of the method, and it is analyzed with various tools. But for tertiary structure prediction, you generally want to create another HMM from the final multiple alignment, in order to search another database (for example, the PDB database), or to score sequences that were selected by some other method. Although you can use `modelfromalign` directly to do the HMM construction, the SAM package includes some simple scripts that set the parameters appropriately.

These scripts do three things: first they remove redundant sequences from the multiple alignment, then calculate a weight for each sequence, and finally build an HMM from the weighted counts at each position in the multiple alignment. The sequence weighting is intended to reduce further the effect of biased sampling of protein sequences, increasing the weight of under-represented sequences in the multiple alignment. One popular weighting scheme, developed by the Henikoffs,[17] works about as well as any other for relative sequence weighting. When we use Dirichlet mixtures to estimate the probabilities of the amino acids, we also need to set the total weight appropriately. For the SAM programs, one specifies how general one wants the model to be (in terms of how many bits more information each match state has on average, compared to the background frequencies). The programs then adjust the total weight to get the desired level of generality. For comparison, the popular BLOSUM50 and BLOSUM62 matrices[29] save about 0.5 and 0.7 bits per position.[30]

We used the `w0.5` script as follows

```
w0.5 hba_human-t2k.a2m hba_human-t2k-w0.5.mod
```

to build an HMM that averages about 0.5 bits of information per column—the sequence logo in Plate 11.1 shows what the match columns of the HMM are looking for. A sequence logo for the match columns of an HMM was built using script `w0.5` from the multiple alignment of almost 1300 sequences found in the non-redundant protein database (NR) by SAM-T2K starting with hemoglobin hba_human as a seed. The height of each column indicates the conservation in each position (expressed as the number of bits of information). The letters in each column are sized according to the probability of each amino acid, and sorted with the most probable amino acid on top. Note the high conservation of the iron-coordinating histidine in position 88. There are only three sequences with residues other than histidine at H88 (probably from sequencing or alignment errors), but histidine is assigned a probability of only 60%, because the prior probability of substituting other aromatics is so high that the conservation here is not fully recognized.

Other highly conserved residues in the heme-binding site include F44, V63, L67, L84, V94, and L102. The lines below the amino-acid sequence show the true secondary structure (as determined by STRIDE[31] on PDB structure 1abwA: H=helix, T=turn, G=3_{10} helix).

This model can then be used for scoring a set of sequences:

```
hmmscore hba_human-scop -i hba_human-t2k-w0.5.mod -db pdb90d -sw 2
```

producing hba_human-scop.dist (the option `-sw 2` specifies that we want local alignment, rather than global alignment). If we take all the SCOP domains from version 1.53, reduced to sequences with 90% or less residue identity,[32] then all the globins have E-values ≤ 4.55e–11, and the best-scoring non-globin has E-value 1.13e–04 providing excellent separation. Note that this HMM does much better than prior reports of HMMs constructed automatically for globin recognition, which did not achieve perfect separation even when given 400 known globins,[20,33] rather than just one (except for SAM-T99[34] which works as well as SAM-T2K on globins). Warning: globins are a fairly easy protein domain to recognize, and this clean a separation cannot be expected for more difficult domains.

The phycocyanins, which are in the same SCOP superfamily as globins, but a different SCOP family, had E-values ranging from 4.71e–02 to 3060, and were emphatically not recognized by the HMM. Thus the model generated from hba_human should be regarded as a family model, not as a superfamily model. If we score the paramecium hemoglobins, which are quite remote from the others and are not in the SCOP test set, we get E-values ranging from 6.16e–03 to 9.72e–02 (using a database size of 4861, to match the SCOP test set). These are not quite recognized by the HMM—there are 5 false positives before the first paramecium hemoglobin and 24 before all the paramecium hemoglobins are found.

For this SCOP test, E-values less than about 3e–5 should be considered roughly equivalent, because the threshold E-value of 0.005 on the last iteration of search in NR (which is about 130 times bigger than the test set) will include sequences scoring better than that into the multiple alignment used to build the HMM. Since all the globins score much better than 3e-5, about all we can say is that they fit the model—there are no sequences being found that are very different from the ones used in creating the HMM.

On a test of 1997 sequences with sequence identity less than 25% (FSSP representative sequences),[35] with homology defined as containing domains classified as similar by SCOP,[36] the E-values can be seen to be somewhat conservative for values larger than 1, but the false-positive rate remains above 0.01 even for very small E-values (see Fig. 11.2).

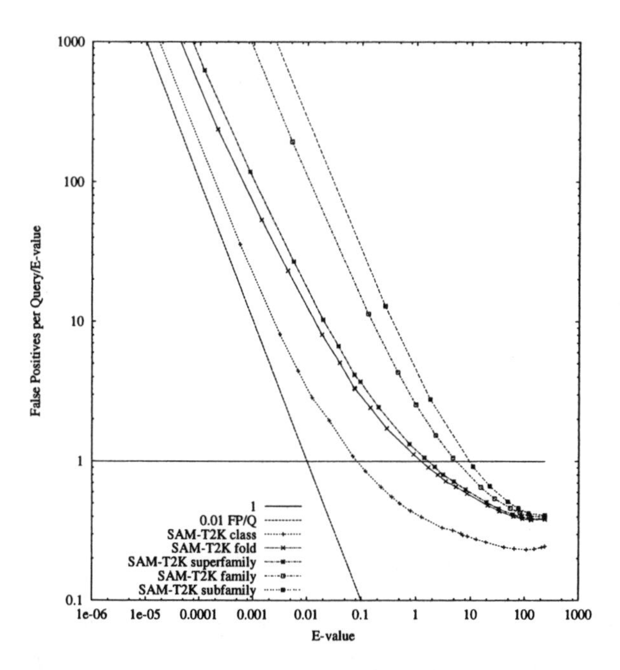

Figure 11.2. This graph compares the theoretically expected number of false positives per query (the E-value), with the observed number of false positives per query in an all-*vs*-all test of 1997 proteins (FSSP representative sequences[35]). The *y*-axis shows the ratio of the observed false positives to those predicted by the E-value computation, which should ideally be near 1. The line for 0.01 false positives/query is included as it appears to be an asymptote for the observed false positives. The different curves reflect the different levels of similarity one can define using the SCOP hierarchy. The superfamily level is most often used as a working definition of homology. The actual number of false positives deviates from the expected number by a factor of 2–10 for E-values larger than 0.01, but is much larger than expected for smaller E-values, staying above 0.01 false positives/query even for quite small E-values. The multiple alignments were built with the default values of the `target2k` script, and the HMMs were built using `w0.5`, and scored with `hmmscore` using local alignment. No calibration was used on the HMMs, so these results are for the default E-value computation $\lambda = 1$.

This "fat-tail" phenomenon occurs not only for SAM-T2K, but for almost all homology-detection methods—there are some statistically significant sequence similarities that do not cause proteins to fold the same way. Most of the strong false positives result from these sequences being admitted into the training set and contaminating the multiple alignment, causing it to model two superfamilies, rather than one. If one has the knowledge and the time to remove these false training sequences by hand, the HMMs can be made much more selective.[37]

At the University of California, Santa Cruz, we have relied heavily on HMM techniques for protein structure prediction in three of the Critical Assessment of Structure Prediction (CASP) experiments. We did well in the fold-recognition category in each of them.[3-5] In CASP4, we had both an automatic server (SAM-T99) and a hand-assisted method (SAM-T2K) evaluated. The automatic server is available for free use on the World-Wide Web.* The methods of the automatic server are primarily what is described in this chapter, though one of the main improvements of SAM-T2K, multitrack HMMs, was discussed in Section 11.3.1.

11.5. Improved Verification of Homology

If we have a conjectured homology between two sequences, perhaps as a result of a search using a SAM-T2K model, we can improve our confidence in the result by combining the results from two different searches, starting with each of the two sequences as a seed. For each sequence, we build a multiple alignment using `target2k`, then create an HMM from the alignment and use it to score the other sequence. The two E-values can be combined by taking their geometric mean, providing a stronger signal if both scores are better than expected by chance. Fig. 11.3 shows the improvement one gets from averaging E-values in this way. There are many other possible ways to combine scores, particularly if we can estimate the relative quality of the target and template HMMs, but this simple averaging technique is surprisingly effective.

As described in Section 11.3.1, we can improve fold recognition using a target model by predicting the secondary structure from the mul-

* http://www.soe.ucsc.edu/research/compbio/HMM-apps/target99-query.html

tiple alignment for the target sequence, and using the predicted probabilities of each of the secondary-structure codes to build a second track for the target HMM. This two-track HMM can be used to score template sequences together with their known secondary structure sequences.

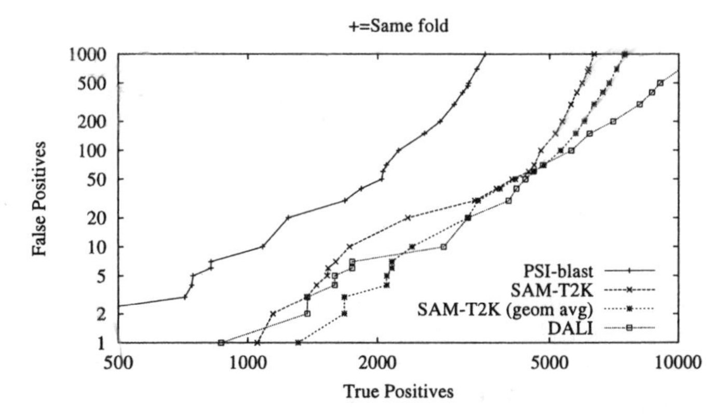

Figure 11.3. This graph plots the number of false positives versus the number of true positives (where "true positives" means containing a SCOP domain from the same fold) in an all-*vs*-all test of 1997 FSSP representative sequences. The multiple alignments were built with the default values of the `target2k` script, and the HMMs were built using `w0.5` and scored after being calibrated. The "geom avg" line is for the method of combining the E-values for HMM A scoring sequence B with HMM B scoring sequence A. The averaging method finds considerably more fold relationships at almost every level of false positives accepted. For comparison, results from using PSI-BLAST[14] and from using the structure-comparison program DALI[38] are also plotted. Note that for fewer than 0.05 false positives/query (100 false positives), the SAM-T2K methods do as well at fold recognition as a method that uses the structural information.

Note that this construction technique does not allow us to construct and score a two-track template HMM, because we do not have a reliable secondary structure for the target sequence. Because of this asymmetry in our knowledge, we cannot use the trick of combining scores of two-track target and two-track template models. We can, however, combine the two-track HMM scores with the scores from the standard amino-acid-only template HMMs to get much more sensitivity (see Fig. 11.4).

Somewhat surprisingly, the two-track HMMs provide increases in selectivity, rather than increases in sensitivity. That is, they increase the number of true positives for low false-positive rates, but not for higher false-positive rates.

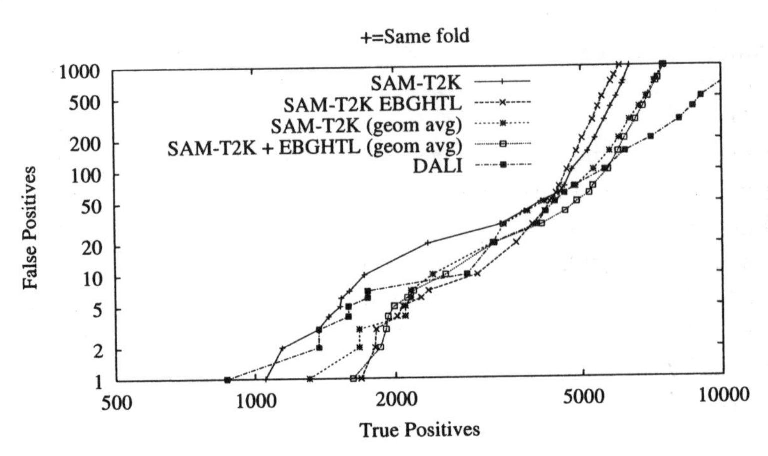

Figure 11.4. This graph shows the number of false positives *versus* the number of true positives (where "true positives" means containing a SCOP domain from the same fold) in an all-*vs*-all test of 1997 FSSP representative sequences. The multiple alignments were built with the default values of the `target2k` script, and the HMMs were built using `w0.5` and scored after being calibrated. The two-track HMMs had a second track predicting a six-state alphabet based on STRIDE's definitions (E=beta strand, B=beta bridge, G=3_{10} helix, H=alpha helix, T=turn, L=other). Although the two-track HMMs are more selective than the amino-acid-only HMMs, combining scores from either style of target HMM with the amino-acid-only HMMs provides similar fold-recognition results. For comparison, results from using the structure-comparison program DALI[38] are also plotted. Note that though DALI has much more information (the complete structures), it does not do better at fold recognition for low false-positive rates.

11.6. Family-Level Multiple Alignments

The default parameters for the `target2k` script have been set for fairly good performance on recognizing SCOP superfamilies, producing HMMs that generalize broadly, though for some seeds (such as hba_human) only SCOP-family-level recognition is achieved. If you want to separate one SCOP family from another (as Pfam[39] does,) the level of generalization in SAM-T2K is usually too much. By specifying the option `-family`, you can request a different set of parameters intended for building family-level multiple alignments from a seed sequence. These parameters have not been tested yet, and are offered only as a starting point for further tweaking.

The Pfam database creates family-level HMMs from hand-curated seed alignments.[39] The SAM-T2K method can also be started with a seed alignment, rather than a single sequence—simply provide the alignment with `target2k`'s `-seed` option. Using the Pfam seed alignments directly with SAM could cause some problems, as the seed alignments often have large regions which would be better modeled as insertions than as alignment columns. The multiple alignment format used for the Pfam seeds cannot express the notion of unaligned residues, so the HMMer package has heuristics to guess which alignment columns should be converted to insertions for better generalization of the HMM. SAM's `buildmodel` program has similar heuristics in the model surgery options of the buildmodel program, but these options are not used by default in the `target2k` script.

The `-seed` option is also useful for superfamily recognition—especially if a structural alignment is available to use as a seed, though a recent study by Julian Gough shows that using multiple HMMs to represent a superfamily works better than trying to build a single HMM from a structural alignment.[37]

11.7. Modeling Non-Contiguous Domains

Some protein domains are formed from non-contiguous pieces of the backbone. For example, the SCOP database[27] has a GroES-like fold for alcohol dehydrogenase (d2ohxa1) consisting of residues 1-174 and 325-374, with a Rossmann-fold domain in the middle.

To model a domain like this, the HMM needs to represent the long insertion in the middle. Although SAM has a special mechanism (the free insertion module, or FIM) for handling arbitrary length insertions, we have found it works just as well to include the insertion as lowercase letters in the seed alignment, and let the insertion probabilities be computed automatically.

If we create a file `d2ohxa1.seq` with the appropriate seed alignment for the multi-helical domain:

```
>d2ohxa1 2.22.1.2.1 (1-174,325-374) Alcohol dehydrogenase [horse]
STAGKVIKCKAAVLWEEKKPFSIEEVEVAPPKAHEVRIKMVATGICRSDDHVVSGTLVTP
LPVIAGHEAAGIVESIGEGVTTVRPGDKVIPLFTPQCGKCRVCKHPEGNFCLKNDLSMPR
GTMQDGTSRFTCRGKPIHHFLGTSTFSQYTVVDEISVAKIDAASPLEKVCLIGC
gfstgygsavkvakvtqgstcavfglggvglsvimgckaagaariigvdinkdkfakake
vgatecvnpqdykkpiqevltemsnggvdfsfevigrldtmvtalsccqeaygvsvivgv
ppdsqnlsmnpmlllsgrtwkgaifggfks
KDSVPKLVADFMAKKFALDPLITHVLPFEKINEGFDLLRSGESIRTILTF
```

then the command

```
target2k -seed d2ohxa1.seq -out d2ohxa1-t2k
```

will build the multiple alignment, and we can create an HMM with the `w0.5` script as before.

The HMM `d2ohxa1-t2k-w0.5.mod` scores the alcohol dehydrogenase family members (in SCOP version 1.53) with E-values ≤ 7.6e–40. The best-scoring sequence outside the superfamily (the first false positive) is 1c17M (subunit A of F1F0 ATP synthase) with E-value 7.17e–04.

We can also look for false negatives—sequences in the GroES-like superfamily that do not score well. The chaperonin-10 sequences, which are in the superfamily but not the alcohol dehydrogenase family, have E-values 133, 211, and 3820 after several hundred unrelated sequences, so this HMM is best viewed as a family-level model, not a superfamily model.

11.8. Building an HMM from a Structural Alignment

The central domain of 2ohxA is an NAD(P)-binding Rossmann-fold domain, a large superfamily of alpha/beta/alpha units with a parallel beta-sheet of six strands (structure 321456). This superfamily is fairly diverse and difficult to capture in a single HMM. Let us start with `d2ohxa2.seq`:

```
>d2ohxa2 3.19.1.1.1 (175-324) Alcohol dehydrogenase [horse]
GFSTGYGSAVKVAKVTQGSTCAVFGLGGVGLSVIMGCKAAGAARIIGVDINKDKFAKAKE
VGATECVNPQDYKKPIQEVLTEMSNGGVDFSFEVIGRLDTMVTALSCCQEAYGVSVIVGV
PPDSQNLSMNPMLLLSGRTWKGAIFGGFKS
```

and build the HMM as before

```
target2k -seed d2ohxa2.seq -out d2ohxa2-t2k w0.5 d2ohxa2-t2k.a2m
d2ohxa2-t2k-w0.5.mod
```

Scoring the SCOP database gives us the first false positive at E-value 1.64e–06 after 10 true positives, at the minimum error point (1.11e–02) there are 34 true positives, 54 false negatives, and 15 false positives.

All eight members of the same family as d2ohxa2 score extremely well (E-values < 8e–27), but several of the other families in the same superfamily also score well, with only occasional false positives, so this can genuinely be viewed as a superfamily model, and not just a family model.

We can try to make a better model for the Rossmann-fold superfamily containing d2ohxa2, by starting from a structural alignment of two of its members. The model built using just d2ohxa2 had particular difficulty recognizing the lactate dehydrogenases and the malate dehydrogenases, so perhaps we could improve the model by including one of them in the alignment. In the FSSP[35] alignment for 2ohxA, the highest Z-score of these is for 2cmd (*E. coli* malate dehydrogenase).

From the FSSP structural alignment, we can extract the aligned section that seems to match the domain we are interested in:

```
              177
               |
2ohxA-domain STCAVFG.LGGVGLSVIMGCKAAG...AARIIGVDINKDKFAKAKEV........GA
2cmd-domain  MKVAVLGaAGGIGQALALLLKTQLpsg-SELSLYDIAPVTPGVAVDLshiptavk--

2ohxA-domain TECVNPQDYKKPIQEVLTEMSNGGVDFSFEVIG...............RLDTMVTAL
2cmd-domain  IKGFSGE-----DATPAL----EGADVVLISAGvrrkpgmdrsdlfnvNAGIVKNLV

2ohxA-domain SCC..QEAYGVSVIVGVP.........PDSQNLSMNP..MLLL..SGRTWKGAIFGG
2cmd-domain  QQVakTCPKACIGIITNPvnttvaiaa------EVLKkaGVYDkn---KLFGVT-TL
                   327
                    |
2ohxA-domain FK.SKDS
2cmd-domain  DIiRSNT
```

If we build the HMM as before

```
target2k -seed 2ohxA-2cmd.a2m -out 2ohxA-2cmd-t2k
w0.5 2ohxA-2cmd-t2k.a2m 2ohxA-2cmd-t2k-w0.5.mod
```

and score the SCOP 90% database, we get the first false positive at E-value 1.18e–05 (DNA polymerase beta) after 31 true positives, and at the minimum-error point (E-value 1.85e–02), there are 51 true positives, 12 false positives, and 37 false negatives.

Although this model scores the superfamily well (58 of the 88 members have E-values less than 1.0), it also scores 97 incorrect domains as well. The HMM built with a single sequence as a seed does not do as well—to get 58 true positives it would need to include 212 false positives.

If we add to the structural alignment one more sequence that was not found with the 2ohxaA-2cmd-t2k HMM (say 2pgd), we can improve the model further. Starting from the alignment

```
2pgd  DIALIG.LAVMGQNLILNMNDHG...F.VVCAFNRTVSKVDDFLANEAKGT....KVLGAH..S
2cmd  KVAVLGaAGGIGQALALLLKTQLpsgS.ELSLYDIA-PVTPGVAVD-LSHIptavKIKGFSgeD
2ohxA TCAVFG.LGGVGLSVIMGCKAAG...AaRIIGVDINKDKFAKAKEV-----ga..-TECVN..P

2pgd  .....LEEMVSKLKKPR.RIILLVK.....AGQAVDNFIEKL....VPLLDIGDIIIDGGNSEY
2cmd  .....ATPALEGA---D.VVLISAGv15nvNAGIVKNLVQQV....AKTCP-KACIGIITNPVN
2ohxA qd7piQEVLTEMSNGGVdFSFEVIG.....----RLDTMVTAlscc---QEAYGVSVIVGVPPD

2pgd  RD.....TMRRCRDLK..DKGI..LFVGSGVS
2cmd  TT.....VAIAAEVLKkaGVYDknKLFGVTTL
2ohxA SQnlsmn-----PMLL..LSGR..TWKGAIFG
```

and building a SAM-T2K alignment as before, the HMM finds 31 true positives before the first false positive (sarcosine oxidase) at E-value 3.93e–05, 47 true positives and 9 false positives at the minimum error point (E-value = 1.01e–2), and 60 true positives with 76 false positives at E-value < 1. To get 58 true positives, we have to accept 30 false positives, so we can see that this model is considerably more selective than either of the first two. We could continue generalizing the model by adding poorly scored members of the superfamily, perhaps 1ofgA next, since it scores very poorly but has a reasonably high Z-score in FSSP.

If we have a structural alignment of dissimilar homologs, it may not seem necessary to run the `target2k` method. If we build a model from just the structural alignment with the command

```
w0.5 2pgd-2cmd-2ohxA.a2m 2pgd-2cmd-2ohxA-w0.5.mod
```

we get only 14 true positives before a false positive at E-value 1.48e–04, 24 true positives with 9 false positives, and 58 true positives with 566 false positives.

For very low false-positive rates, our best results on this domain were with the SAM-T2K method applied to a structural alignment of multiple sequences, with 31 true positives before a false positive. Also, if we want to find the remote homologs, accepting a few false positives, we again get the best results by using a structural alignment as the seed for SAM-T2K (58 true positives with only 30 false positives). The SAM-T2K method does get better results on this example than using just a structural alignment without searching for more homologs.

In tests, we have found using many HMMs, each built with SAM-T2K from a single seed, simpler and as effective at finding remote homologs as building HMMs from structural alignments.[37] We believe that combining information from diverse seeds would be useful, but that building an HMM from a structural alignment is not as valuable as one might expect.

11.9. Improving Existing Multiple Alignments

You can clean up an existing multiple alignment by using it as a seed and specifying the `-tuneup` option. This option turns off the search for similar sequences and turns off the `-force_seed 1` option that normally copies the seed alignment without modification.

The seed alignment will be used to create an HMM, then the sequences in the alignment will be used as the set of potential homologs to search and align. The output alignment may contain only a subset of the original sequences, if some of the sequences score too poorly to meet the thresholds. If you want to include all the sequences, set the `-thresholds` variable to have a very large final threshold.

The `-tuneup` option can also be used to add unaligned sequences to an existing multiple alignment with the -homologs option. For example,

```
target2k -seed hba_human.seq -homologs globins50.seq -tuneup  \
        -db_size 600000 -out hba_human-50
```

adds 50 globin sequences to the single hba_human globin alignment, creating a multiple alignment of 51 sequences. The `-db_size` option says what size database should be assumed for computing E-values. If it is omitted, the size of the non-redundant database that would normally be searched is used.

11.10. Creating a Multiple Alignment from Unaligned Sequences

It is not really necessary to specify a seed for the SAM-T2K method | if the `-close` or `-homologs` option is used, then an initial multiple alignment can be created by `buildmodel` from the specified unaligned sequences. If both `-close` and `-homologs` are given, then only the close set are used for the initial alignment, but the full set of homologs is used for subsequent iterations.

For example,

```
target2k -homologs globins50.seq -tuneup -db_size 600000 \
        -out globins50-tuned
```

will align the 50 globin sequences without using a seed sequence, leaving the alignment in globins50-tuned.a2m.

We have tested this multiple alignment method using the BAliBASE benchmark[40] and the SAM-T99 tuneup option, and found that SAM-T99 produces multiple alignments of about the same quality as ClustalW.[41,42] These tests showed us that there was little to gain from realigning SAM alignments using CLUSTAL, but also that SAM-T99 was a slow way to align small numbers of sequences. Although the SAM multiple alignment algorithm is linear in the number of the sequences, while CLUSTAL is quadratic, the constants for SAM are large enough that CLUSTAL is faster if there are fewer than 500 sequences to align.

11.11. Conclusions

Hidden Markov models have been an effective method for predicting protein structure through fold recognition and alignment, but there is still a lot to do.

We expect the next few years to bring several improvements in fold recognition, not only from the constantly increasing sequence and template databases, but also from improvements in the methods. For example, we have only just started examining multitrack HMMs, and have not yet determined what local properties of protein structures are best to use as auxiliary tracks. Improved computation of statistical significance will let us set our thresholds more precisely, giving better sensitivity and selectivity. We could try more sophisticated ways of combining results from multiple HMMs for the same fold, such as the product-of-p-values method of Bailey and Grundy.[43]

We could try adding "sanity checks" to select fold-recognition predictions that produce compact, protein-like structures—the current alignment methods do not examine the predicted 3D structure at all, and often result in obviously bad predictions. Although almost all hand-assisted predictions include some of these sanity checks, automating the procedure has been surprisingly difficult to do well.

We can also combine our fold-recognition methods with other methods for predicting protein structure, such as the fragment-packing methods of Rosetta[44] or the keyword matching of SAWTED.[45]

References

1. Altschul S, Gish W, Miller W, Myers E, Lipman D: **Basic local alignment search tool.** *J Mol Biol* 1990, **215**:403–410.

2. Pearson W, Lipman D: **Improved tools for biological sequence comparison.** *Proc Natl Acad Sci USA* 1988, **85**:2444–2448.

3. Karplus K, Sjölander K, Barrett C, Cline M, Haussler D, Hughey R, Holm L, Sander C: **Predicting protein structure using hidden Markov models.** *Proteins* 1997, **Suppl 1**:134–139.

4. Karplus K, Barrett C, Cline M, Diekhans M, Grate L, Hughey R: **Predicting protein structure using only sequence information.** *Proteins* 1999, **Suppl 3**(1):121–125.

5. Karplus K, Karchin R, Barrett C, Tu S, Cline M, Diekhans M, Grate L, Casper J, Hughey R: **What is the value added by human intervention in protein structure prediction?** *Proteins* 2002, in press.

6. Lathrop R, Smith T: **Global optimum protein threading with gapped alignment and empirical pair score functions.** *J Mol Biol* 1996, **255**(4):641–665.

7. Panchenko AR, Marcher-Bauer A, Bryant SH: **Combination of threading potentials and sequence profiles improves fold recognition.** *Proteins* 1999, **Suppl 3**(1):133–140.

8. Miyazawa S, Jernigan R: **Residue-residue potentials with a favorable contact pair term and unfavorable high packing density term, for simulation and threading.** *J Mol Biol* 1996, **256**(3):623–44.

9. Lemer CM, Rooman MJ, Wodak SJ: **Protein structure prediction by threading methods: evaluation of current techniques.** *Proteins* 1995, **23**(3):337–355.

10. Sippl M: **Knowledge-based potentials for proteins.** *Curr Opin Struc Biol* 1995, **5**:229–235.

11. Mirny LA, Shakhnovich EI: **Protein structure prediction by threading. Why it works and why it does not.** *J Mol Biol* 1998, **283**(2):507–526.

12. Westhead DR, Collura VP, Eldridge MD, Firth MA, Li J, Murray CW: **Protein fold recognition by threading: comparison of algorithms and analysis of results.** *Protein Eng* 1995, **8**(12):1197–1204.

13. Cline M: *Protein Sequence Alignment Reliablity: Prediction and Measurement.* Ph.D. thesis. University of California, Computer Science, UC Santa Cruz, CA 95064, 2000.

14. Altschul S, Madden T, Schaffer A, Zhang J, Zhang Z, Miller W, Lipman D: **Gapped BLAST and PSI-BLAST: A new generation of protein database search programs.** *Nucleic Acids Res* 1997, **25**:3399–3402.

15. Karplus K, Barrett C, Hughey R: **Hidden Markov models for detecting remote protein homologies.** *Bioinformatics* 1998, **14**(10):846–856.

16. Durbin R, Eddy S, Krogh A, Mitchison G: *Biological Sequence Analysis: Probabilistic Models of Proteins and Nucleic Acids.* Cambridge, U.K.: Cambridge University Press, 1998.

17. Henikoff S, Henikoff JG: **Position-based sequence weights.** *J Mol Biol* 1994, **243**(4):574–578.

18. Altschul SF, Carroll RJ, Lipman DJ: **Weights for data related by a tree.** *J Mol Biol* 1989, **207**:647–653.

19. Vingron M, Sibbald PR: **Weighting in sequence space: a comparison of methods in terms of generalized sequences.** *Proc Natl Acad Sci USA* 1993, **90**:8777–8781.

20. Karchin R, Hughey R: **Weighting hidden Markov models for maximum discrimination.** *Bioinformatics* 1998, **14**(9):772–782.

21. Sjölander K, Karplus K, Brown MP, Hughey R, Krogh A, Mian IS, Haussler D: **Dirichlet mixtures: A method for improving detection of weak but significant protein sequence homology.** *Comput Appl Biosci* 1996, **12**(4):327–345.

22. Eddy S: *HMMER WWW site.* http://hmmer.wustl.edu/

23. Eddy SR: **Profile hidden Markov models.** *Bioinformatics* 1998, **14**(9):755–763.

24. Altschul SF, Gish W, Miller W, Myers EW, Lipman DJ: **A basic local alignment search tool.** *J Mol Biol* 1990, **215**:403–410.

25. Smith TF, Waterman MS: **Identification of common molecular subsequences.** *J Mol Biol* 1981, **147**:195–197.

26. Park J, Teichmann S, Hubbard T, Chothia C: **Intermediate sequences increase the detection of homology between sequences.** *J Mol Biol* 1997, 273:349–354.

27. Murzin AG, Brenner SE, Hubbard T, Chothia C: **SCOP: A structural classification of proteins database for the investigation of sequences and structures.** *J Mol Biol* 1995, **247**:536–540.

Information on SCOP is available at http://scop.mrc-lmb.cam.ac.uk/scop

28. *NR.* http://www.ncbi.nlm.nih.gov/BLAST/blast_databases.html

All non-redundant GenBank CDS translations+PDB+SwissProt+PIR+PRF Database. Distributed on the Internet via anonymous FTP from ftp://ftp.ncbi.nlm.nih.gov/blast/db

29. Henikoff S, Henikoff JG: **Amino acid substitution matrices from protein blocks.** *Proc Natl Acad Sci USA* 1992, **89**:10915–10919.

30. Altschul SF: **Amino acid substitution matrices from an information theoretic perspective.** *J Mol Biol* 1991, **219**:555–565.

31. Frishman D, Argos P: **Knowledge-based protein secondary structure assignment.** *Proteins* 1995, **23**:566–579.

32. Brenner SE, Koehl P, Levitt M: **The ASTRAL compendium for sequence and structure analysis.** *Nucl Acids Res* 2000, **28**:254–256; http://astral.stanford.edu/.

33. Haussler D, Krogh A, Mian IS, Sjölander K: **Protein modeling using hidden Markov models: Analysis of globins.** In: *Proceedings of the Hawaii International Conference on System Sciences*, vol. 1. Los Alamitos, California: IEEE Computer Society Press, 1993:792–802

34. Hughey R, Karplus K, Krogh A: *SAM: Sequence Alignment and Modeling Software System, version 3. Technical Report UCSC-CRL-99-11*. University of California, Santa Cruz, Computer Engineering, Santa Cruz, CA, October 1999. Available from http://www.cse.ucsc.edu/research/compbio/sam.html.

35. Holm L, Sander C: **Mapping the protein universe.** *Science* 1996, **273**(5275):595–603.

36. Hubbard T, Murzin A, Brenner S, Chothia C: **SCOP: A structural classification of proteins database.** *Nucleic Acids Res* 1997, **25**(1):236–239.

37. Gough J, Karplus K, Hughey R, Chothia C: **Assignment of homology to genome sequences using a library of hidden Markov models that represent all proteins of known structure.** *J Mol Biol* 2001, **313**:903–319.

38. Holm L, Sander C: **Protein structure comparison by alignment of distance matrices.** *J Mol Biol* 1993, **233**(1):123–138.

39. Sonnhammer ELL, Eddy SR, Durbin R: **Pfam: A comprehensive database of protein families based on seed alignments.** *Proteins* 1997, **28**:405–420.

40. Thompson JD, Plewniak F, Poch O: **BAliBASE: A benchmark alignment database for the evaluation of multiple alignment programs.** *Bioinformatics* 1999, **15**(1):87–88.

41. Thompson JD, Higgins DG, Gibson TJ: **CLUSTALW: Improving the sensitivity of progressive multiple sequence alignment through sequence weighting, position-specific gap penalties, and weight matrix choice.** *Nucleic Acids Res* 1994, **22**(22):4673–4680.

42. Karplus K., Hu B: **Evaluation of protein multiple alignments by SAM-T99 using the BaliBASE multiple alignment test set.** *Bioinformatics* 2001, **17**:713–720.

43. Bailey TL, Grundy WN: **Classifying proteins by family using the product of correlated p-values.** In: *Int. Conf. Computational Molecular Biology (RECOMB99)*, April 11-14 1999. New York: ACM Press, 1999:10–14.

44. Simons KT, Bonneau R, Ruczinski I, Baker D: ***Ab initio* protein structure prediction of CASP III targets using ROSETTA.** *Proteins* 1999, **Suppl 3**(1):171–176.

45. MacCallum RM, Kelley LA, Sternberg MJE: **SAWTED: Structure assignment with text description-enhanced detection of remote homologues with automated SWISS-PROT annotation comparisons.** *Bioinformatics* 2000, **16**(2):125–129.

12

Local Genome Organization, Gene Expression, and Structural Genomics: Evolution at Work

Wayne Volkmuth and Nickolai Alexandrov

Abstract

Discovering distant evolutionary relationships between proteins is a fundamental problem in modern computational biology. This relationship often allows us to transfer the results of expensive and time consuming experiments from one protein to another, and frequently leads to productive speculations about the biological role of newly sequenced genes. It is widely believed that protein three-dimensional structure is much more conservative than its amino acid sequence, making structural similarity a reliable indication of the evolutionary link. Sequence-based methods of detecting homologous proteins are not always able to distinguish between true homologs and false positive hits in the twilight zone of sequence similarity. We should pay attention to other, sequence independent clues about a protein's ancestry that evolution might offer. Here we describe two such evolutionary clues that could be used to help infer the evolutionary relationship via structural similarity and improve the ability to predict the biochemical function. The first such clue is a positional conservation along the genome, i.e., nearby genes tend to be structurally related more often than expected by chance alone. The second such clue is present in expression data: genes that are correlated in expression are more apt to share a common fold than two randomly chosen genes.

12

*Local Genome Organization, Gene Expression, and Structural Genomics: Evolution at Work**

Wayne Volkmuth and Nickolai Alexandrov

Ceres Inc., Malibu, California

12.1. Introduction

The most powerful method of protein structure and function prediction—homology search—is limited by difficulties in finding remote sequence similarities. The function of about 30% of *Arabidopsis thaliana* genes cannot be predicted by sequence similarity search methods.[1-5] About 40% of the identified genes in human chromosomes 21 and 22 do not have detectable homology to known genes.[6,7] Even a small improvement in our ability to identify distant homologs can help us make functional predictions for a large number of newly discovered genes.

However, there are completely different, sequence independent evolutionary traces which can be exploited to extend our predictive power. These traces can be especially helpful given the huge amount of biological information being accumulated from various large-scale scientific projects. An easy and commonly acceptable method of verifying correctness of the prediction of evolutionary link is to check the similarity between three-dimensional protein structures.[8,9] Various fold clas-

* Parts of this chapter are reproduced from reference[24]. Courtesy of World Scientific Publishing Co. Pte. Ltd. (Singapore.)

sifications exist; for our purposes we use that of the Structural Classification of Proteins (SCOP).[10]

Progress in fold recognition (prediction of the protein structure via establishing an evolutionary link) is assessed at the regular CASP conferences (http://predictioncenter.llnl.gov/). The recent CASP4 meeting clearly showed that an expert, knowledge-based approach is superior to a purely computational, fully automated approach. The advantage of experts is that they use not only biochemical or biophysical information on protein sequence and structural properties, but also knowledge of protein function derived from biochemical experiments. In support of the utility of expert knowledge, it has, in fact, already been demonstrated that even something as simple as a key-word comparison of protein descriptions in the database increases the accuracy of fold recognition programs.[11] Clearly then, incorporation of additional data will improve the accuracy of fold prediction.

Gene duplication is an important evolutionary event, resulting in higher frequency of neighboring genes sharing a common ancestor. It is not surprising then to observe a fold enrichment among neighboring genes. However, even if we exclude all neighboring homologs identified through sequence comparison, the fold enrichment still holds. This means that the positional conservation in many cases conveys more evolutionary information than the sequences alone.

We will distinguish *biochemical* function of a protein from the gene's *biological* function. By biological function we mean what the organism accomplishes with the gene and other co-expressed genes, for example a signal transduction cascade. Just as sequence conservation during evolution can be used to infer structural homology and biochemical function, one might wonder if Nature has left a trace of structural information during its evolution of biological (as opposed to biochemical) function. Could it be that as evolution made the pathways more complex, genes were adopted from these same pathways to accomplish the biological function more efficiently?

Recently, results of many high-throughput gene expression experiments have become available. Analysis of these data allows us to identify co-expressed genes, which are most likely involved in the same pathway. As in the case of the fold enrichment seen between nearby genes on the genome, the fold enrichment between pairs of co-expressed genes conveys evolutionary information over and above that from sequence similarity alone. Hence, *biological* function of a gene says something about the gene's corresponding protein fold, and via the fold something about the *biochemical* function of that protein.

In the following, then, we show that both physical distance on the genome and correlation in expression show traces of evolution manifested in structural homology. Those weak signals could, in principle, be used to improve prediction of structural homology in the twilight zone of sequence similarity.

12.2. Methods

12.2.1. Genomes

To investigate the evolutionary trace of fold enrichment along the genome, we used the genomes of the model organisms *Saccharomyces cerevisiae*[12] and *Arabidopsis thaliana*.[1-5] The yeast genome consists of 16 chromosomes, with 6,310 identified ORFs, available from the *Saccharomyces* Genome Database (SGD) at http://genome-www.stanford.edu/Saccharomyces/. The smallest chromosome, chromosome I, is ~0.23 Megabases and has 107 ORFs. The largest chromosome, chromosome IV, is ~1.53 Mb and encodes 819 ORFs. The recently finished *Arabidopsis* genome consists of five chromosomes, with 25,498 genes predicted. The smallest chromosome is chromosome IV and is ~17.5 Mb in length, containing 3,825 protein encoding genes. The largest chromosome is chromosome I, approximately 29.1 Mb in length, with 6,543 genes. We downloaded *Arabidopsis* genes from the NCBI web site (http://www.ncbi.nlm.nih.gov).

Only chromosomes II and IV were available as one contig at the time we made our analysis, so we used only 7,852 protein-coding genes from these two chromosomes in our analysis of fold enrichment in the genome neighborhood.

12.2.2. Microarray Expression Data

Yeast and *Arabidopsis* expression data were downloaded from the Stanford Microarray Database.[13] A subset of experiments assessing the performance of microarrays was excluded from our analysis. The resulting dataset contained expression data from 308 yeast and 162 *Arabidopsis* microarray experiments.

The similarity in expression across all experiments between a pair of genes was measured using the Spearman rank correlation coefficient on the normalized ratios.[14] The Spearman *r* was chosen because it is a robust statistic that will capture any monotonic relationship between a pair of variables, as opposed to the commonly used Pearson correlation coefficient which is suitable for detecting linear relationships between pairs of variables. Missing data points were handled by pairwise deletion of observations from the Spearman *r* calculation and any pair of genes having fewer than 10 experiments in common were ignored. No special attempt was made to account for the redundancy due to experimental replicates or similarities in subsets of experiments. We note, however, that for the purposes of a global correlation analysis at least, it would be more desirable to have a larger number of distinct, diverse experiments than to have experimental replicates, since the correlation coefficient implicitly takes inherent experimental variation into account.

Spurious significant correlations might be introduced between a pair of genes that are not actually co-expressed if the two genes are sufficiently similar that cross-hybridization occurs.[15,16] No large scale experimental study of cross-hybridization on microarrays has been done, so we assessed the degree of cross-hybridization indirectly as follows. Pairwise similarity was measured using Wash-U BLASTN, version 2.0 with M = 2 and all other parameters set to their defaults. We compared the overall distribution of correlation coefficients between pairs of genes to the distribution between non-identical chip features showing similarity of ≥85% over ≥50 nt. There is a clear shift towards one for the similar sequences as seen in Fig. 12.1. The distribution for clones having 70–85%, ≥50 nt similarity is shifted towards 1 as well, but the shift is less pronounced (data not shown). To be conservative and minimize the possibility of cross-hybridization, we discarded any chip feature having a sequence with an HSP showing 70% or greater similarity over at least 50 nt to some other gene in the genome. Applying this approach for yeast is straightforward since the microarray features are PCR fragments of the ORFs, the complete sequence of the features is known,[17] and there are essentially no introns in yeast. In the case of *Arabidopsis*, the full sequence of the clones serving as source material for the microarrays is not available; so we mapped ESTs from the clones to the cDNAs from the annotation of the *Arabidopsis* genome[1-5] and assumed that the entire cDNA sequence was present in the microarray feature, then screened against all other cDNAs in *Arabidopsis*. While conservative, our approach cannot guarantee complete exclusion of similar features because the genomic annotations often lack full UTRs or have other errors.

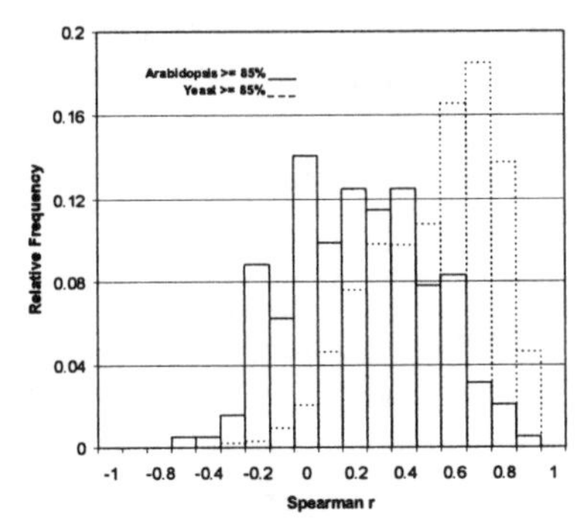

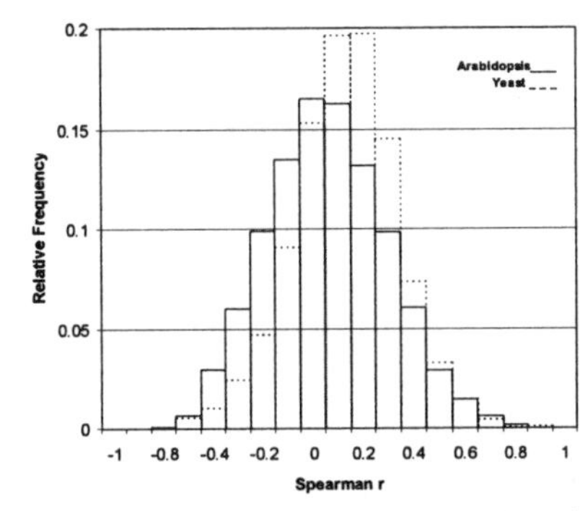

Figure 12.1. (**a**) Histogram of Spearman *r* distribution for clones showing ≥ 85% similarity over ≥ 50 nt. (**b**) Histogram of Spearman *r* distribution overall for clones included in analysis, i.e., those that show no similarity to any other clone at the 70%, 50 nt level. It should be pointed out that Fig. 12.1a strongly suggests that the *Arabidopsis* data is of inferior quality to the Yeast data, see comment in Section 12.3 below. Courtesy of World Scientific Publishing Co. Pte. Ltd. (Singapore).[24]

The cross-hybridization filtering resulted in 4,280 yeast features and 3,011 *Arabidopsis* features. The smaller number of filtered features for *Arabidopsis* was primarily a consequence of feature redundancy on the chip (more than one clone for a given cDNA) and a higher amount of gene duplication in *Arabidopsis*. A correlation and cluster analysis was performed with the biological sensibility of results conforming to those from the literature, though the *Arabidopsis* results were less compelling.[18-20]

12.2.3. Fold Assignment

For each gene we assigned fold(s) according to the SCOP-1.55 classification of protein structures.[10] Assignment was done by WU-Blastp[21] search against the Astral database of non-redundant SCOP domains at the 95% identity level.[22] We considered all matches with a P-value < 0.001. At that level of significance approximately 2% of our assignments are wrong.[8] One protein may consist of more than one domain; in those cases multiple folds were assigned to the corresponding gene. Out of 6,310 yeast genes, we assigned folds to 1,839 genes (29%). Out of 27,469 *Arabidopsis* genes from the TIGR gene index, we assigned folds to 9,147 genes (33%). The distribution of different SCOP folds in the two genomes is shown in Fig. 12.2, with the most frequent folds summarized in Table 12.1. More advanced methods of fold assignments, e.g., PSI-BLAST, the profile-profile technique, and threading, increase the coverage, but overall do not change the statistical observations.

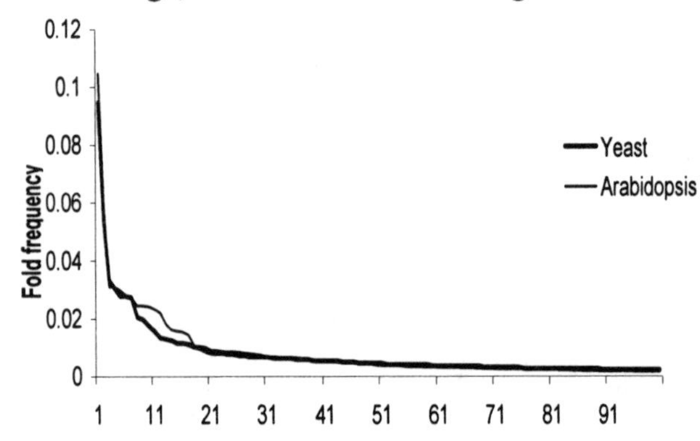

Figure 12.2. Fold distributions in yeast and *Arabidopsis* genomes. Courtesy of World Scientific Publishing Co. Pte. Ltd. (Singapore).[24]

Table 12.1. Ten most frequent folds in the yeast and *Arabidopsis* genomes. Seven folds belong to the ten most frequent folds in both genomes. Courtesy of World Scientific Publishing Co. Pte. Ltd. (Singapore).[24]

SCOP_1.55 fold	Description	Yeast		*Arabidopsis*	
		Rank	Frequency	Rank	Frequency
c.37	P-loop containing nucleotide triphosphate hydrolases	1	0.095	2	0.059
d.144	Protein kinase-like (PK-like)	2	0.054	1	0.105
c.1	TIM beta/alpha-barrel	3	0.033	5	0.030
b.69	7-bladed beta-propeller	4	0.030	10	0.024
a.118	alpha-alpha superhelix	5	0.028	7	0.027
c.2	NAD(P)-binding Rossmann-fold domains	6	0.028	8	0.025
d.58	Ferredoxin-like	7	0.027	3	0.031
g.38	Zn2/Cys6 DNA-binding domain	8	0.021	> 279	0
f.2	Membrane all-alpha	9	0.020	18	0.011
c.55	Ribonuclease H-like motif	10	0.018	21	0.009
a.4	DNA/RNA-binding 3-helical bundle	11	0.016	4	0.031
g.44	RING finger domain, C3HC4	22	0.008	6	0.028
a.104	Cytochrome P450	149	0.001	9	0.024

12.2.4. Non-Redundant Set of Proteins

Since our intent is to determine if we can detect distant homologs, we created a non-redundant set of proteins from the overall set of proteins that had folds assigned to them. To create the non-redundant set, the following procedure was applied: for each protein, beginning with the longest, all shorter proteins were removed from the list if they matched the first protein with a P-value < 1.0e–3.

12.2.5 Fold Enrichment along the Genome

The relative enrichment of folds along the genome was defined as the ratio of the probability of finding the same fold between pairs of genes a given distance apart in the genome, to the probability of finding the same fold between two randomly selected pairs of genes in the genome. The ratio is therefore a function of the distance in nucleotides between gene pairs. At a given distance, a ratio greater than one implies that more similar folds are occurring than one would expect if folds were distributed randomly over that distance. A ratio of one indicates that the folds are distributed randomly.

12.2.6. Fold Enrichment for Genes with Similar Patterns of Expression

The relative enrichment for co-expressed genes was defined by calculating the ratio of the probability of having a matching fold at or above a given level of correlation coefficient to that expected by randomly choosing pairs of genes. Fold enrichment for correlated genes is therefore a function of the Spearman r, with a ratio greater than one indicating that a pair of correlated genes is more likely to share fold than expected from chance. Error bars were estimated from counting error.

12.3. Results

12.3.1. Fold Enrichment Along the Genome

One of the most frequent evolutionary events is gene duplication, with the *Arabidopsis* genome being especially rich in tandemly repeated genes,[1-5] therefore, it is not surprising to see enrichment of homologous genes in the chromosomal neighborhood for both organisms. The enrichment can, however, still be observed even for the set of non-redundant proteins (see Fig. 12.3).[1-5]

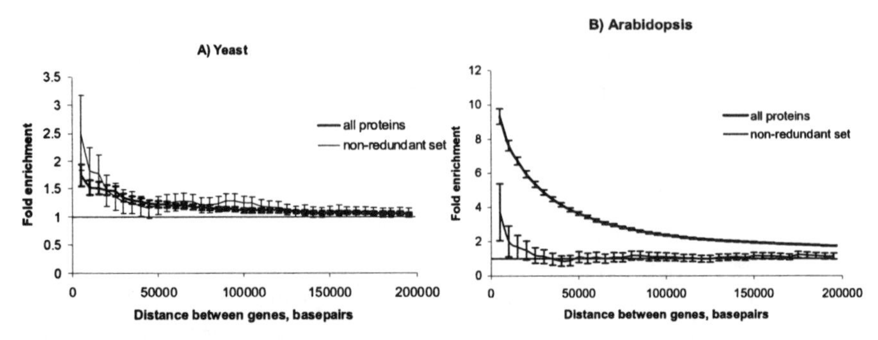

Figure 12.3. We examined two sets of genes for fold enrichment along the genome. The first set was all genes that had a fold assignment. The second set was a subset of the first, consisting of genes whose proteins showed no significant homology to one another. For each set, fold enrichment was measured as the ratio of the frequency of same fold for genes within *d* nucleotides of each other, to the frequency of the same fold in randomly selected genes. The enrichment ratio is then plotted as a function of distance *d*. Courtesy of World Scientific Publishing Co. Pte. Ltd. (Singapore).[24]

12.3.2. Fold Enrichment for Genes with Similar Patterns of Expression

Plate 12.1a shows a plot of the fold enrichment in yeast as a function of *r*. Plate 12.1b shows the corresponding plot for *Arabidopsis*. Both organisms show enrichment that is significantly elevated from the baseline of 1.0, although the difference is more pronounced for yeast. The enrichment is maintained even when redundant proteins are removed. We examined two sets of genes for fold enrichment among co-expressed genes. The first set contains all genes that had a fold assignment and showed no significant sequence homology to other genes at the nt level; see methods description for details on selection. The second set was a subset of the first, consisting of genes whose proteins showed no significant homology to one another. For each set, fold enrichment was measured as the ratio of the frequency of same fold for genes correlated at *r* or better to each other, relative to the frequency of the same fold in randomly selected genes. The enrichment ratio is then plotted as a function of r. Error bars are counting statistics only.

We suspect that the extremely weak signal in the case of *Arabidopsis* is in fact real, and that it is weak as compared to yeast, at least in part, because of dataset size and in part because of overall data quality. A power calculation shows that for the set of yeast data, we can reliably detect correlations down to ~0.38 (significance 0.01, power 80%, Bonferonni correction, power calculation assumed Pearson r rather than Spearman r). For *Arabidopsis* the threshold is roughly 0.50. This weaker detection ability is confounded by relatively poor quality data for *Arabidopsis*. The poorer quality *Arabidopsis* data means that the threshold for a biologically significant correlation is higher than for the yeast data. The difference in quality is evident from the much smaller overall shift towards a correlation of 1 in the distribution of Fig. 12.1a). With more and better quality data, the peak should become cleaner.

12.4. Summary and Conclusions

Folds among nearby genes in the genome and among co-expressed genes are enriched relative to that expected by chance alone. We examined the distributions of enriched folds and were unable to explain the enrichment through a bias in folds in either case. From this we conclude that the enrichment we see is a more-or-less general feature of folds in organisms. Therefore one should, in principle, be able to incorporate such information into fold prediction for proteins whose fold is unknown. Plate 12.1 shows the results of combining sequence similarity with co-expression data to improve fold prediction.

The mechanism behind the enrichment of folds along the genome seems clear. Gene duplications lead to pairs of genes with similar ancestors, and even after substantial divergence results in no *sequence* homology between nearby genes on the genome, there is still a remnant of structural similarity. It is that remnant which accounts for the observed enrichment.

The mechanism behind the enrichment of folds among co-expressed genes is less clear. One hypothesis is that during the course of evolution of a particular metabolic pathway, a newly duplicated gene is created.

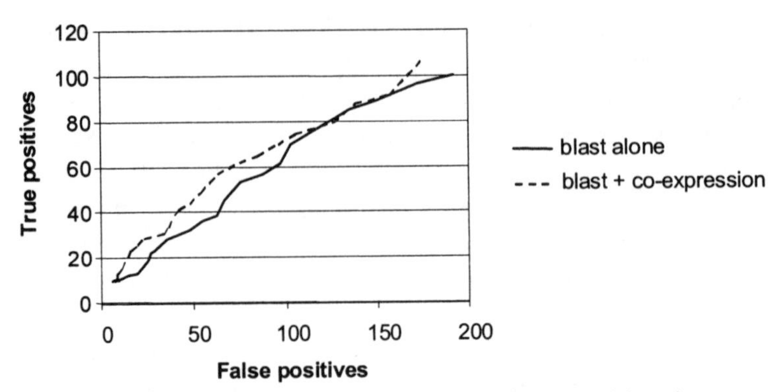

Figure 12.4. True-positive *vs.* false-positive plot comparing fold assignments made with blast alone to fold assignment based on a linear combination of blast score and correlation coefficient. To generate the plot, all pairs of proteins in our non-redundant set were compared to one another by blast, then the pairs were listed in rank order by blast score (blast alone) or a linear combination of blast score and correlation coefficient (blast + co-expression). A pair on the list was counted as a true positive if they had the same fold assignment and a false positive otherwise. The plots shows the cumulative number of true *vs.* false positives at any point on the list for the two scores. For a given number of false positives, the number of correctly assigned folds is higher for the combined blast and correlation coefficient score than for the blast score alone.

For the sake of illustration let us say that the duplicated gene is an enzyme. Since that newly duplicated gene, which includes the promoter region of the original gene, is now redundant, one of two things must happen. Either one of the duplicated genes will disappear or the pair will diverge apart in sequence, with one retaining the original function (by function here we mean both biochemical function as well as biological role), and the other taking on a new function. Since both originally operate on the same substrate, there is a structural constraint to how the pair diverges in sequence, and this constraint tends to cause the fold to be maintained.

The extent to which the behavior described above actually explains how Nature evolves pathways remains to be demonstrated. It is interesting to note, however, that we made a correct, blinded prediction of protein fold for a recent target in the CASP4 competition, using, in part, exactly the above reasoning. The target in question was pectin methylesterase, which is co-expressed with its metabolic pathway neighbor pectate lyase.[23] Both enzymes share exactly the same SCOP fold, the single-stranded right-handed beta-helix.

References

1. Theologis A, Ecker JR, Palm CJ, Federspiel NA, Kaul S, White O, Alonso J, Altafi H, Araujo R, Bowman CL, Brooks SY, Buehler E, Chan A, Chao Q, Chen H, Cheuk RF, Chin CW, Chung MK, Conn L, Conway AB, Conway AR, Creasy TH, Dewar K, Dunn P, Etgu P, Feldblyum TV, Feng J, Fong B, Fujii CY, Gill JE, Goldsmith AD, Haas B, Hansen NF, Hughes B, Huizar L, Hunter JL, Jenkins J, Johnson-Hopson C, Khan S, Khaykin E, Kim CJ, Koo HL, Kremenetskaia I, Kurtz DB, Kwan A, Lam B, Langin-Hooper S, Lee A, Lee JM, Lenz CA, Li JH, Li Y, Lin X, Liu SX, Liu ZA, Luros JS, Maiti R, Marziali A, Militscher J, Miranda M, Nguyen M, Nierman WC, Osborne BI, Pai G, Peterson J, Pham PK, Rizzo M, Rooney T, Rowley D, Sakano H, Salzberg SL, Schwartz JR, Shinn P, Southwick AM, Sun H, Tallon LJ: **Sequence and analysis of chromosome 1 of the plant *Arabidopsis thaliana*.** *Nature* 2000, **408**:816–820.

2. Lin X, Kaul S, Rounsley S, Shea TP, Benito MI, Town CD, Fujii CY, Mason T, Bowman CL, Barnstead M, Feldblyum TV, Buell CR, Ketchum KA, Lee J, Ronning CM, Koo HL, Moffat KS, Cronin LA, Shen M, Pai G, Van Aken S, Umayam L, Tallon LJ, Gill JE, Venter JC, *et al.*: **Sequence and analysis of chromosome 2 of the plant *Arabidopsis thaliana*.** *Nature* 1999, **402**:761–768.

3. Salanoubat M, Lemcke K, Rieger M, Ansorge W, Unseld M, Fartmann B, Valle G, Blocker H, Perez-Alonso M, Obermaier B, Delseny M, Boutry M, Grivell LA, Mache R, Puigdomenech P, De Simone V, Choisne N, Artiguenave F, Robert C, Brottier P, Wincker P, Cattolico L, Weissenbach J, Saurin W, Quetier F, Schafer M, Muller-Auer S, Gabel C, Fuchs M, Benes V, Wurmbach E, Drzonek H, Erfle H, Jordan N, Bangert S, Wiedelmann R, Kranz H, Voss H, Holland R, Brandt P, Nyakatura G, Vezzi A, D'Angelo M, Pallavicini A, Toppo S, Simionati B, Conrad A, Hornischer K, Kauer G, Lohnert TH, Nordsiek G, Reichelt J, Scharfe M, Schon O, Bargues M, Terol J, Climent J, Navarro P, Collado C, Perez-Perez A, Ottenwalder B, Duchemin D, Cooke R, Laudie M, Berger-Llauro C, Purnelle B, Masuy D, de Haan M, Maarse AC, Alcaraz JP, Cottet, A, Casacuberta E, Monfort A, Argiriou A, Flores M, Liguori R, Vitale D, Mannhaupt G, Haase D, Schoof H, Rudd S, Zaccaria P, Mewes HW, Mayer KF: **Sequence and analysis of chromosome 3 of the plant *Arabidopsis thaliana*. European Union Chromosome 3 *Arabidopsis* Sequencing Consortium, The Institute for Genomic Research & Kazusa DNA Research Institute.** *Nature* 2000, **408**:820–822.

4. Mayer K, Schuller C, Wambutt R, Murphy G, Volckaert G, Pohl T, Dusterhoft A, Stiekema W, Entian KD, Terryn N, Harris B, Ansorge W, Brandt P, Grivell L, Rieger M, Weichselgartner M, de Simone V, Obermaier B, Mache R, Muller M, Kreis M, Delseny M, Puigdomenech P, Watson M, McCombie WR, *et al.*: **Sequence and analysis of chromosome 4 of the plant *Arabidopsis thaliana*.** *Nature* 1999, **402**:769–777.

5. Tabata S, Kaneko T, Nakamura Y, Kotani H, Kato T, Asamizu E, Miyajima N, Sasamoto S, Kimura T, Hosouchi T, Kawashima K, Kohara M, Matsumoto M, Matsuno A, Muraki A, Nakayama S, Nakazaki N, Naruo K, Okumura S, Shinpo S, Takeuchi C, Wada T, Watanabe A, Yamada M, Yasuda M, Sato S, de la Bastide M, Huang E, Spiegel L, Gnoj L, O'Shaughnessy A, Preston R, Habermann K, Murray J, Johnson D, Rohlfing T, Nelson J, Stoneking T, Pepin K, Spieth J, Sekhon M, Armstrong J, Becker M, Belter E, Cordum H, Cordes M, Courtney L, Courtney W, Dante M, Du H, Edwards J, Fryman J, Haakense., Lamar, E, Latreille P, Leonard S, Meyer R, Mulvaney E, Ozersky P, Riley A, Strowmatt C, Wagner-McPherson C, Wollam A, Yoakum M, Bell M, Dedhia N, Parnell L, Shah R, Rodriguez M, See LH, Vil D, BakerJ, Kirchoff K, Toth K, King L, Bahret A, Miller B, Marra M, Martienssen R, McCombie WR, Wilson RK, Murphy G, Bancroft I, Volckaert G, Wambutt R, Dusterhoft A: **Sequence and analysis of chromosome 5 of the plant *Arabidopsis thaliana*. The Kazusa DNA Research Institute, The Cold Spring Harbor and Washington University in St. Louis Sequencing Consortium & The European Union *Arabidopsis* Genome Sequencing Consortium.** *Nature* 2000, **408**:823–826.

6. Dunham I., Shimizu N, Roe BA, Chissoe S, Hunt AR, Collins JE, Bruskiewich R, Beare DM, Clamp M, Smink LJ, Ainscough R, Almeida JP, Babbage A, Bagguley C, Bailey J, Barlow K., Bates KN, Beasley O, Bird CP, Blakey S, Bridgeman AM, Buck, D, Burgess J, Burrill WD, O'Brien, KP, *et al.*: **The DNA sequence of human chromosome 22.** *Nature* 1999, **402**:489–495.

7. Hattori M, Fujiyama A, Taylor TD, Watanabe H, Yada T, Park HS, Toyoda A, Ishii K, Totoki Y, Choi DK, Soeda E, Ohki M, Takagi T, Sakaki Y, Taudien S, Blechschmidt K, Polley A, Menzel U, Delabar J, Kumpf K, Lehmann R, Patterson D, Reichwald K, Rump A, Schillhabel M, Schudy A: **The DNA sequence of human chromosome 21. The chromosome 21 mapping and sequencing consortium.** *Nature* 2000, **405**:311–319.

8. Brenner SE, Chothia C, Hubbard TJ: **Assessing sequence comparison methods with reliable structurally identified distant evolutionary relationships.** *Proc Natl Acad Sci USA* 1998, **95**:6073–6078.

9. Park J, Karplus K, Barrett C, Hughey R, Haussler D, Hubbard T, Chothia C: **Sequence comparisons using multiple sequences detect three times as many remote homologues as pairwise methods.** *J Mol Biol* 1998, **284**:1201–1210.

10. Murzin AG, Brenner SE, Hubbard T, Chothia C: **SCOP: A structural classification of proteins database for the investigation of sequences and structures.** *J Mol Biol* 1995, **247**:536–540.

11. MacCallum RM, Kelley LA, Sternberg MJ: **SAWTED: Structure assignment with text description–enhanced detection of remote homologues with automated SWISS-PROT annotation comparisons.** *Bioinformatics* 2000, **16**:125–129.

12. Goffeau A, Barrell BG, Bussey H, Davis RW, Dujon B, Feldmann H, Galibert F, Hoheisel JD, Jacq C, Johnston M, Louis EJ, Mewes HW, Murakami Y, Philippsen P, Tettelin H, Oliver SG: **Life with 6000 genes.** *Science* 1996, **274**:546,563–567.

13. Sherlock G, Hernandez-Boussard T, Kasarskis A, Binkley G, Matese JC, Dwight SS, Kaloper M, Weng S, Jin H, Ball CA, Eisen MB, Spellman PT, Brown PO, Botstein D, Cherry JM: **The Stanford microarray database.** *Nucleic Acids Res* 2001, **29**:152–155.

14. Press WH, Teukolsky SA, Vetterling WT, Flannery BR: *Numerical Recipes in C.* Cambridge, U.K: Cambridge University Press, 1992:639–640.

15. Kane M., Jatkoe TA, Stumpf CR, Lu J, Thomas JD, Madore SJ: **Assessment of the sensitivity and specificity of oligonucleotide (50mer) microarrays.** *Nucleic Acids Res* 2000, **28**:4552–4557.

16. Girke T, Todd J, Ruuska S, White J, Benning C, Ohlrogge J: **Microarray analysis of developing *Arabidopsis* seeds.** *Plant Physiol* 2000, **124**:1570–1581.

17. DeRisi JL, Iyer VR, Brown PO: **Exploring the metabolic and genetic control of gene expression on a genomic scale.** *Science* 1997, **278**:680–686.

18. Eisen MB, Spellman PT, Brown PO, Botstein D: **Cluster analysis and display of genome-wide expression patterns.** *Proc Natl Acad Sci USA* 1998, **95**:14863–14868.

19. Heyer LJ, Kruglyak S, Yooseph S: **Exploring expression data: Identification and analysis of coexpressed genes.** *Genome Res* 1999, **9**:1106–1115.

20. Marcotte EM, Pellegrini M, Thompson MJ, Yeates TO, Eisenberg D: **A combined algorithm for genome-wide prediction of protein function.** *Nature* 1999, **402**:83–86.

21. Gish W: WU Blast, Personal Communication, 1994.

22. Brenner SE, Koehl P, Levitt M: **The ASTRAL compendium for protein structure and sequence analysis.** *Nucleic Acids Res* 2000, **28**:254–256.

23. Tierny Y, Bechet M, Joncquiert JC, Dubourguier HC, Guillaume JB: **Molecular cloning and expression in *Escherichia coli* of genes encoding pectate lyase and pectin methylesterase activities from *Bacteroides thetaiotaomicron*.** *J Appl Bacteriol* 1994, **76**:592–602.

24. Volkmuth W, Alexandrov N: Evidence for sequence-independent evolutionary traces in genomic data. *Proc Pac Symp Biocomput*, 2002, in press.

13

Protein Structure Prediction on the Basis of Combinatorial Peptide Library Screening

Igor Tsigelny, Yuriy Sharikov, Vladimir Kotlovyi, Michael J. Kelner, and Lynn F. Ten Eyck

Abstract

A scientist who has the molecular structure of a drug and is trying to find the drug target has three general ways to proceed. The first option is to run *in vivo* tests trying to define proteins and/or DNA (RNA), which will show decrease or increase when cells interact with the drug, or which can be identified by other biochemical techniques. A second option is to run *in silico* calculations to create pharmacophore models of the drug. A third option is to run an *in vitro* screening of combinatorial libraries of peptides or DNA fragments to find which ones will bind to the drug. This leaves one with the problem of reconstructing possible protein or DNA targets from the set of relatively small peptides or DNA pieces. Peptides that show good binding do not necessarily match any piece of real protein, and in any case only part of the peptide may be responsible for the binding activity. This presents a difficult problem in recovering information from noisy data. We have developed new methods for attacking this problem. These methods are implemented in a set of programs that together help find a short list of possible drug target proteins and predict their tertiary structure. The first of these programs is HMM-ELONGATOR, which predicts putative protein targets based on a set of peptides shown to bind a drug molecule from combinatorial libraries. If the three dimensional structure of the putative targets is not known, the sequences of the target proteins are processed by the program HMM-SPECTR (see Chapter 4), which generates possible structures, These structures than are processed by the program DOT,[1] which evaluates possible modes of interaction and identifies possible binding sites. Predicted protein is than compared with the set of proteins proposed by the program ANALOG-SEARCH on the basis of *in-vivo* experiments.

13

Protein Structure Prediction on the Basis of Combinatorial Peptide Library Screening

Igor Tsigelny, Yuriy Sharikov, Vladimir Kotlovyi, Michael J. Kelner, and Lynn F. Ten Eyck

University of California, San Diego; San Diego Supercomputer Center, La Jolla, California

13.1. Concept of the Comprehensive System

Rational drug design depends on identification of a drug target, determination of its structure, and design of candidate drug molecules that are designed to modify the action of the drug target. This is contrasted to the traditional approach in which large numbers of compounds are screened for biological activity. Once a promising activity was identified, a laborious process of testing variations on the structure of the drug was begun to optimize the activity and minimize unwanted side effects. More recently high throughput methods of screening very large libraries of known compounds have been developed which greatly enhance this process. It is still necessary to identify the drug target and mode of action to fully understand the medical properties of the treatment, to improve the drug, and to evaluate possible risks. Discovering the target protein or DNA that binds tightly to a given drug can be compared to finding a needle in a haystack. Contemporary drug design uses a number of computer based strategies to find this needle. In these problems, the best use of computational methods is to provide experimentalists with a list of the most promising possibilities to test. Computational methods are not yet sufficiently advanced that they can be relied upon to produce definitive answers to difficult problems in biology, but they can drastically reduce the number of possibilities that the experimentalists need to consider.

Plate 13.1 shows that a scientist who has the molecular structure of a drug and is trying to find the drug target has three general ways to proceed. The first option is to run *in vivo* tests trying to define proteins and/or DNA (RNA) which will show decrease or increase when cells interact with the drug, or which can be identified by other biochemical techniques. These methods often work well when the drug can be easily labeled and is not metabolized too drastically, the binding is strong, and the target is a relatively abundant protein. They may also work well if the biological effect clearly implicates some known metabolic or signal transduction pathways.

A second option is to run *in silico* calculations to create pharmacophore models of the drug. This requires preparation and testing of a number of variations on the drug structure. This may still be inadequate for determining the target, because the pharmacophore model may not correspond to any known three-dimensional structure of a protein. These studies will be necessary at some point in the drug development process, but are easier to carry out with knowledge of a target structure.

A third option is to run an *in vitro* screening of combinatorial libraries of peptides or DNA fragments to find which ones will bind to the drug. This leaves one with the problem of reconstructing possible protein or DNA targets from the set of relatively small peptides or DNA pieces. Peptides that show good binding do not necessarily match any piece of real protein, and in any case only part of the peptide may be responsible for the binding activity. This presents a difficult problem in recovering information from noisy data.

At the San Diego Supercomputer Center, we have developed two new methods for attacking this problem. These methods are part of a set of programs that together help find a short list of possible drug target proteins and predict their tertiary structure. The first of these programs is HMM-ELONGATOR, which predicts putative protein targets based on a set of peptides shown to bind a drug molecule from combinatorial libraries. This chapter describes HMM-ELONGATOR and provides some examples of the use of the other programs.

If the three dimensional structure of the putative targets is not known, the sequences of the target proteins may be processed by the program HMM-SPECTR (see Chapter 4), which generates possible structures. Finally, these structures can be used in our program DOT,[1] which evaluates possible modes of interaction and identifies possible binding sites Predicted protein is then compared with the set of proteins proposed by the program ANALOG-SEARCH on the basis of *in vivo* experiments. This program instantly shows if these molecules have any homology to known proteins from the database of HMMs for specific disease related proteins families and gives a set of homologous sequences (unpublished results).

In the best case scenario these are the same or the similar proteins and the task of the entire experimental-theoretical system is fulfilled. In the medium case both systems HMM-ELONGATOR and ANALOG-SEARCH propose the lists of proteins and the solution can be found on the crossing of these two lists. In the worst case scenario these two lists contain completely different proteins. In such a case the experimental part *in vivo* or *in vitro* or both have to be continued with somewhat different initial experimental parameters.

13.2. HMM-ELONGATOR

13.2.1. Problem Description

Several techniques such as phage display, etc., can screen combinatorial peptide libraries for properties such as binding to a drug or protein of interest. The results of these experiments are lists of short peptides that show the desired binding activity. These peptides are assumed to be related to fragments of an active protein. Reconstructing the putative protein can be a difficult task because of the statistical uncertainty associated with aligning many of short, poorly overlapping sequences. Furthermore, the libraries often contain peptides that are not parts of any protein. Many of these bind to the probe molecule. Use of these peptides in multiple alignments increases the noise level in the data. The statistical significance of the alignments is already weak because the peptides are short (7-12 residues). The challenge of these experiments is to obtain longer peptides that can be more reliably used to identify or predict the sequences of proteins that will show the desired activity.

The program HMM-ELONGATOR addresses these problems through use of a more discriminating probabilistic model. Recently this program identified two probable target proteins for the anticancer drug Irofulven (HMAF) from phage display binding data. Both targets have since been experimentally confirmed *in vivo*.[2-4]

13.2.2. Elongation Strategies

Peptide elongation without substitution. In this case short peptides are aligned with all proteins from a databank and the results are used to find overlapping and adjacent peptides that can be substituted for a longer peptide. Such a process is iterative and leads eventually to construction of an HMM featuring all short and elongated peptides (Fig. 13.1).

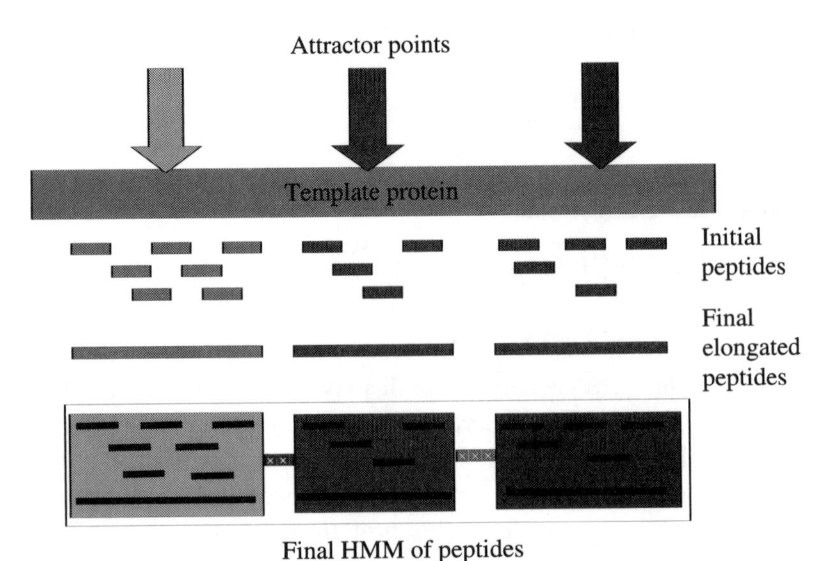

Figure 13.1. Construction of HMM on the basis of initial and elongated peptides.

Peptide elongation with substitution. The program selects the highest scoring peptides for all chosen proteins. These peptides are substituted by fragments of protein sequences, including a core fragment of sequences that directly aligns with the peptides, and additional *N* residues are added from the protein sequence from both left and right sides of the core sequence. The value of N can range from 10 residues to 50–60 residues. The resulting polypeptide chains extracted from real protein sequences are now aligned, and an HMM is built from the aligned sequences. This HMM is used to search the PIR database for the best fitting proteins.

We tested both strategies on the set of peptides that bind to the HMAF molecule. This set was obtained using phage libraries. Three different ER2537-derived phage libraries (New England Biolabs, Beverly, MA) were used for biopanning studies: a 7-mer, a 12-mer, and a C7C. The latter consisted of a 7-mer library flanked by two cysteines to allow disulfide formation. We received 157 sequences of peptides that bind to HMAF.

The first step of the algorithm is common for both strategies (with and without substitution).

Each sequence is aligned with each of proteins in the PIR database. Proteins having fragments that align with the peptides within a specified cut-off value of identity (or specific score) are stored in our data warehouse. An extract from one of the data files is shown in Table 13.1. After the search we count the number of possible high score alignments for all 157 peptides with each of the stored proteins. All proteins are sorted on the number of 'hits'—number of alignments of peptides, which were found to have significant scores when aligned to each of specified proteins.

Table 13.1. Extract from the data storage of proteins showing high scores when aligned to the peptide **AETVEGSCLAKS**.

Peptide →	**AETVEGSCLAKS**	**Pairwise alignment score**
Protein ID	**Protein name**	
Z3BPM3	coat protein A precursor – phage M13	12.8
T14288	DNA (cytosine-5-)-methyltransferase	9.6
I49564	polycystic kidney disease-related protein	9.4

On the next step the algorithm can proceed in two different ways, with and without substitution of amino acids.

Elongation 'with substitution'

We select a peptide that shows the highest score for all chosen aligned proteins and prepare a table (Table 13.2) including as columns the ID of a protein, the sum of BLAST scores for alignments of all peptides for this protein; and the ID of peptide with the maximum score for this protein. Table 13.2 is sorted by the sums of scores. Selected peptides are then substituted by fragments of the protein sequence.

Table 13.2. Extract from the file with the results of hit and score alaysis: AVG—sum of scores for each hit devided to the number of hits.

Number of files	⟶ 30		
Protein ID	Protein name	Number of hits	AVG score
I38344	titin, cardiac muscle – human	3	10.0
T12651	NADH dehydrogenase (ubiquinone)	2	6.0
YGBSG2	gramicidin S synthetase	2	6.7
S08436	pol polyprotein - human immunodeficiency	2	6.2

The resulting substitution sequence includes: a core sequence of a protein, which is directly aligned with the peptide, and additional *N* residues form the protein adjacent to the left and right borders of this core sequence. A value of *N* can be from 10 to 50 residues. The resulting long polypeptide chains, which represent fragments of real protein sequences, now are aligned and an HMM is built on the aligned sequences. This HMM is used for the subsequent search of PIR database for the best fitting protein. This algorithm (with substitution) found topoisomerase I as one of the best fitting proteins for the 157 peptides set. To evaluate this prediction possible interactions of Irofulven with topoisomerase I were computed to see if these occur in regions that can disrupt the function of topoisomerase. Docking was performed on 256×256×256 Å grid with the program DOT[1] using the crystal structure of Topoisomerase I.[5]

Plate 13.2 shows the predicted possible positions of Irofulven in a complex with topoisomerase. These positions have the minimum energies according to the results of docking program. There are three principal clustering positions of Irofulven (marked by arrows). These positions correspond to the crystallographic position of DNA within the topoisomerase—DNA complex. Irofulven thus would prevent normal positioning of DNA in the DNA-topoisomerase complex and consequently disrupt the topoisomerase machinery. All three Irofulven positions of are close to points defined by HMM-ELONGATOR as sites in topoisomerase that correspond to the short peptides from the combinatorial libraries (yellow on Plate 13.2).

Elongation 'without substitution' The algorithm without substitution is applied cyclically until the elongation stops. It starts with the selection of peptide sequences that can be overlapped partly or completely compliment each other when aligned to the sequence of any protein target (Fig. 13.1). Results from the data stored in the form shown in Table 13.3 are used for peptide elongation.

Table 13.3. Residues in protein sequences T33245, T42207, JQ1190, S59398, T00777, T21267, T19553, which align with peptides from the set of 157. Overlapping alignments are shown in bold.

Protein ID	Regions of proteins aligned with peptides
T33245	84-101 23-40 138-155
T42207	**481-497 210-224 2354-2372**
JQ1190	**203-224 213-233**
S59398	1-20 8-29
T00777	282-301 293-306
T21267	**823-837 74-95 1059-1076**
T19553	**355-375** 463-479 470-490

An example of progressive elongation using this algorithm is shown in Table 13.4. First, all proteins with fewer than three 'hits,' or for which the scores for the alignments are low, are eliminated from consideration for elongation. Regions in which peptides overlap on a single protein are used to generate a new peptide sequence, which is a consensus sequence based on the sequence of the protein and of each overlapping peptide. The consensus sequence is *added* to the list of short peptides from the current cycle. The new list of peptides is again aligned against the entire PIR sequence database and the 'hits and scores' analysis is repeated as illustrated in Table 13.2. This search produces all of the proteins originally found, because all of the original peptides are still present, but it also produces additional proteins because the longer peptides may raise the statistical significance of previously poor alignments. The process stops when no further elongations are discovered.

A number of proteins from the same family may have high of scores when being aligned to elongated peptides. If this effect is ignored, the results are significantly biased. In such a case we create an HMM for the family and use the consensus sequence of this HMM as a template protein sequence for the entire family.

Table 13.4. Examples of peptide elongation.

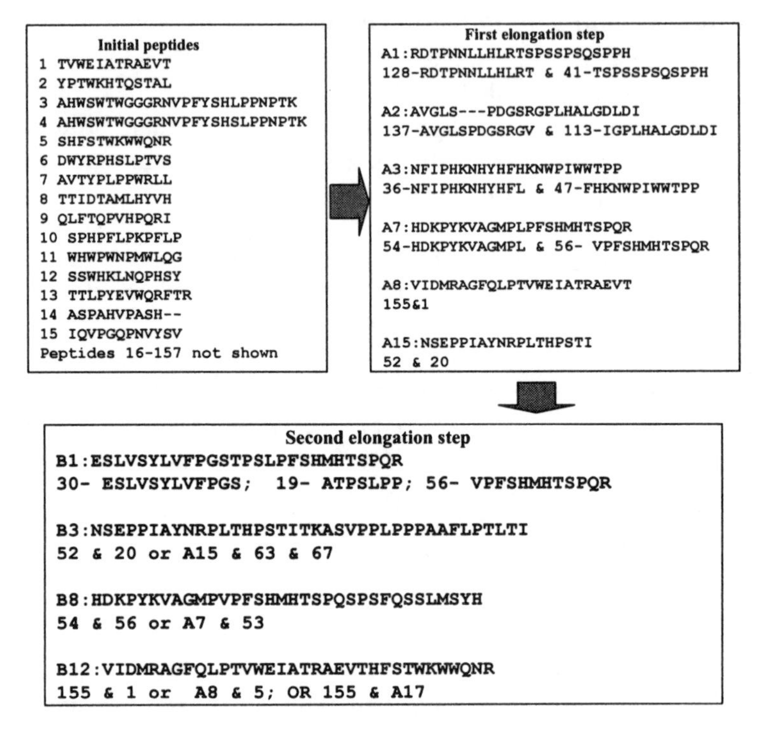

Initial peptides

1 TVWEIATRAEVT
2 YPTWKHTQSTAL
3 AHWSWTWGGGRNVPFYSHLPPNPTK
4 AHWSWTWGGGRNVPFYSHSLPPNPTK
5 SHFSTWKWWQNR
6 DWYRPHSLPTVS
7 AVTYPLPPWRLL
8 TTIDTAMLHYVH
9 QLFTQPVHPQRI
10 SPHPFLPKPFLP
11 WHWPWNPMWLQG
12 SSWHKLNQPHSY
13 TTLPYEVWQRFTR
14 ASPAHVPASH--
15 IQVPGQPNVYSV
Peptides 16-157 not shown

First elongation step

A1:RDTPNNLLHLRTSPSSPSQSPPH
128-RDTPNNLLHLRT & 41-TSPSSPSQSPPH

A2:AVGLS---PDGSRGPLHALGDLDI
137-AVGLSPDGSRGV & 113-IGPLHALGDLDI

A3:NFIPHKNHYHFHKNWPIWWTPP
36-NFIPHKNHYHFL & 47-FHKNWPIWWTPP

A7:HDKPYKVAGMPLPFSHMHTSPQR
54-HDKPYKVAGMPL & 56- VPFSHMHTSPQR

A8:VIDMRAGFQLPTVWEIATRAEVT
155&1

A15:NSEPPIAYNRPLTHPSTI
52 & 20

Second elongation step

B1:ESLVSYLVFPGSTPSLPFSHMHTSPQR
30- ESLVSYLVFPGS; 19- ATPSLPP; 56- VPFSHMHTSPQR

B3:NSEPPIAYNRPLTHPSTITKASVPPLPPPAAFLPTLTI
52 & 20 or A15 & 63 & 67

B8:HDKPYKVAGMPVPFSHMHTSPQSPSFQSSLMSYH
54 & 56 or A7 & 53

B12:VIDMRAGFQLPTVWEIATRAEVTHFSTWKWWQNR
155 & 1 or A8 & 5; OR 155 & A17

A number of proteins from the same family may have high scores when being aligned to elongated peptides. If this effect is ignored, the results are significantly biased. In such a case we create an HMM for the family and use the consensus sequence of this HMM as a template protein sequence for the entire family.

Once the process is complete we take *N* proteins with the best total alignment scores (where N is a small number) as templates. The templates are used to align all of the peptides. Using this alignment (excluding the sequences of the template protein and leaving only the sequences of peptides incorporated in the 'final model' of polypeptide chain) we construct the final HMMs. We then search the PIR databank for proteins having the best score versus each of the HMMs from this set. The resulting list of proteins is sorted by score. The template proteins often appear as one of the top scores, but frequently there are other proteins found which have higher scores.

The final step is to align the selected proteins with the initial set of short peptide sequences. The proteins that give the best scores and number of 'hits' are chosen as results of the search. This algorithm (without substitution) found c-myc as one of the best fitting proteins for the 157 peptide set. Plate 13.3 shows the 3D structure of c-myc that was modeled using the Homology program[6] from a protein structure prediction made by HMM-SPECTR (see Chapter 4).

Acknowledgments

This research was supported by funds from the
Tobacco-Related Disease Research Program (award 9RT-0057) and from the
National Science Foundation (grant DBI 9911196).

References

1. Mandell JG, Roberts VA, Pique ME, Kotlovyi V, Mitchell JC, Nelson E, Tsigelny I, Ten Eyck LF: **Protein docking using continuum electrostatics and geometric fit.** *Protein Eng* 2001, **14**:105–113.

2. McMorris TC, Yu J, Hu Y, Estes LA, Kelner MJ: **Design and synthesis of antitumor acylfulvenes.** *J Org Chem* 1997, **62**:3015–3018.

3. McMorris TC, Yu J, Lira R, Dawe R, MacDonald JR, Waters SJ, Estes LA, Kelner MJ: **Structure-activity studies of antitumor agent irofulven (hydroxymethylacylfulvene) and analogues.** *J Org Chem* 2001, **66**:6158–6163.

4. Kelner MJ, Tsigelny I, Sharikov Y, Ten Eyck LF, Elyadi AN, McMorris TC: **Phage display panning of irofulven identifies potential intracellular targets.** Proc. AACR-NCI-EORTC Intl Conf. Molecular Targets & Cancer Therapeutics, Miami Beach, FL, 2001.

5. Stewart L, Redinbo MR, Qui X, Hol WGJ, Champoux JJ. **A model for the mechanism of human Topoisomerase I.** *Science* 1998, **279**:1534–1540.

6. *Homology Program.* Accelrys, Inc., 2000.

Consensus Structure Prediction

- **A User's Guide to Fold Recognition**
- **Structure Prediction Meta Server**

14

A User's Guide to Fold Recognition

Naomi Siew and Daniel Fischer

Abstract

As the gap between the number of available protein sequences and protein structures widens, the field of structure prediction is proliferating. One of the current approaches of structure prediction is fold recognition, and a large variety of fold recognition methods are already available. The CAFASP and LiveBench experiments have shown that fully automated fold recognition is becoming increasingly powerful.

Here we describe a number of lessons learned from these experiments and show how they can be applied by a non-expert user for particular prediction targets. We illustrate three cases where the use of fold recognition helped biologists in planning and devising experiments and in generating verifiable hypotheses. In addition, we describe one of the latest advances in the field, the meta-predictor. As of today, meta-predictors are the most powerful automated fold recognition approach.

14

A User's Guide to Fold Recognition

Naomi Siew and Daniel Fischer

Ben Gurion University, Beer-Sheva, Israel

14.1. Introduction

Fold recognition, also known as fold assignment or threading, is now over ten years old and has already become a well established research sub-area of the protein structure prediction field.[1,2] Fold recognition research is a most challenging and difficult scientific quest, which has drawn considerable interest because of its enormous applicability. Despite the fast growth in the number of solved 3D protein structures, the majority of protein sequences show no significant sequence similarity to proteins of known structure.[3,4] However, it is estimated that most of these proteins will eventually turn out to have a three-dimensional (3D) structure which is similar to one of the already seen folds.[3,5,6] Because researchers interested in a particular protein cannot always wait until its 3D structure is solved, they often attempt to generate an *in silico* model of its structure. Thus, in the absence of sequence similarity to known structures, fold recognition is increasingly becoming a powerful tool. The various structural genomics projects will result in larger applicability of prediction methods because these projects generate only representative structures from each protein family.[7-11] Fold recognition will play a critical role in translating the information on the relatively small fraction of those proteins whose structures were solved, into accurate models for all other proteins. *In silico* 3D models of most proteins will thus be available years before their experimental structures are determined. Therefore, fold recognition is gaining great importance, and advances in this field result in a stronger impact of the structural genomics approach.

Since the pioneer works in fold recognition,[12-16] dozens of methods have been developed (for reviews see, e.g., references.)[2,17,18] A number of these methods are available to the community as programs to be installed at the user's site. A number of them can be purchased as commercial packages, and others are made available via the internet as computer servers. Modern fold recognition methods have matured considerably and are becoming increasingly more powerful. The large number of research groups developing fold recognition methods indicates that the structure prediction problem is far from being solved, and obviously there is still a lot of room for improvement.

The variety of available fold recognition methods has resulted in many biologists feeling at a loss. Why are there so many programs? Why should I use a new, scarcely known method? Which one should I trust? Maybe none? Maybe all? Even if a biologist has chosen a fold recognition method, (s)he has yet to face a series of other questions: is the result I obtained reliable? Now that I have a 3D model, what can I do with it? Can I use the model to obtain biological insights? What if the model is wrong? Why should I trust an *in silico* result?

In this paper we attempt to address some of these questions. Rather than surveying the field (good reviews are listed above), we concentrate on how to use fold recognition methods; how to better interpret their results; and what can be done in order to assess the reliability of the resulting model. We begin with three examples illustrating how fold recognition aided researchers/biologists in arriving at novel biological insights, and how the *in silico* derived hypotheses helped design experiments to test these new ideas. We then describe the fold recognition process from a user's perspective, with emphasis on the interpretation of the results.

14.2. Examples of Using Fold Recognition for Biological Research

Structure prediction methods can be helpful tools for the biologist in planning and preparing experiments for studying a protein or a biological system, and in arriving at biological hypotheses and insights. Here we present three examples of how the use of structure prediction methods helped biologists in their research. Other examples can be found at the bioinbgu web site (http://www.cs.bgu.ac.il/~bioinbgu). In the first example,[19] structure prediction helped the researchers to approach the studied protein from a whole different perspective and arrive at a new hypothesis regarding the protein's function, thus breaking through a pattern of thinking that had shown no experimental evidence. In the second example,[20] structure prediction aided in planning additional site-specific mutagenesis experiments and in proposing a binding site, leading to a better understanding of the studied protein's structure-function activity. In the third example,[21] structure prediction led to construction of a detailed molecular model of a protein which is part of a complex biological system involving a few proteins that interact together. This provided insights facilitating the design of new experiments aimed at identifying potential protein-protein interaction sites and helped in rationalizing functional data.

At the time the above works were carried out, no clear homologues with known structure could be identified. Due to the fast growth of the structural and sequence databases, it is possible that what in the past appeared to be a difficult prediction target, can now be considered an easier target.

14.2.1. Plant Resistance Gene Products

Rigden *et al.* studied plant resistance gene products (RGPs).[19] These are proteins that are involved in the protection of plants against pathogenic attacks. RGPs comprise a number of classes, whose structures are made of various combinations of different domains.[22] Based on motifs found along the sequences of these proteins,[23] it was previously hypothesized that one of the RGP domains includes a nucleotide-binding site (NBS), although this idea was not supported by experiment. Extensive sequence analysis of the NBS motifs in RGPs showed that these sequence motifs are also found in proteins that do not exhibit nucleotide binding properties.[19] Therefore, the researchers used structure prediction tools in looking for an alternative hypothesis for the function of the predicted NBS regions in RGPs.

The sequences of NBS RGP domains from a few plants were sent to three fold recognition servers,[24-26] in search for structural homologues. The CheY family domain fold received high scores by all methods.

A closer analysis revealed that the secondary structure prediction for the RGP domains is highly similar to the secondary structure of the CheY-like domains. In addition, sequence alignment revealed two identical positions of conserved residues in both the RGP and the CheY-like fold family, known to be functionally important in the CheY-like family. A mutation in one of these positions in the tomato RGP lead to complete loss of activity,[27] a fact that could not be explained by the NBS hypothesis. Further sequence motif conservation was found despite overall poor sequence similarity between these two families of proteins. Based on this fold recognition prediction, the researchers built detailed models for the RGP domains.[19]

In addition to residue location similarities, functional analogies were found between the His-Asp phosphotransfer pathway involving the CheY-like folds, and the signaling pathways involving RGPs. For example, both systems' membrane or cytosolic signals are transduced by branched, multistep pathways involving kinases and phosphatases.[19] Based on these similarities, further analogous functional characteristics were noticed in the signal transduction pathways of the two systems; as well as similar positions of hydrophobic residues in the models and other possible functional sites. This led to suggesting a different function for the RGB putative NBS domain than that previously hypothesized, and enabled designing experiments that can confirm or refute the new theory.[19]

14.2.2. Acetohydroxyacid Synthase

Mendel *et al.* studied the enzyme acetohydroxyacid synthase (AHAS).[20] Valine inhibition of AHASs plays an important role in the regulation of biosynthesis of branched-chain amino acids in bacteria,[28] as well as in eukarya.[29] Despite intensive research over decades, this step has not been well understood as a biochemical phenomenon.[20] AHAS III is one of three isozymes in enterobacteria. It is a holoenzyme of the structure α_2/β_2 constructed of two large subunits and two small ones.[30] The large subunit has low catalytic activity and is not sensitive to valine inhibition, while the small subunit (SSU) has no catalytic activity but binds valine.[31] SSU associates with the large subunit and is required for full catalytic activity and valine sensitivity. The other isozymes are AHAS I, which is also inhibited by valine, in the required presence of SSU,[32] and

AHAS II, which is insensitive to valine regulation, although it contains a small subunit.[33]

The researchers concentrated on the *E. coli* AHAS III. Its large catalytic subunit is homologous to several thiamin diphosphate-dependent enzymes, while the small, regulatory subunit shows no homology to any protein of known structure. In the lack of a 3D structure for SSUs, it is not possible to explain its inhibition by valine and its other properties.

Three approaches were used here in order to study the valine-regulatory site in SSU: studying spontaneous mutants; designing site specific mutants, based on comparison of AHAS SSU sequences that are valine-sensitive (AHAS I and AHAS III) and those that are not (AHAS II); and searching for structural homologues, in order to study structure-function relationship.

Fold recognition[24] found that the N-terminus of SSU is compatible to the structure of the C-terminal regulatory domain of 3-phosphoglycerate dehydrogenase (3PGDH). The N-terminus of SSU is also the region with the highest sequence homology between various SSUs. Based on this structure prediction and on the properties of the SSU mutants, a structural model was built for SSU. From the model, a valine binding site in the protein, homologous to the serine binding site in 3PGDH, can be proposed. The researchers report that the model can help explain some of the mutants' experimental characteristics. The model also aided in designing another mutant, which, as predicted, caused the enzyme to be resistant to valine.[20]

On a higher molecular level, the SSU molecular model also serves as a starting point for a hypothesis on how the enzyme's large and small subunits pack together, and can serve as a basis for designing further experiments for structure-function activity studies. Thus, fold recognition clearly aided in deriving verifiable hypotheses, some of which were already corroborated by experiment.

14.2.3. Endothelial Cell Protein C/Activated Protein C Receptor

Villoutreix *et al.* studied the endothelial cell protein C/activated protein C receptor (EPCR), a transmembrane glycoprotein whose physiological function is still not fully characterized.[21] EPCR binds both the zymogen, or proenzyme, protein C (PC) and the activated protein C (APC),[34] a key coagulant multi-modular serine-protease. EPCR and PC/APC are components of a complex coagulation pathway,[35,36] whose cascade of events is still not fully understood. The pathway and its regulation involve many protein-protein and protein-membrane interactions, many of which are not clearly characterized at the moment. In order to better understand this system, Villoutreix and colleagues have previously built several 3D models of coagulation proteins which shed new light on parts of this system and helped design experiments (see, e.g., references.)[37,38] Here they further study EPCR.[21]

Based on fold recognition prediction scores[24] and sequence comparisons, mouse CD1 was chosen as a template for building the EPCR model. Since it is known that CD1 and MHC class I molecules are highly structurally similar, despite their low sequence similarity, structural and experimental data known for MHC class I molecules could be used during the building and analysis of the EPCR model.

Three models were built, for the human, bovine, and mouse EPCR. EPCR contains three domains, two of which are similar to α_1 and α_2 domains of CD1 and MHC class I (see Plate 14.1). The 3D molecular model of human EPCR (hEPCR) was built based on the structure of the α_1 and α_2 domains of mouse CD1. These domains are composed of an eight-stranded antiparallel β-pleated sheet and two long anti-parallel α-helices. Mouse CD1 was chosen as a template based on fold recognition results. Two other regions in CD1/MHC class I do not have parallels in EPCR. Instead, EPCR has a transmembrane region, which was predicted based on known structures of transmembrane proteins.[21] The modeling of the α_1 and α_2 domains took into account the locations of conserved key residue positions, a predicted disulfide bond, the 3D distribution of hydrophobic, polar and charged residues, and potentially glycosylated residues. The EPCR transmembrane region was modeled based on the structure of known transmembrane proteins.[21]

The models can serve as a basis for understanding previously obtained biochemical results, and can serve for deducing hypotheses concerning the receptor's function and its interactions with other molecules. Based on the model, a binding site was suggested, a hypothesis which was later confirmed by site-directed mutagenesis. In addition, based on the model, it was suggested that EPCR does not bind calcium. This information is of importance for experimental design, for instance, when interactions are evaluated in the presence and absence of calcium ions.

In this example, fold recognition helped in analyzing the structure of two interacting proteins thus enabling us to gain new insights on the system's structure and function.

To conclude, fold recognition methods can enable the building of approximate 3D models that can serve as the basis for constructing verifiable new hypotheses about the function and structure of proteins not previously fully characterized. These hypotheses can direct further experimental work, and thus, fold recognition can help in planning new experiments, such as site directed mutagenesis. Finally, fold recognition aids in exploring functional similarities between families of proteins, which are otherwise non-evident due to lack of significant sequence similarity.

14.3. How to Fold Recognize?

A few years ago only a handful of fold recognition methods, which required significant user's expertise, were available. The CAFASP[39,40] (http://www.cs.bgu.ac.il/~dfischer/CAFASP2) and LiveBench[41,42] (http://BioInfo.PL/LiveBench) experiments have catalyzed the development of new, improved, automated programs, often incorporating aspects of human expertise that were applied manually in the older methods. This has resulted in the wide applicability of the new methods, and today there are over a dozen different programs. These modern methods have become more automated and user friendly, and require less user expertise. For the "easier" prediction targets, fully automated methods can be used. However, for harder prediction targets, significant human expertise is still required. Using these programs, users of fold recognition methods have developed their own prediction protocols, as illustrated in the examples above.

The CAFASP and LiveBench experiments have provided a wealth of information about the available methods. These experiments have shown that there are a number of top fold recognition methods, and that it is hard to point to one best method. LiveBench2 showed that the top performing servers are able to correctly predict the fold of about one third of the "difficult" prediction targets. However, the servers' specificities were not as good.[42] The CAFASP2 results lead to similar conclusions, identifying five top servers,[24,25,43-45]

Both experiments showed that a meta-predictor, which utilizes the predictions of several servers, performs better than any of the individual servers. In addition, the CAFASP2 experiment showed that only 6 human predictors out of 103 performed better than the consensus prediction based on the automatic servers' predictions. This demonstrates that automatic predictions can be very helpful for human predictors. CAFASP and LiveBench are also useful in assessing the reliability of the servers. Each fold recognition method scores the results using different statistical measures. Consequently, a user needs to understand what these scores mean, and when they are likely to indicate a confident prediction. In particular, the LiveBench experiments allow the user to estimate the likelihood that a given score produced by a particular method represents a reliable prediction. LiveBench's specificity analysis lists the scores of each method's ten strongest false positive predictions. With this information one could estimate how reliable the results of a specific server are.

Based on lessons learned in the above experiments, the following suggests a fold recognition procedure. We refer to the protein sequence to be modeled as the query.

14.3.1. Searching for Homologues of Known Structure

The first step when building a 3D model of a protein, using any modeling approach, is to search for homologous sequences of known structure, using, for example, fast sequence comparison techniques such as FASTA[46] and BLAST.[47] If at least one such homologue with high enough sequence similarity is found (usually, sequence identity above 30-40% is sufficient) then a 3D model can be built directly, using Homology Modeling methods (see, e.g., references[2,48,49] for recent reviews). If no protein of known structure shows significant sequence similarity to the query, then more sensitive methods are needed. Distant homologies can often be detected using, for example, the fast and widely used PSI-BLAST.[50]

Even when PSI-BLAST (or another sequence-based method) identifies a distant homologue of known structure, it is still advisable to use fold recognition as well, in order to verify and refine the results. The

LiveBench experiments have demonstrated that a correct PSI-BLAST prediction is often corroborated by strong fold recognition results.[41] In addition, fold recognition methods usually produce alignments which are 10 to 15% more accurate than the alignments produced by PSI-BLAST.[41] This is of great importance for building a 3D molecular model for the query protein, since it has been observed that the quality of a model is strongly dependent on the accuracy of the sequence-structure alignment (see, e.g., references.)[1,51]

Another reason it is advisable to verify PSI-BLAST's results is that PSI-BLAST is not error free, and can occasionally produce false positive results, especially when no good PDB hits were found in the first iteration. For example, in LiveBench1 PSI-BLAST reported structurally unrelated proteins with scores better than 10E-06, the highest scoring false positive having a score below 10E-32 (see http://bioinfo.pl/LiveBench/1/ and reference.)[41] PSI-BLAST's false positives can be easily detected by running fold recognition methods. If fold recognition does not identify the same fold with a confident score, then the user should be careful, as this often occurs when PSI-BLAST's result is incorrect.

When PSI-BLAST cannot identify a homologous sequence with a known fold, fold recognition can be the next step in the effort to find a compatible 3D structure for the query sequence. In order to illustrate the fold recognition process, the following shows actual results for two queries for which PSI-BLAST could not identify similar sequences of known structure.

The first query is a novel apolipoprotein sequence of unknown function, apoM, discovered by Villoutreix and colleagues.[52] Based on the results obtained from several methods, the researchers proposed that apoM adopts the lipocalin fold. Fold recognition methods provide a highly confident prediction for this target, and thus it can be considered a relatively easy prediction target, probably at the difficulty level of the easier CAFASP2 fold recognition targets (see http://www.cs.bgu.ac.il/~dfischer/CAFASP2/ and reference.)[40] The second query is the ORF Vng2399h from Halobacterium species NRC-1.[53] Vng2399h is a new ORF of unknown function, with no homologous sequences in the databases (i.e., it is an ORFan.)[4] Fold recognition methods do not give confident predictions for this query. On the contrary, the results indicate that either this protein has a novel fold, or its 3D structure is non-trivially different from any of the known folds. In this case, negative fold recognition results can be used as a basis for identifying attractive structure determination targets.[54]

14.3.2. Running Your Favorite Fold Recognition Method

For illustration purposes, we chose to first run our queries on the bioinbgu server,[24] which has been a top performer in the LiveBench and CAFASP experiments. The bioinbgu server runs five different methods and returns a consensus prediction (for a detailed description of each of the five components and the consensus computation see reference.)[24] The server's top results for the apoM query are shown in Table 14.1.

Table 14.1. Bioinbgu fold recognition results for apoM.

gonp		gonpm		seqpprf		seqpmprf		prfseq		consensus	
1a3y	12.2	1ew3	6.8	1a3y	13.3	1a3y	11.8	2apd	5.9	1a3ya	42.0
1ew3	10.8	1mup	6.6	1ew3	11.5	1mup	10.7	1ew3	5.1	1ew3a	22.4
1bj7	10.5	1dfv	5.4	1mup	11.3	1bj7	10.5	1beb	4.4	1mup	16.3
1mup	10.2	2apd	5.2	1bj7	10.9	2apd	10.3	1mup	4.1	2apd	11.5
1obp	10.2	1a3y	4.6	1bbp	10.2	1ew3	10.3	2prf	3.8	1bj7	11.2
1dfv	9.5	1f3u	4.0	1obp	10.2	1dfv	10.2	1cqa	3.6	1dfva	6.4
1beb	8.9	1beb	4.0	1dfv	10.1	1bbp	9.9	1bj7	3.5	1bbpa	5.7
1epa	8.8	1bj7	3.8	1beb	9.2	1obp	9.8	1bbp	3.4	1beba	5.6
1bbp	8.6	1elc	3.5	2apd	8.9	1beb	9.2	1c05	3.4	1obpa	5.4
2apd	7.5	1bbp	3.3	1epa	7.7	1epa	8.6	1elc	3.4	1epaa	3.2

The first five columns in Table 14.1 correspond to each of the five methods used in bioinbgu: gonp, gonpm, seqpprf, seqpmprf and prfseq. For each method, the PDB codes of the top scoring folds are listed along with the score. The last column corresponds to the consensus prediction (see below).

Can 1a3ya be used as a template for building a 3D model for our query? Is 42.0 a reliable score? From the information given in the LiveBench2 and LiveBench3 "spec" tables (see http://bioinfo.pl/LiveBench/2/ and http://bioinfo.pl/LiveBench/current/) one can conclude that for bioinbgu, a top hit with a score higher than 17 gives a correct prediction in approximately 90% of the cases, whereas a score above 28 is almost certainly correct. Thus, we can conclude that the bioinbgu prediction for apoM is highly reliable.

However, additional care should be taken. The quality of the alignment should be checked. Does it cover all the sequence? How many

gaps are there and where? How does the predicted secondary structure of the query match the observed secondary structure of the fold? The bioinbgu server returns a url where the detailed results can be browsed, including the alignments. In our case, we observe that both the alignment and the secondary structure match are very good, concluding that for the apoM query we do obtain a strong fold recognition result.
In contrast to apoM, no strong results were obtained for the second query, Vng2399h (see Table 14.2.)

Table 14.2. Bioinbgu fold recognition results for Vng2399h.

gonp		gonpm		seqpprf		seqpmprf		prfseq		consensus	
1quq	4.3	5hvp	4.3	5hvp	3.9	NA	0.0	1b34	5.3	5hvpa	11.8
5hvp	4.2	1bkb	3.6	1hgx	3.8	NA	0.0	5hvp	4.3	1quqa	5.7
1cmv	3.6	1eal	3.6	1quq	3.6	NA	0.0	1com	3.6	1b34b	4.6
2mip	3.5	1vfa	3.6	2mip	3.5	NA	0.0	1eal	3.5	1hgxa	3.6
1eje	3.5	1c3k	3.5	1d3b	3.5	NA	0.0	1wit	3.2	2mipa	2.2
1a2z	3.0	1a3y	2.7	1bvq	3.3	NA	0.0	1eft	3.2	1d3bb	1.3
1hgx	2.9	1b34	2.7	1acx	3.1	NA	0.0	1sfp	3.0	1eal	1.1
1dhy	2.8	1eif	2.6	1kjc	2.8	NA	0.0	1lic	2.9	1bvqa	1.0
1xgs	2.7	1cgt	2.5	1b34	2.8	NA	0.0	1kja	2.8	1cmva	1.0
1qor	2.6	1bmv	2.4	1csq	2.7	NA	0.0	1a2v	2.8	1coma	0.8

Here we see that the consensus top hit has a relatively low score of 11.8. In addition, notice the column corresponding to seqpmprf. This component of bioinbgu applies a method similar to seqpprf, but uses additional information from sequences that are homologous to the query (when available). When no such sequences exist, bioinbgu reports "NA" in the seqpmprf column. Targets with no homologous sequences are harder to predict, because most fold recognition methods exploit the evolutionary information of neighboring sequences in the form of multiple alignments, profiles[25,43] or HMMs.[26]

The sequence-structure alignments of the top hits for Vng2399h contain a number of gaps, some of them large, and most of the alignments do not cover the full length of the query. Thus, we conclude that no reliable fold recognition prediction can be obtained for this target using the bioinbgu server. This could mean that the method is not sensitive enough to identify a compatible fold for Vng2399h (if one exists) or that the 3D structure of Vng2399h may correspond to a new fold or to a fold which is significantly different from those currently available.

When the top ranking fold recognition result is not strong enough to allow a confident prediction, additional evidence can be gathered by observing the first 3 to 5 ranking folds. Are there a number of top results having the same fold? If so, this is reinforcing evidence. In the apoM case, all the PDB entries at the top ranks correspond to the lipocalin fold. Thus, even if the top ranking score would not have been as high, obtaining several top hits having the same fold may still allow one to reach a confident prediction. In the Vng2399h case, there is no consistency among the top hit folds. This further indicates that a prediction is not possible in this case.

When sequences homologous to the target exist, another good practice is to verify whether additional runs, using the homologous sequences as queries, give stronger results. If a strong result is obtained for one of the homologues, then it is a positive sign. If no strong result is obtained for the homologues, then one could check whether the top ranking results obtained with the homologues consistently find the same fold. This procedure was applied by Rigden *et al.*, as described above.[19] However, this step is not applicable for Vng2399h since it is an ORFan.

In addition, it is advisable for the user to verify that the query does not contain significant low-complexity regions, transmembrane helices, signal peptides, or more than one domain. Targets with these characteristics are problematic and difficult to predict. Vng2399h contains a number of low-complexity regions, a fact which probably contributes to the difficulty in predicting its structure. It is advisable that the user remove potentially problematic regions from the query, partition the query into separate domains (if known), and in general, avoid prediction of queries with significant low-complexity regions or a number of transmembrane helices.

14.3.3. Running Other Methods

For the difficult cases it is recommended to use additional fold recognition methods to check whether a particular method gives a strong result, or whether the different methods give consistent (albeit weak) results.

For the apoM query, practically all the available fold recognition servers, including all five bioinbgu's components, give both a strong and a consistent result, reinforcing even more our confidence in this prediction. For the Vng2399h query, we clearly observe an inconsistency in the types of folds predicted by the various servers, in accordance with bioinbgu's results.

To facilitate running a number of methods, and to organize the results for analysis, Leszek Rychlewski has created a meta-server (see Chapter 15 in this book.) The query is submitted only once to the meta-server, which automatically sends it to each of a dozen or so servers, collects the results and presents them in a single summary page, along with additional information.

For the apoM query, all the meta-server methods returned consistent and strong results. Thus, we can predict with the highest certainty that the correct fold for apoM is the lipocalin fold, as suggested by Villotreux *et al.*[52] The meta-server results for the Vng2399h query are available at http://bioinfo.pl/meta/target.pl?id=4307. A quick look at the results shows that almost every method predicts a different fold, and that all predictions have low scores. This confirms our previous conclusions, namely, that no reliable *in silico* prediction can be produced for this query, and that Vng2399h can be identified as an attractive target for further experimental studies.

14.3.4. Why Run More Than One Method?

As noted above, when no strong results are obtained by one method, one can check whether a fold is consistently predicted by a number of methods. Such consistency can be regarded as reinforcing evidence that each server has a number of correct predictions for certain targets, for which the other servers perform poorly. The reason for this may be that each method exploits different types of information in different (and sometimes incompatible) ways, and consequently, given a particular target, one method may find the correct answer, whereas others may not.

The bioinbgu consensus was built with this rationale in mind. The consensus is based on the five methods run by bioinbgu, each exploiting evolutionary information in different ways. Tests showed that each component on its own had its strengths, but that the best performance was obtained by producing a consensus prediction of the five methods.[24] The consensus idea was carried even further in CAFASP2. There, a semi-automated consensus prediction of all the participating automated servers was created using semi-automatic methods, and was named CAFASP-CONSENSUS.[40] Among the automated servers that participated in CAFASP2, the CAFASP-CONSENSUS ranked first, and only six human participants in CASP4 ranked higher (see http://PredictionCenter.llnl.gov/casp4/ and reference.)[40] Thus, CASP4 and CAFASP2 clearly demonstrated the potential of consensus predictions. After CAFASP2, Arne Elofsson created a fully automated meta-predictor, PCONS (see chapter 16 in this book). PCONS feeds the results of the top available 5 to 7 servers to a neural network, and determines the most likely consensus result. Tests on LiveBench2 and LiveBench3 have shown that PCONS is a powerful prediction method (see http://bioinfo.pl/LiveBench/2/ and http://bioinfo.pl/LiveBench/current.)

14.3.5. 3D-Shotgun Meta-Predictor

A different meta-predictor, currently under development in our group, is named 3D-shotgun. This method computes a "meta-prediction" based on all the structurally similar models that are produced by bioinbgu's five components. Scores are computed using both the structural similarities between the individual models and their reported scores. Preliminary tests on LiveBench3 targets show that when applied on the five bioinbgu components, 3D-shotgun is significantly more sensitive and selective (see http://bioinfo.pl/LiveBench/current) than the current bioinbgu consensus, placing 3D-shotgun among the top performing methods tested in LiveBench.

3D-shotgun can also be applied to the results of other servers. Preliminary tests based on the top hits of three servers, namely, bioinbgu's consensus itself, FFAS[43] and 3D-PSSM,[45] have been performed. This combined meta-prediction of three servers is named 3DS3. Preliminary results show a significant improvement over bioinbgu's 3D-shotgun. The 3D-shotgun method (both the version using the five bioinbgu components, and the version using results from a number of servers) will soon be made available as a server.

14.4. Summary

The field of fold recognition is growing fast. At the moment, there are a number of good methods available. None of them can be considered the "best" one. Each one of these methods has its strengths and its weaknesses. Consistency in results can be enhanced through consensus predictions. A number of consensus approaches were described here, and new ones will definitely be developed in the future.

As fold recognition matures, it will continue to aid biologists in designing experiments and arriving at hypotheses regarding the structure-function activity of their studied proteins. These hypotheses can then be readily checked in the laboratory. We have shown here a few examples of published results with insights from fold recognition. We also described the fold recognition procedure, including two detailed examples of queries that gave two different types of results. The results for the first query, apoM, gave a confident prediction: apoM is most likely compatible to the lipocalin fold. No function can be straightforwardly deduced from this result, since the lipocalin fold class contains members with a wide variety of functions. However, a 3D molecular model, built based on this prediction, may help design experiments aiming at characterizing this protein. The second query, Vng2399h, is an ORFan sequence for which no prediction is possible. The contradictory, low-scoring results hint at the possibility that this sequence adopts either a novel fold, or a fold that is non-trivially different from any known structure. This example demonstrates how a "negative" fold recognition result can be used in order to identify attractive targets for experimental work, such as structure determination.

Fold recognition methods are slowly but steadily incorporating aspects of human expertise, thus becoming easier to use and freeing humans from some tedious aspects of the prediction process. In addition, the availability of automated methods allows their application on genomic scales. Thus, fold recognition is becoming an essential complement of structural genomics.[11] However, automated methods will never totally replace human expertise,[2] in particular, when biological knowledge exists for the prediction target, or when human intuition can be applied to interpreting the results.

For updates on new methods and on improvements of the current methods, please visit the CAFASP and LiveBench web-pages.

Acknowledgments

We thank Bruno Villoutreix for kindly providing us with the color plate, and for discussions. We acknowledge Arne Elofsson and Leszek Rychlewski for developing PCONS and the meta-server. CAFASP, LiveBench and the CAFASP-CONSENSUS are the result of our close collaboration with Leszek Rychlewski and Arne Elofsson. N.S. is partly supported by a grant from the Ministry of Science, Israel, and by the Kreitman Foundation Fellowship. D. F. research was partially supported by Grant # 9900032 from the United States-Israel Binational Science Foundation (BSF), Jerusalem, Israel.

References

1. Moult J, Hubbard T, Fidelis K, Pedersen JT: **Critical assessment of methods of protein structure prediciton (CASP): Round III.** *Proteins* 1999, **Suppl 3**:2–6.
2. Siew N, Fischer D: Convergent evolution of CAFASP and computer chess tournaments: **CASP, Kasparov and CAFASP.** *IBM Systems J* 2001, **40**(2):410–425.
3. Fischer D, Eisenberg D: **Assigning folds to the proteins encoded by the genome of *Mycoplasma genitalium*.** *Proc Natl Acad Sci USA* 1997, **94**:11929–11934.
4. Fischer D, Eisenberg D: **Finding families for genomic ORFans.** *Bioinformatics* 1999, **15**:759–762.
5. Chothia C: **One thousand folds for the molecular biologist.** *Nature* 1992, **357**:543–544.
6. Orengo CA, Jones DT, Thornton JM: **Protein superfamilies and domain superfolds.** *Nature* 1994, **372**:631–634.
7. Montelione GT, Anderson S: **Structural genomics: Keystone for a human proteome project.** *Nature Struct Biol* 1999, **6**:11–12.
8. Kim SH: **Shining a light on structural genomics.** *Nature Struct Biol* 1998, **5**:643–645.

9. Gaasterland T: **Structural genomics taking shape.** *Trends Genet* 1998, **14**:135.

10. Abbott A: **Computer modellers seek out ten most wanted proteins (news).** *Nature* 2001, **409**(6816):4.

11. Fischer D, Baker D, Moult J: **We need both computer models and experiments (correspondence).** *Nature* 2001, **409**:558.

12. Bowie JU, Luthy R, Eisenberg D: **A method to identify protein sequences that fold into a known 3-dimensional structure.** *Science* 1991, **253**:164–170.

13. Sippl MJ, Weitckus S: **Detection of native like models for amino acid sequences of unknown three dimensional structure in a database of known protein conformations.** *Proteins* 1992, **13**:258–271.

14. Jones D, Thornton J: **Protein fold recognition.** *J Comput-Aided Mol Des* 1993, **7**:439–456.

15. Godzik A, Kolinski A, Skolnick J: **Topology fingerprint approach to the inverse folding problem.** *J Mol Biol* 1992, **227**:227–238.

16. Bryant SH, Lawrence CE: **An empirical energy function for threading protein sequences through folding motifs.** *Proteins* 1993, **16**:92–112.

17. Fischer D, Rice D, Bowie JU, Eisenberg D: **Assigning amino acid sequences to 3-dimensional protein folds.** *FASEB J* 1996, **10**:126–136.

18. Fischer D, Eisenberg D: **Predicting structures for genome proteins.** *Curr Opin Struct Biol* 1999, **9**:208–211.

19. Rigden DJ, Mello LV, Bertioli DJ: **Structural modeling of a plant disease resistance gene product domain.** *Proteins* 2000, **41**:133–143.

20. Mendel S, Elkayam T, Sella C, Vinogradov V, Vyazmensky M, Chipman D, Barak Z: **Acetohydroxyacid synthase: A proposed structure for regulatory subunits supported by evidence from mutagenesis.** *J Mol Biol* 2001, **307**:465–477.

21. Villoutreix BO, Blom AM, Dahlback B: **Structural Prediction and analysis of endothelial cell protein C/activated protein C receptor.** *Protein Eng* 1999, **10**(12):833–840.

22. Hammond-Kosack KE, Jones JDG: **Plant disease resistance genes.** *Annu Rev Plant Physiol Plant Mol Biol* 1997, **48**:575–607.

23. Traut TW: **The functions and consensus motifs of nine types of peptide segments that form different types of nucleotide-binding sites.** *Eur J Biochem* 1994, **222**:9–19.

24. Fischer D: **Hybrid fold recognition: Combining Sequence derived properties with evolutionary information.** *Pac Symp Biocomput* 2000:119–130.

25. Jones DT: **GenTHREADER: An efficient and reliable protein fold recognition method for genomic sequences.** *Protein Sci* 1996, **5**:947–955.

26. Karplus K, Barrett C, Hughey R: **Hidden Markov models for detecting remote protein homologies.** *Bioinformatics* 1998, **14**:846–856.

27. Salmeron JM, Oldroyd GED, Rommens CMT, Scofield SR, Kim HS, Lavelle DT, Dahlbeck D *et al.* **Tomato Prf is a member of the leucine-rich repeat class of plant resistance genes and lies embedded within the pto kinase gene cluster.** *Cell* 1996, **86**:123–133.

28. Umbarger HE: **The study of branched chain amino acid biosynthesis—Its roots and its fruits.** In: *Biosynthesis of Branched Chain Amino Acids*. Barak Z, Chipman DM, Schloss JV, eds. Weinham, Germany: VCH, 1990:1–24.

29. Pang SS, Duggleby RG: **Expression, Purification, characterization, and reconstitution of the large and small subunits of yeast acetohydroxyacid synthase.** *Biochemistry* 1999, **38**:5222–5231.

30. Barak Z, Clavo JM, Schloss JV: **Acetolactate synthase isozyme III from *Escherichia coli*.** *Methods Enzymol* 1988, **166**:455–458.

31. Vyazmensky M, Elkayam T, Chipman DM, Barak Z: **Isolation of subunits of acetohydroxy acid synthase isozyme III and reconstitution of the holoenzyme.** *Methods Enzymol* 2000, **324**:95–103.

32. Eoyang L, Silverman PM: **Role of small subunit (IlvN polypeptide) of acetohydroxyacid synthase I from *Escherichia coli* K-12 in sensitivity of the enzyme to valine inhibition.** *J Bacteriol* 1986, **166**:901–904.

33. Schloss JV, VanDyke DE, Vasta JF, Kutny RM: **Purification and properties of *Salmonella typhimurium* acetolactate synthase isozyme II from *Escherichia coli* HB101/pDU9.** *Biochemistry* 1985, **24**:4952–4959.

34. Fukudome K, Esmon CT: **Identification, cloning, and regulation of a novel endothelial cell protein C/activated protein C receptor.** *J Biol Chem* 1994, **269**(42):26486–26491.

35. Dahlback B: **The protein C anticoagulant system: inherited defects as basis for venous thrombosis.** *Thromb Res* 1995, **77**(1):1–43.

36. Aiach M, Borgel D, Gaussem P, Emmerich J, Alhenc-Gelas M, Gandrille S: **Protein C and protein S deficiencies.** *Semin Hematol* 1997, **34**(3):205–216.

37. Villoutreix BO, Teleman O, Dahlback B: **A theoretical model for the Gla-TSR-EGF-1 region of the anticoagulant cofactor protein S: From biostructural pathology to species-specific cofactor activity.** *J Comput-Aided Mol Des* 1997, **11**:293–304.

38. Villoutreix BO, Dahlback B: **Structural investigation of the A domains of human blood coagulation factor V by molecular modeling.** *Protein Sci* 1998, **7**(6):1317–1325.

39. Fischer D, Barret C, Bryson K, Elofsson A, Godzik A, Jones D, Karplus KJ, Kelley LA, MacCallum RM, Pawlowski K, Rost B, Rychlewski L, Sternberg M: **CAFASP-1:**

Critical assessment of fully automated structure prediciton methods. *Proteins* 1999, **Suppl 3**:209–217.

40. Fischer D, Elofsson A, Rychlewski L, Pazos F, Valencia A, Rost B, Ortiz AR, Dunbrack R Jr: **CAFASP2: The second critical assessment of fully automated structure prediction methods.** *Proteins* 2001, **45 Suppl 5**:171–183.

41. Bujnicki JM, Elofsson A, Fischer D, Rychlewski L: **LiveBench-1: Continuous benchmarking of protein structure prediction servers.** *Protein Sci* 2001, **10**(2):352–361.

42. Bujnicki J, Elofsson A, Fischer D, Rychlewski L: **LiveBench-2: Large-scale automated evaluation of protein structure prediction servers.** *Proteins* 2001, **45 Suppl 5**:184–191.

43. Rychlewski L, Jaroszewski L, Li W, Godzik A: **Comparison of sequence profiles. Strategies for structural predictions using sequence information.** *Protein Sci* 2000, **9**(2):232–241.

44. Shi J, Blundell TL, Mizuguchi K: **Fugue: Sequence-structure homology recognition using environment-specific substitution tables and structure-dependent gap penalties.** *J Mol Biol* 2001, **310**(1):243–257.

45. Kelley RM, MacCallum LA, Sternberg MJ: **Enhanced genome annotation using structural profiles in the program 3D-PSSM.** *J Mol Biol* 2000, **99**(2):499–520.

46. Pearson WR, Lipman DJ: **Improved tools for biological sequence comparison.** *Proc Natl Acad Sci USA* 1988, **85**(8):2444–2448.

47. Altschul SF, Gish W, Miller W, Myers EW, Lipman DJ: **Basic local alignment search tool.** *J Mol Biol* 1990, **215**(3):403–410.

48. Sali A: **Modeling mutations and homologous proteins.** *Curr Opin Biotechnol* 1995, **6**(4):437–451.

49. Johnson MS, Srinivasan N, Sowshamini R, Blundell TL: **Knowledge-based protein modeling.** *CRC Crit Rev Biochem Mol Biol* 1994, **29**:1–68.

50. Altschul SF, Madden TL, Schaffer AA, Zhang J, Zhang Z, Miller W, Lipman DJ: **Gapped BLAST and PSI-BLAST: A new generation of protein database search programs.** *Nucleic Acids Res* 1997, **25**(17):3389–3402.

51. Fischer D: **Modeling three-dimensional protein structures for amino acid sequences of the CASP3 experiment using sequence-derived predictions.** *Proteins* 1999, **Suppl 3**:61–65.

52. Duan J, Dahlback B, Villoutreix BO: **Proposed lipocalin fold for apolipoprotein M based on bioinformatics and site-directed mutagenesis.** *FEBS Lett* 2001, **499**(1-2):127–132.

53. Ng WV, Kennedy SP, Mahairas GG, Berquist B, Pan M, Shukla HD, Lasky SR, Baliga NS, Thorsson V, Sbrogna J, Swartzell S, Weir D, Hall J, Dahl TA, Welti R, Goo YA, Leithauser B, Keller K, Cruz R, Danson MJ, Hough DW, Maddocks DG,

Jablonski PE, Krebs MP, Angevine CM, Dale H, Isenbarger TA, Peck RF, Pohlschroder M, Spudich JL, Jung KW, Alam M, Freitas T, Hou S, Daniels CJ, Dennis PP, Omer AD, Ebhardt H, Lowe TM, Liang P, Riley M, Hood L, DasSarma S. **Genome sequence of Halobacterium species NRC-1.** *Proc Natl Acad Sci USA* 2000, **97**(22):12176–12181.

54. Fischer D: **Rational structural genomics: Affirmative action for ORFans and the growth in our structural knowledge.** *Protein Eng* 1999, **12**(12):1029–1030.

15

Structure Prediction Meta Server

Leszek Rychlewski

Abstract

Various fold recognition servers are accessible on the internet. All of them use different algorithms, scoring functions, and formats for presenting the results. All of them have different performance characteristics, different sensitivity, specificity and reliability cutoffs. All of them are known to produce in rare cases false positive predictions with scores above the reliability cutoff provided by the authors. A common procedure to increase the confidence of predictions is to confirm them with independent predictions obtained from unrelated servers. The Structure Prediction Meta Server facilitates this procedure. It collects models from many high quality services and translates them into standard formats enabling convenient analysis of the results. In addition, the Meta Server offers an infrastructure for the creation of automated jury algorithms, which analyze the set of results for the user and calculate the reliability score for a consensus prediction.

15

Structure Prediction Meta Server

Leszek Rychlewski

BioInfoBank Institute, Poznan, Poland

15.1. Introduction

The 3D structure of the protein provides the biologist with a potentially very powerful source of information about the function of the protein. The knowledge of the structure of the target protein is a major requirement for the understanding of the effects of mutations, for the rational modification of the target, and for the design of proteins with altered or novel functions. The experimental methods used to determine the structure of proteins are costly and slow. They have little chance of keeping up with the growing demand for structures caused by the massive genome sequencing efforts, and the resulting flood with amino acid sequences of proteins. The prediction of the structure using computational methods presents a complementary alternative to the experimental determination. The main obstacle of the utilization of the computational approach is the limited quality of produced models. This problem has been addressed for over 30 years and little success has been achieved in the struggle to improve significantly the reliability and performance of the prediction procedures.

The most robust way of predicting the structure of a protein based on its sequence alone is by inference, using the structure of an evolutionary related cousin, a homologous protein.[1] The difficulty of this procedure depends on the level of similarity between both representatives of the family. In some cases the similarity is very profound and standard sequence alignment shows extended similarity, i.e., many identical residues on large parts of the aligned amino acid sequence of both proteins. In other cases, when the similarity moves into the twilight zone, the alignment is much more complicated, resulting in the ambiguity of the prediction. Finally, for some proteins, standard sequence comparison methods will not detect any proteins with known structure in available databases. Such cases represent an application area for programs aimed to detect distant sequence similarity, fold recognition methods based on sequence-structure fitting functions (threading[2] or hybrid methods)[3] or "*ab initio*"[4,5] folding algorithms. The main difference between the first two (fold recognition methods) and the last method (*ab initio*) is the set of available structures. The fold recognition methods provide alignments to templates in the databases and extract large pieces of the structure corresponding to parts of the proteins, which seem to be conserved between the target and the template. The *ab initio* methods generate *de novo* structures and evaluate the fitness between the sequence of the target and the generated structure in a long iterative process. Despite visible progress, their reliability is currently below the threshold allowing simple practical application.

Current fold recognition procedures follow several standard approaches and utilize various recurring components including initial secondary structure prediction,[6] the incorporation of evolutionary information transformed into sequence profiles[7,8] or use scoring functions (potentials) developed for threading procedures based on propensities of amino acids to engage in contacts with each other and with the solvent.[9] The main differences between most methods are not new fundamental ideas about the prediction process, but much more the details of implementation and coupling of the components. The correct way to combine all modules remains however an art, which is being developed by leading scientific groups in an exhausting manual labor.

The potentially large amount of combinations of components, which can be utilized by new fold recognition procedures, represents the biggest obstacle in the development plans. One of the possible solutions of this problem would be the automation of this process. A preliminary draft of a machinery aimed at this task could include three devices. The first device would provide new modules and procedures for the prediction of partial structural information. The modules would be then tested and their individual utility would be evaluated by the second device. The third device would combine the modules using statistical procedures, neural networks or jury systems into a superior prediction engine. All devices would be interconnected using standard data transmission formats.

Except for the first, all other devices are already in the development stage. The structure prediction Meta Server[10] represents the framework for all of them. The second device is the subject of several evaluation programs, including CAFASP,[11] ToolShop[12] or LiveBench.[13] An attempt to design the third device resulted in the development of Pcons,[14] a consensus structure prediction server. The first device is still missing for obvious reasons, and currently the prediction procedures are provided by the research teams in the form of servers. It will probably take a long time before computer programs will achieve the creativity of human experts.

15.2. The Meta Server

The main goal of the Meta Server (see http://bioinfo.pl/meta) is to facilitate data exchange between prediction servers, evaluation programs and visualization modules. The general model of the information flow and the components of the Meta Server are presented in Fig. 15.1.

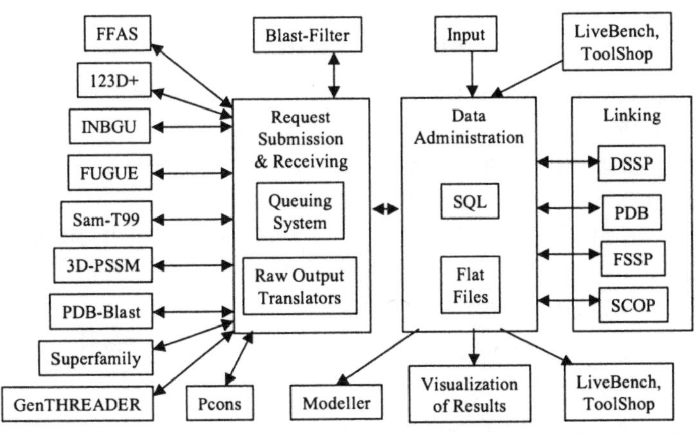

Figure 15.1. The protein structure prediction Meta Server. The components are described in details in the text.

The prediction process starts with the submission of the sequence of the target protein using an HTML form provided by the WWW server. Initiation of the prediction process by automated programs is also implemented and used by the evaluation programs LiveBench and ToolShop. The processing of the structure prediction is conducted using the following components:

15.2.1. User Input and Job Status Display

The input page of the Meta Server requests several basic parameters including the E-mail of the user, the name of the job and the sequence of the target proteins. The user also has the option of skipping several of the coupled structure prediction servers. The input is collected by a PERL script, which sends the information via E-mail to the main server. The script adds information about the remote host (IP and hostname) to the data sent to the main server. The E-mail based communication allows the mirroring of the input script on many machines.

The WWW server also provides a page for the administration of previously submitted prediction jobs. The page enables regular expression queries on several fields maintained in the dedicated SQL table. The available fields include the E-mail address of the user, the host name of the computer from which the request was initiated and the job name of the request assigned by the user. The search is conducted using an SQL server, which is a part of the data administration component.

15.2.2. Job Deposition and Administration

The job manager is responsible for the administration of the data stored for all predictions. The manager is based on a combination of a flat file repository and an SQL-driven database engine. The databases engine conveniently solves several problems. It uses only one table for the administration of all jobs. The engine is responsible for the assignment of unique identifiers to new jobs based on an auto-increment field in the table. Other fields include:

- 'dir': a directory name where all job-related files can be found;
- 'len': the length of the sequence;
- 'born': the date of the submission of the target;
- 'host': the host name (or host IP if the host name is not available);

- 'job': the job provided by the user;
- 'seq': the sequence of the target protein;
- and a set of fields describing the status of the request for all coupled server.

The SQL server enables a relatively fast lookup of some basic information about each job, which is used by the job status display script. Other information is stored in the form of flat files in a sub-directory tree, wherein each sub-directory contains the data related to one target. The information collected from prediction servers and their translation requires a relatively large amount of disk space. Because of this, the data management is divided between the SQL database engine and a directory structure. The incorporation of the data stored in flat files into the SQL database would greatly decrease the speed of access to the fileds in the SQL tables. It would also quite quickly exhaust the table size limits on the database server.

The raw results obtained from the prediction servers are stored in the job-specific directory using a server-specific filename (see Table15.1) without a suffix. The files with suffixes contain the following information specific for each prediction server:

- '.align': an alignment of the template sequence to the target sequence (one line per template);
- '.start': the position of the first aligned residue from the template (one line per template);
- '.name': the PDB code of the template (one line per template);
- '.scop': the SCOP ID of the template or more IDs if the aligned part of the template contains several domains (one line per template);
- '.score': the alignment score of each hit returned by the prediction server (one line per template);
- '.ss': the secondary structure pattern of the aligned template (one line per template);
- '.stime': the time and log of the request submission to the prediction server;
- '.casp': the returned models translated into CASP[15] "pfrmatal" format;
- '.pdb': the returned models translated into PDB (CASP "pfrmatts") format.

The files are created after the raw output was received from the prediction server and after it was translated.

15.2.3. Request Submission Queuing

As described earlier, the SQL table contains a set of fields specific for each prediction server describing the status of the prediction request. These fields are set automatically when a new target arrives based on the selection of the user. For example, if the user wishes to skip some services, the status field for this service will be updated accordingly. The user can also modify the fields manually in case the Blast-Filter stopped the submission of the request to prediction servers (see below). The administrator also has the option of resubmitting previous requests in case of errors or inaccessibility of the prediction servers, or if translation problems of the raw output have been encountered.

The submission of the requests to structure prediction servers is handled by a CRON job initiated every 15 minutes. The 15 minute delay is introduced as a security measure which prevents the flooding of remote servers with prediction requests as a result of erroneous submission loops that have occurred in the past. This delay results in a upper limit of submission frequency of four requests per hour.

The periodic CRON job verifies the status of all target submission requests separately for all prediction servers. The server-specific status can have several values:

- 'skipped,' the submission to this server was skipped;
- 'wait,' the request is still being processed by the Blast-Filter;
- 'queued,' the request is queued and will be send to the prediction server;
- 'busy,' the request was send to the prediction server and no correct result was received yet;
- 'error,' an error occurred in the submission, in the translation of the raw output or the prediction timed out after 24 hours of waiting for the results;
- 'ready,' the prediction by this server is ready.

For each server only one 'busy' request is allowed, except for servers which are known to be able to operate in parallel fashion, i.e., Pcons2 enables the parallel processing of up to three requests.

The queuing system sets a higher priority to requests submitted be users than by automated programs, like the LiveBench program or the ToolShop evaluation.

15.2.4. Blast-Filter

The newly inserted sequence of the target protein is sent to the Blast-Filter. The Blast-Filter is responsible for the filtering of trivial predictions. The filter uses a simple sequence comparison (using Blast)[16] between the target sequence and proteins with known structure (PDB database)[17] downloaded weekly from RCSB (http://rcsb.org). If a hit was found with an E-value below 0.001 then the prediction is aborted. The Blast-Filter notifies the user via provided E-mail about the trivial hit. The sequence is nevertheless left in the job database. The user has the option of overwriting the decision of the filter and of submitting the target sequence to the coupled prediction servers, using buttons (or links) displayed on the job result page.

15.2.5. Local and Remote Prediction Services

The submission of the request to a prediction service is formatted according to the requirements of the remote server. In some cases the submission is skipped because a target sequence is too long (several servers impose a limit of 800 amino acids) or due to access restrictions introduced by the servers (the GenTHREADER family[18] which block requests from non academic sites if no license was obtained). The current list of independent (non-consensus) fold recognition servers is shown in Table 15.1.

Table 15.1. List of fold recognition servers coupled to the Meta Server. The left column shows the name of the server and the file name (in brackets) used to store the raw results in the data depository and as field in the SQL table.

NAME (filename)	DESCRIPTION
PDB-Blast (pdbblast)	PDB-Blast is based on the PSI-Blast[19] program. Before the fifth iteration the sequence profile is saved and used to scan the database of proteins with known structure (proteins in PDB from last week). A locally installed version of the service is being used.

Continued on next page

NAME (filename)	DESCRIPTION
FFAS (burnham)	FFAS[8] is based on comparing sequence profiles with each other. Profiles are generated for protein families in a different way than by PSI-Blast but PSI-Blast is used to collect the proteins of a family. The official version at the Burnham Institute is used.
OLD-FFAS (ffas)	The old, not official, local version of FFAS is also coupled to the Meta Server for historical reasons.
Sam-T99 (samt99)	Sam-T99[20] builds a multiple alignment (the Sam-T99 alignment) by iterated search using hidden Markov models. Uses the alignment to predict secondary structure (using various methods) and to build an HMM used for searching PDB for similar proteins. Also, a library of HMMs built by similar methods from PDB sequences is used to score the target sequence.
3D-PSSM (foldfit)	3D-PSSM[21] is based on a threading approach using 1D and 3D profiles coupled with secondary structure and solvation potential. The current version of this server uses a fold library, which is automatically updated every week.
GenTHREADER (genthreader) mGenTHREADER (mgenthreader)	GenTHREADER[18] is using a combination of various methods including sequence alignment with structure based scoring functions and it uses a neural network based jury system to calculate the final score for the alignment. mGenTHREADER takes as input an automatically generated PSI-BLAST profile instead of the sequence.
INBGU (inbgu)	INBGU[22] is a combination of five methods, which exploit sequence and structure information in different ways and produces one consensus prediction of the five. It uses predicted versus observed secondary structure and sequence profiles for both the target and for the folds in the library. The evaluated results are taken from the consensus prediction only.
ShotGun-INBGU (shgu)	This server uses the same prediction components as INBGU. The consensus of the prediction results is built in a novel way.
FUGUE (fugue)	Environment-specific substitution tables were derived from the structure-based alignments in the HOMSTRAD database. Each alignment in HOMSTRAD was converted into a scoring template (profile) using the environment-specific substitution tables and environment-dependent gap penalties. The program FUGUE[23] searches a sequence or sequence alignment against the library of profiles.
123D+ (123d)	123D+[24] is a program, which combines substitution matrix, secondary structure prediction, and contact capacity potentials to thread a sequence through the set of structures. Contact capacity potentials reflect mainly hydrophobicity of the amino acids. Hydrophobicity is the major driving force for protein folding. Without pair-wise contact potentials it can use a simple and fast dynamic programming algorithm to align a sequence with a structure.

Continued on next page

NAME (filename)	DESCRIPTION
SUPERFAMILY (superfamily)	The search method[25] uses a library (covering all proteins of known structure) consisting of 820 (SCOP) superfamilies, each of which is represented by a group of hidden Markov models. Each model is created from a seed sequence which is aligned to many superfamily homologues; the model is built from the alignment. A hit to a model is not a hit to the seed but is a hit to the superfamily, which represents the model.

Additional set of servers include several secondary structure prediction services, i.e., PSI-Pred,[26] Jpred,[2,27] and the secondary structure prediction module from SAM-T99. They are handled in a similar fashion as fold recognition servers. Other more sophisticated and popular meta services are available for local structure prediction. Nevertheless the local structure prediction was included to enable the comparison of predicted secondary structure patterns with patterns extracted from aligned structural templates on the results pages produced by the Meta Server.

The Meta Server offers special treatment for remote consensus fold recognition servers. Pcons, the only currently available consensus server, requires the output of several independent (non-consensus) fold recognition server to create a consensus prediction. The newer version of Pcons (Pcons2) is able to utilize for example PDB-Blast, FFAS (the old and the official version), 3D-PSSM, GenTHREADER (both versions), INBGU, SAM-T99, and FUGUE. The Meta Server places a request to Pcons into its queue if at least three of the utilized individual servers have returned results and they were translated into PDB format. Each new result received from a prediction server, which can be utilized by Pcons, triggers the queue placement again. A repetitive queue placement modifies the same field, thus if the request was not processed before it will be submitted only once.

15.2.6. Raw Output Converters

Most prediction servers coupled to the Meta Server return results in their own specific format (raw format). The authors of the Meta Server did not succeed yet in the wide promotion of a unified form of presenting prediction results. But they have succeeded in convincing the prediction server developers to send all necessary results in one E-mail message. This greatly simplified the communication between prediction providers and the Meta Server (for which they are very thankful.)

The raw format converters are responsible for the translation of the results of predictions received as an E-mail message into a uniform presentation. There are currently two types of results, consisting of alignments to templates or structures of models:

- The first, and most popular, result presentation includes alignments to templates available from the public database of proteins with known structure (PDB). This type of result includes the alignments, their scores and the PDB codes of the templates. The Meta Server uses only a maximum of 10 hits for further processing. The main problem encountered in the processing is due to ambiguous presentation of protein sequences in PDB files and in the resulting discrepancies in the sequences of templates stored in template databases of the prediction servers. The alignment to the template reported by the server has usually the correct target sequence, but the sequence of the template may differ between the sequence used by the Meta Server and by the prediction server. Because of this problem, the sequence of the template provided by the server has to be mapped onto the sequence of the PDB protein stored in the Meta Server databases. After this mapping additional analysis available for the PDB proteins can be conducted. The PDB template can be linked with external databases providing secondary structure assignment, structural classification or additional functional annotation.
- The second type of results returned by servers includes the coordinates of the models. Only one coupled server, ShotGun-INBGU, provides this information as the only available output. On one hand this format would greatly simplify the process of obtaining the structure of the protein based on its sequence. On the other hand it makes additional processing much more complex. Using coordinates of C-alpha alone it is difficult to say what PDB template was used, if any. This makes it more difficult to add annotation of the secondary structure or structural classification of the hit. Although ShotGun-INBGU helps in finding the template by providing this information in the output, it often uses multiple templates. To extract an alignment between the target protein and the template, a structural alignment of the model with the template

structure would be necessary. This is currently not implemented. If no information about the template is provided in the prediction output, the Meta Server could still try to find a matching structure by running a structural comparison between the returned model and all known structures. This procedure is unfortunately very time consuming. Because of the mentioned problems, the further processing of this type of server output is currently very limited.

The alignment to the template structure is very helpful for the verification of the hits using biological hints. In many cases for example, the conservation of few crucial active site residues can provide additional arguments for the correctness of the model. On the other hand, '*ab initio*' methods do not use any templates in their structure prediction procedures. They can however provide useful models in cases where fold recognition methods fail, as was shown during the last CASP4 meeting. Because of this, further development of the procedures for the analysis of predicted models is planned.

15.2.7 Visualization and Linking

The visualization of prediction results uses HTML as viewing language and is provided in the same fashion for all prediction servers. For each, up to 10 (some return less) alignments to PDB templates are displayed. If a secondary structure assignment is available for the PDB template, the alignment is displayed in a form where residues in helices are colored red and residues in beta sheets are colored blue. The alignments displayed currently for servers returning only model coordinates contain the target sequence instead of the template sequence and no secondary structure assignment is provided.

For each hit the score and the PDB code of the template is shown (if available). The PDB code is used by a PDB-FSSP[28] translator to display the structural classification of the PDB template based on FSSP, if it is available for the template. A second translator maps the PDB code to the SCOP[29] structural classification of protein domains. In many cases the mapping is ambiguous due, for example, to the fact that the template protein has several domains with different SCOP classification. In such cases the aligned sequence of the template is used to conduct a comparison with the sequences of domains stored in the SCOP databases. If the target protein has several domains it can still result in distinct hits. In such cases the strongest hit to a SCOP domain is reported first and displayed. The others are stored and displayed in the status of the browser when the user points the mouse on the visible SCOP classification.

The produced HTML result page includes links to the raw output of the server and to the translation of the output into PDB ('pfrmatts') and CASP ('pfrmatal') format. In addition, for each hit a link to the alignment in FASTA format and to the PDB model is displayed. Both are generated on-the-fly if requested.

15.2.8. Interfaces

The Meta Server provides interfaces for the communication with various other programs or services, which use stored partial results. The Meta Server interacts with the LiveBench program, with a Consensus Prediction Methods (Pcons) and with a modeling service based on Modeller-3.[30]

- LiveBench interaction involves the submission of new LiveBench jobs to the Meta Server. The Meta Server submits the weekly selected targets to the prediction services and stores the results locally. The results are further submitted to three model evaluation services: MaxSub,[31] LGscore,[32] and Touch (L. Rychlewski, unpublished). The results are collected again and presented to the evaluation procedures in LiveBench.
- Pcons, as mentioned above, relies on predictions obtained by other prediction methods. The Pcons interface collects the predictions of selected methods, translates them into PDB format and sends them to Pcons. The processing of Pcons results is conducted in a way similar to the other prediction servers. The difference is that Pcons does not return alignments but ranks the obtained models using its own criteria providing new scores for each. The result of Pcons is only a list of new scores. The original alignments and other related information is taken from the original results obtained from the independent prediction servers.
- The Modeller interface prepares input data for the Modeller-3 program. The program is aimed at improving crude models generated by extracting aligned parts from the structures of the templates. Modeller-3 is unfortunately a commercial program and this inter-

face is not available for public use outside the institution hosting the Meta Server.

15.3. Discussion

The Meta Server has various applications:

- The Meta Server provides an infrastructure for the data exchange between various components of the development process of proteins structure prediction methods. It provides interfaces for automated evaluation programs and enables the construction of accurate jury systems.
- The Meta Server is a powerful and very convenient tool for biologists interested in structural analysis of individual proteins. The biologists can obtained an overview of the predictions obtained with the most accurate methods in the field. The current benchmarking results obtained by closely related evaluation programs (LiveBench or ToolShop) help in the estimation of the reliability of the predictions.
- The Meta Server is completely automated and can be used as a screen for proteins intended to be placed in the Structural Genomics pipeline. Proteins, for which the Meta Server can not detect reliable fold classification, could be treated with priority in the structure determination process.

The Meta Server offers wide application possibilities and visible benefits. The standardized interface structure represents an example. Remote service providers are able to use the Meta Server infrastructure to offer their services as additional processing options. This will promote the development of prediction procedures, which provide significant added value or complementary expertise. The bioinformatics research teams will be able to focus on method development: while the complex communication infrastructure used to transmit data between the user and the central server will be available from the Meta Server development team. The team can also provide standardized benchmarks with training and testing sets and predictions results obtained from other methods, which is currently happening in the framework of the LiveBench and

ToolShop program. The team could also offer continuously updated template sets for remote prediction runs. Thus, even if the Meta Server is based on simple software modules, which solve rather trivial problems, it can make significant impact on the development process of structure prediction methods and contribute this way to the scientific progress in the structure prediction community.

The advantage of the Meta Server for the biologist is straightforward. No time consuming downloading and local installation of prediction methods is necessary. No local updates are needed. The prediction service is available from any computer connected to the global Internet and the large computer-time requirements are covered by specialized centers. Another important contribution of the Meta Server is that it may change the attitude of users toward online processing of biological information. Currently most of the companies strongly oppose sending proprietary sequence information outside internal firewalls. The blurring differences between non-profit research centers and for-profit institutions will result in an increasing fraction of academic scientists concerned with data security. The future plans for the Meta Server include secure access and secure processing of information. The secure version of the Meta Server, will promote the remote, on-line biological data processing by attracting users with the most accurate prediction services and with high security of data transmission. Simplified and convenient access to the latest achievements in protein structure prediction will certainly result in a wider application of concepts of structural biology in daily biological labor.

References

1. Wood TC, Pearson WR: **Evolution of protein sequences and structures.** *J Mol Biol* 1999, **291**:977–995.

2. Jones DT: **Theoretical approaches to designing novel sequences to fit a given fold.** *Curr Opin Biotechnol* 1995, **6**:452–459.

3. Jaroszewski L, Rychlewski L, Zhang B, Godzik A: **Fold prediction by a hierarchy of sequence, threading, and modeling methods.** *Protein Sci* 1998, **7**:1431–1440.

4. Bonneau R, Strauss CE, Baker D: **Improving the performance of Rosetta using multiple sequence alignment information and global measures of hydrophobic core formation.** *Proteins* 2001, **43**:1–11.

5. Ortiz AR, Kolinski A, Rotkiewicz P, Ilkowski B, Skolnick J: ***Ab initio* folding of proteins using restraints derived from evolutionary information.** *Proteins* 1999, **Suppl 3**:177–185.

6. Rost B: **Predicting one-dimensional protein structure by profile based neural networks.** *Methods Enzymol* 1996, **266**:525–539.

7. Gribskov M, McLachlan AD, Eisenberg D: Profile analysis: detection of distantly related proteins. *Proc Natl Acad Sci USA* 1987, 84:4355-4358.

8. Rychlewski L, Jaroszewski L, Li W, Godzik A: **Comparison of sequence profiles. Strategies for structural predictions using sequence information.** *Protein Sci* 2000, **9**:232–241.

9. Jones DT, Thornton JM: **Potential energy functions for threading.** *Curr Opin Struct Biol* 1996, **6**:210–216.

10. Bujnicki JM, Elofsson A, Fischer D, Rychlewski L: **Structure prediction MetaServer.** *Bioinformatics* 2001, **8**:750–751.

11. Fischer D, Barret C, Bryson K, Elofsson A, Godzik A, Jones D, Karplus KJ, Kelley LA, MacCallum RM, Pawlowski K, Rost B, Rychlewski L, Sternberg M: **CAFASP-1: Critical assessment of fully automated structure prediction methods.** *Proteins* 1999, **Suppl 3**:209–217.

12. Rychlewski L: **ToolShop: Prerelease inspections for protein structure prediction servers.** *Bioinformatics* 2001, **17**(12):1240–1241.

13. Bujnicki JM, Elofsson A, Fischer D, Rychlewski L: **LiveBench-1: Continuous benchmarking of protein structure prediction servers.** *Protein Sci* 2001, **10**:352–361.

14. Lundstrom, Rychlewski L, Bujnicki JM, Elofsson A: **Pcons: A neural network based consensus predictor that improves fold recognition.** *Protein Sci* 2001, **10**(11):2354–2362.

15. Moult J, Hubbard T, Fidelis K, Pedersen JT: **Critical assessment of methods of protein structure prediction (CASP): Round III.** *Proteins* 1999, **Suppl 3**:2–6.

16. Altschul SF, Gish W, Miller W, Myers EW, Lipman DJ: **Basic local alignment search tool.** *J Mol Biol* 1990, **215**:403–410.

17. Berman HM, Westbrook J, Feng Z, Gilliland G, Bhat TN, Weissig H, Shindyalov IN, Bourne PE: **The Protein Data Bank.** *Nucleic Acids Res* 2000, **28**:235–242.

18. Jones DT: **GenTHREADER: An efficient and reliable protein fold recognition method for genomic sequences.** *J Mol Biol* 1999, **287**:797–815.

19. Altschul SF, Madden TL, Schaffer AA, Zhang J, Zhang Z, Miller W, Lipman DJ: **Gapped BLAST and PSI-BLAST: A new generation of protein database search programs.** *Nucleic Acids Res* 1997, **25**:3389–3402.

20. Karplus K, Barrett C, Cline M, Diekhans M, Grate L, Hughey R: **Predicting protein structure using only sequence information.** *Proteins* 1999, **Suppl 3**:121–125.

21. Kelley LA, McCallum CM, Sternberg MJ: **Enhanced genome annotation using structural profiles in the program 3D-PSSM.** *J Mol Biol* 2000, **299**:501–522.

22. Fischer D: **Hybrid fold recognition: combining sequence derived properties with evolutionary information.** *Pac Symp Biocomput* 2000:119–130.

23. Shi J, Blundell TL, Mizuguchi K: **Fugue: Sequence-structure homology recognition using environment-specific substitution tables and structure-dependent gap penalties.** *J Mol Biol* 2001, **310**:243–257.

24. Alexandrov NN, Nussinov R, Zimmer RM: **Fast protein fold recognition via sequence to structure alignment and contact capacity potentials.** *Pac Symp Biocomput* 1996:53–72.

25. Gough J, Karplus K, Hughey R, Chothia C: **Assignment of homology to genome sequences using a library of Hidden Markov Models that represent all proteins of known structure.** *J Mol Biol* 2001, **313**:903–319..

26. Jones DT: **Protein secondary structure prediction based on position-specific scoring matrices.** *J Mol Biol* 1999, **292**:195–202.

27. Cuff JA, Clamp ME, Siddiqui AS, Finlay M, Barton GJ: **JPred: A consensus secondary structure prediction server.** *Bioinformatics* 1998, **14**:892–893.

28. Holm L, Sander C: **Dali/FSSP classification of three-dimensional protein folds.** *Nucleic Acids Res* 1997, **25**:231–234.

29. LoConte L., Ailey B, Hubbard TJ, Brenner SE, Murzin AG, Chothia C: **SCOP: A structural classification of proteins database.** *Nucleic Acids Res* 2000, **28**:257–259.

30. Sanchez R, Sali A: **Evaluation of comparative protein structure modeling by MODELLER-3.** *Proteins* 1997, **Suppl 1**:50–58

31. Siew N, Elofsson A, Rychlewski L, Fischer D: **An automated measure to assess the quality of protein structure prediction.** *Bioinformatics* 2000, **9**:776–85.

32. Cristobal S, Zemla A, Fischer D, Rychlewski L, Eloffson A: **A study of quality measures for protein threading models.** *BMC Bioinformatics* 2001, **2**:5

PART II

Methods of Structure and Sequence Alignment

- **Pcons: Consensus Fold Recognition**
- **Protein Fold Space & Sequence-Structure Relationships**
- **A Flexible Method for Structural Alignment**
- **Multiple Structure Alignment**
- **Protein Kinase Structural Alignments**

16

Improved Fold Recognition by Using the PCONS Consensus Approach

Huisheng Fang, Björn Wallin, Jesper Lundström, Christer von Wowern, and Arne Elofsson

Abstract

In the CASP and CAFASP processes, it has been shown that manual experts are better at predicting the fold of an unknown protein than fully automated methods. The best manual predictions seem to be performed by authors using a wide-range of different methods; and the most obvious similarity between them is that they have worked on fold recognition for years. Several of these experts also develop methods, however, these methods do not perform as well as the experts themselves. What are the secrets that the manual experts possess, but are not able to put into a computer?

Here we show that one such secret is the use of a "consensus" approach in fold recognition. Using several different methods, the same method with different parameters, or searching using several homologous sequences, allows a "consensus" prediction to be made. The consensus analysis can also be done, using only a single sequence and a single method, by searching for similar hits among the top-scoring hits. In contrast, most automatic methods only use a single sequence or a single set of parameters and do not use the top-scoring hits to search for "consensus" predictions.

Here, we will describe a new method for fold recognition, Pcons, that utilizes the "consensus analysis" to improve automatic fold recognition. We will describe the process behind the development of Pcons, starting with the use of a "semi-automatic" method in CASP4, and the later development of the fully automated Pcons method. We will show some results from large scale benchmarking that show the advantages of Pcons. Finally we will describe some recent developments that have improved the performance of Pcons further.

16

Improved Fold Recognition by Using the PCONS Consensus Approach

Huisheng Fang, Björn Wallin, Jesper Lundström, Christer von Wowern, and Arne Elofsson

Stockholm Bioinformatic Center, Stockholm University, Stockholm, Sweden

16.1. Introduction

As the genome projects proceed, we are presented with an exponentially increasing number of protein sequences, but with only a very limited knowledge of their structure or function. Since the experimental determination of the structure or the function is a nontrivial task, the quickest way to gain some understanding of these proteins and their genes is by relating them to proteins or genes with known properties.

From other examples in this book it is clear that there are many different fold recognition methods. Different methods are based on single sequences,[1,2] multiple sequence alignments or profiles,[3-6] and predicted[7-10] or experimentally determined[11] structures. It is not clear which features are most important and how they should best be combined. However, what is noticed is that it seems to be of great importance, with detailed choices of parameters, to get the best performance.

Several different means of benchmarking the performance of these methods has been developed, including large scale benchmarks,[12-15] blind-predictions,[16-18] and automatic benchmarking of all newly solved protein structures.[19,20] Several groups have also benchmarked the alignment quality for different fold recognition methods.[19,21-23]

The large scale benchmarks were based on databases of structurally related proteins, such as SCOP[24] and CATH.[25] In these benchmarks it was concluded that the use of fraction identity as a measure of similarity between two proteins should be abandoned.[12] It is much better to use statistical measures, such as E- and P-values used in BLAST[4] and FASTA.[26] In these studies the performance of several different methods was compared; and methods that use multiple sequence information perform better than methods using only single sequence information.[14,23] Different studies gave slightly different results on how to utilize the multiple sequence information best. However, it was clear that PSI-BLAST[4] performed very well and was one of the computationally most efficient ways to use multiple sequence information.

It was also observed that different methods perform best at different levels of relationship.[23] For proteins that only share the same fold, i.e., do not have a common ancestor, it is better to use methods that completely ignore the sequence; while for protein that are more closely related it is better to utilize the sequence information.

In later studies other fold recognition methods were also included in large-scale benchmarks. Here, it was observed that several fold recognition methods perform better than PSI-BLAST when detecting distantly related proteins.[19,20,23] Several, but not all, of these methods use structural information. The structural information can be included in several different ways. INBGU[27] and 3D-PSSM[10] use predicted secondary structures, 3D-PSSM and FUGUE[28] use structural alignments, and GenTHREADER uses a threading potential. In addition, at least two methods that only use sequence information, FFAS[6] and SAM-T99,[5] seem to perform better than PSI-BLAST. It was also noted that no method could reliably distinguish between weak correct hits and wrong hits.[19]

In the LiveBench studies it was observed that a correct prediction often is obtained from one method.[19] Further, some studies have also focused on the quality of the generated alignment.[19,21-23] An important conclusion from these studies is, that for different targets, the best predictions are often made by different methods. It is quite common for a sin-

gle method and pairs of distantly related proteins, to find that the optimum choice of alignment parameters differs from case to case. Thus, when evaluating the accuracy of a structure prediction protocol on a large set, it is quite clear that it's performance could be increased if different, best suited approaches could be applied in appropriate cases.
The exact choice of parameters such as gap-penalties is of great importance for the performance of a method. Therefore, the development of better prediction methods is an art that only a few groups master.

16.2. Why are Manual Predictions Better?

How can these observations be used for making automatic fold recognition prediction as good as predictions of experts? To answer these questions, we need to try to understand what knowledge experts use that is not used by the servers (developed by them and others). We believe that there are three important contributions by the experts. These are biological knowledge, structural verifications and consensus analysis. Our recent studies indicate that, of these three factors, the last one is most important; and that incorporating it into an automated method provides a significant improvement.

16.2.1. Biological Knowledge

If a protein is known to be a DNA-binding protein, any high scoring hit to a DNA-binding domain would get the attention of an expert. Due to the current limitations in computer-readable classifications of protein functions this knowledge is hard to automatize. However, some attempts have been made. The SAWTED algorithm[29] searches for related keywords in SWISS-PROT[30] between two proteins. SAWTED is utilized by 3D-PSSM and was shown to increase its performance by a few percentages (MacCallum, personal communication). It is difficult to judge how much biological knowledge actually improves fold recognition, but certainly in some cases it can be important.

16.2.2. Structural Analysis

As good as biological knowledge can be, manual experts can also use structural knowledge. For instance, it is known that secondary structure prediction algorithms are better at predicting secondary structures than fold recognition methods. Therefore, if a model clearly has four helixes, a hit to an immunoglobulin can easily be disregarded.

Structural information can be used either directly in the alignment algorithm, as in INBGU, or as a post-processing filter, as in GenTHREADER.[31] The difference is that, when used in the alignment algorithm, a different alignment is obtained; but when used as a post-processing method, only the score of a particular alignment is influenced by the structural information. Two different types of information have successfully been used in fold recognition methods, predicted secondary structures and residue contact information.

Intuitively it seems as if a post-processing filter might be most useful to deselect false positives. This also seems to be correct as GenTHREADER, the only method that use a post-processing filter, has a very good specificity.[20] The inclusion of structural information in the alignment procedure might instead improve the alignment of distantly related proteins.

16.2.3. Consensus Analysis

A common trick used by fold recognition experts is to use what could be referred to as a consensus analysis. Here, not only one prediction for each target is considered. Instead, models from different predictions with similar scores, using different parameters, from different methods, or for homologous sequences, are taken into account. In contrast, an automatic fold recognition method returns a list of hits and, when the performance is measured, only the single highest scoring hit is used. Until the introduction of Pcons we were not aware of any method that utilized this type of information.

What is an expert actually doing when he examines several hits, and why can it be used to increase the performance? One obvious feature is that using several parameters increases the possibility to create at least one good model. A method can create several predictions for each target-template pair and then use the one with the highest (normalized) score. Some groups have already used a simple form of consensus predictors, where several models are created for each sequence-template pair. The INBGU method performs five alignments using combinations of single sequence and profile data.[27] The 3D-PSSM method performs three alignments for each sequence-template pair.[10] In both INBGU and 3D-PSSM all alignments are made using predicted secondary structure information for the query sequence and the experimentally determined secondary structure of the template protein. The alignments of INBGU are made using either single sequence or multiple sequence information of the query and the template. In 3D-PSSM, two alignments are made using the query sequence and two different template profiles; one is derived from a superfamily-wide structural alignment, the third alignment uses the template sequence and a profile obtained from the query sequence. For each query-template pair, these methods choose one alignment. 3D-PSSM chooses the highest scoring one, while INBGU also takes the rank into account. However, these methods only consider the different templates individually, while manual experts often examine multiple hits. Therefore, if hits 2 to 9 all are Tim-barrels, but the first hit is something else, then the manual expert would guess on a Tim-barrel structure while an automatic method would not.

Here we describe the process leading to the development of a consensus method that tries to mimic the work of an expert, Pcons. We show recent results from the LiveBench process that benchmarks the performance of Pcons and other fold recognition methods.[19,20] Finally, we will describe some current attempts to increase the performance of Pcons.

16.3. Consensus Predictions in CASP4

During the CASP4 process it was realized that the meta-server, described in another chapter, gave all participants in CASP4 the possibility of easily using a large set of fold recognition methods. Several of the top-performing groups in CASP4 utilized the results from the meta-server. However, they also used manual knowledge and other methods. In contrast, together with Daniel Fisher and Leszek Rychlewski, we wanted to examine if an automatic consensus prediction would perform better than the single servers and possibly as good as the manual predictors.

During CAFASP we were not able to create a fully automated consensus server; therefore, we used a semi-automatic procedure to perform the consensus predictions. These were submitted to CASP as the CAFASP-CONSENSUS semi-automatic predictions. Since the end of CAFASP we have developed the fully automatic consensus predictor Pcons. The process of both the automatic and manual consensus predictions can be described in three steps:

- First, predictions are obtained from a set of web-servers using the meta-server.[19]
- The second step is to identify related predictions. In the manual predictions, this was simply done by listing the number of predictions for a particular SCOP fold. In the automated predictions, the number of other models that were similar to a particular model was listed.
- In the third and last step the goal is to select one model out of all the models from the servers. In the automated consensus predictor, this is done by a neural network that combines the particular score of a model with the information about the number of other similar models. For the manual consensus predictions, we used the same information as an input to our human neural networks.

In CASP4 we detected three scenarios: The first scenario consists of trivial predictions, where most methods predict the same fold with significant scores. In this case we picked one of these predictions more or less randomly. The second scenario was when no servers gave any significant hits, but that a particular SCOP fold was clearly most frequent among the top hits. In this case we selected one prediction from this fold. In the last scenario there were no fold selected significantly more frequently than others. Here we tried to use additional information. It was noted that in the first two scenarios the most frequent fold was almost always the correct one.

According to the official ranking the CAFASP-CONSENSUS predictor performed better than all other automatic methods, and only six manual groups managed to perform better; although all of the consensus data and our predictions were publicly available. Post-predicting the CASP4 targets showed that the automatic Pcons predictor performed as well as the best individual methods, but not significantly better.

16.4. Pcons

The Pcons approach presented here differs significantly from earlier consensus predictors. It follows the approach described above for the semi automatic CAFASP-CONSENSUS predictor used in CASP4. Pcons follows the following steps, see Figs. 16.1 and 16.2.

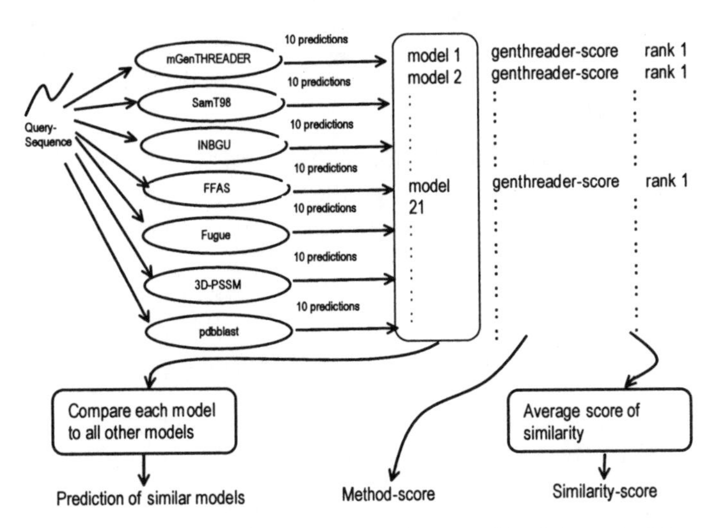

Figure 16.1. Description of the Pcons method used in this study. First up to 70 models are collected from seven different web servers. The structure of these models is compared to the structure of all other models. Three data points are fed into the network, the score, the fraction similar models and the average similarity to other top ranked models. A separate network is trained for each server. For each model obtained from one server the LGscore is predicted.

1. Firstly, a set of publicly available protein fold recognition web servers is utilized to produce input data. This data is converted into possible models for the query target.
2. Secondly, all collected models being compared used structural superposition and evaluation algorithms.
3. Thirdly, similarity between models, plus the particular score for this model, is used to predict the quality of the model.

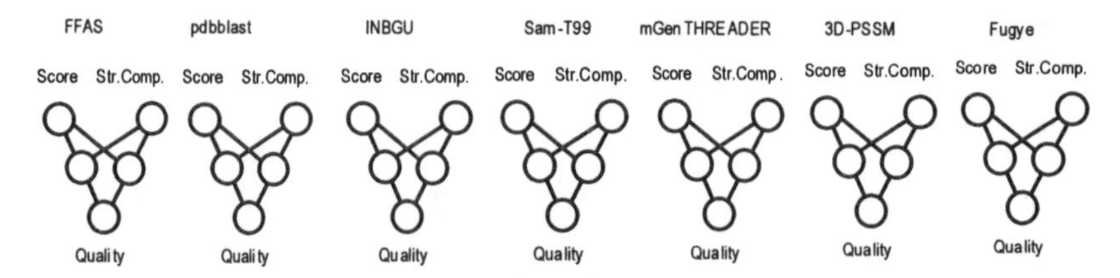

Figure 16.2. Description of the neural networks used by Pcons. The structure of these models, and the related templates are compared to the structure of all other models and templates. Two types of data points are given to the network, the score and the fraction of similar models. A separate network is trained for each server. For each model obtained from one server, the LGscore is predicted by a neural network.

Following the protocol sketched above, there are several possible choices where an automated consensus predictor can be implemented. In the following sections we will discuss how some of these choices affect the performance of Pcons.

16.4.1. Collection of Publicly Available Models

The current version of Pcons utilizes seven servers (PDB-Blast, FFAS, INBGU, mGenTHREADER, SAM-T99, 3D-PSSM, and FUGUE). The top 10 hits from each server were converted into models by the meta-server. To create the optimal consensus predictor, it is necessary to use the optimal number of models from the best servers. We have used the servers that performed best in LiveBench[20] and some limited experimentation. The conversion uses the simplest possible method, i.e., using the backbone coordinates from the template. The resulting model might include gaps and if there are insertions there will be some residues missing. In a future we plan to replace this process by using a homology modeling method to build a full model of the query protein for each predicted model. This will allow the use of additional terms that can be used as an input to the network.

16.4.2. Structural Comparison

The main idea about the consensus predictor is to use the observation that the correct fold often does not occur alone among the top hits. For instance if many of the top hits are immunoglobulins, it is quite likely that the query protein actually is an immunoglobulin. But, what is the best method of examining how many immunoglobulins are found among the top hits? One possible method could be to use one of the databases over protein folds, such as FSSP, SCOP or CATH. However, this involves several problems. These databases all use different domain definitions, while the different servers could use any of these or another set of domain definitions. Another problem is that these protein fold databases are updated less frequently than some of the servers. To avoid these problems we use structural comparisons of the models in Pcons.

Structural comparisons can be done in two different ways. Either the structure of the template or the structure of the models can be used. Structural comparisons of the models can be done either by taking the alignment into account (alignment dependent) or ignoring this (alignment independent), while the template comparison needs to use alignment independent structural superpositions. In our first studies, we used both template and model comparisons[32] in an alignment independent way. However, the structural alignment method is too time-consuming for a server version. Therefore, all versions of Pcons discussed in this study only utilize alignment dependent model comparisons. In our experience the difference in performance between the different structural superposition methods is not very significant.

Besides the decision to use alignment dependent structural superpositions, it is also possible to choose a measure for the similarity between two models. One simple measure of similarity could be RMSD. However, it is possible that the models are only partly correct, in which case other measures are better. We choose to use the same type of measures as used for the evaluation of the model quality; see below.

16.4.3. Prediction of Quality of the Models

Pcons predicts the quality and accuracy of all collected models, and if several servers predict one particular fold, Pcons will assign a high score to it. Pcons also differs from most earlier methods in the way in which correct predictions are defined. Pcons is trained to predict the quality of a model, while most other methods are optimized to detect if a correct fold is recognized or not. This might be advantageous, as it is not trivial to uniquely define folds;[33] and even if the correct fold is found, the alignment could potentially be wrong. The problems with domain definitions and updating of fold recognition methods, described above, make it very difficult to train Pcons to predict the correct fold.

Several different measures have been developed to measure the quality of a protein model. These can be divided into four groups: global, such as RMSD, which uses the whole model to measure the quality, alignment dependent, which uses the most similar segment between the model and the correct structure, alignment independent measures which are identical to alignment dependent measures but allow a shift in the alignment, and finally, template based measures. We have recently reviewed these measures and concluded that "alignment dependent" measures were best at identifying the best models, while "alignment independent measures" were better at detecting fold recognition.[34]

In the first Pcons version we used the alignment independent measure LGscore2 to evaluate model accuracy.[34] In the more current version of Pcons we have used an updated version of the alignment dependent LGscore. The updated version provides better statistics for short segments. Since the first Pcons was released we have used several different measures of the quality, including MaxSub,[35] new versions of LGscore and Touch.[20] Here we observed that by using the LGscore, we obtained the best correlation between the predicted and the real quality measures. However, the difference was quite small. It was also observed that if Pcons was trained to predict a particular measure, it performed slightly better using that measure for the evaluation. Surprisingly, the scoring method used for the structural comparisons was of greater importance than the scoring used as the goal for the predictions. The best predictions of MaxSub quality were obtained if MaxSub was used as the measure of structural similarity, etc.

16.5. Performance of Pcons

As we have claimed above, Pcons performs better than any single automatic method and the performance might even rival manual experts. Here, we will describe some different tests done to analyze the performance of different fold recognition servers. These are based on the LiveBench benchmarking system.[19,20] For simplicity we have mainly used MaxSub[35] as the evaluation method, but the results are similar using any other evaluation method.

16.5.1. Performance in LiveBench2

LiveBench2 is based on a large set of 203 proteins. Each week all new structures in PDB that do not show any significant sequence similarity to an already known protein are collected. A meta-server submits a sequence to all participating servers, collects the predictions, and analyzes the results automatically. Pcons-I was trained on data from LiveBench1 and was thus included in LiveBench2. Results have been divided into easy and hard target, depending on the best score obtained by PDB-Blast. The LiveBench process is continuously proceeding and can be found at http://bioinfo.pl/livebench/.

There are several possible methods of analyzing the results from LiveBench. In the complete analysis, several methods have been used, and only results that are consistent throughout all methods, are considered significant. For simplicity, here we choose to analyze the results using MaxSub.[35] In Table 16.1 it can be seen that Pcons-I does not detect significantly more correct hits for the easy targets, while for the hard targets the number of correct hits is increased by about 10% from 50 (for 3D-PSSM) to 56. This indicates a small but significant improvement for the hard targets.

Table 16.1. Number of correct first ranked models by servers and consensus predictors.

Method	Easy (44)	Hard (155)	All (199)
FFAS	36	37	75
pdbblast	38	19	57
inbgu	39	39	83
mGenTHREADER	39	35	79
3D-PSSM	40	50	93
Sam-T99	31	29	61
fugue	35	36	74
Pcons	38	56	99
Pcons-II	42	65	107
Pcons-II-MLR	42	75	117

Another important feature of an automatic fold recognition method is to be able to distinguish between correct and incorrect hits. In Fig. 16.3 we plot the number of correct predictions at a given number of incorrect predictions. It can be seen that the consensus predictor identifies significantly more correct hits than any single server does. None of the individual servers detect more than the "easy" targets before a number of false predictions are detected; while Pcons detects more than 20 of the "hard" targets before any significant number of false positive predictions.

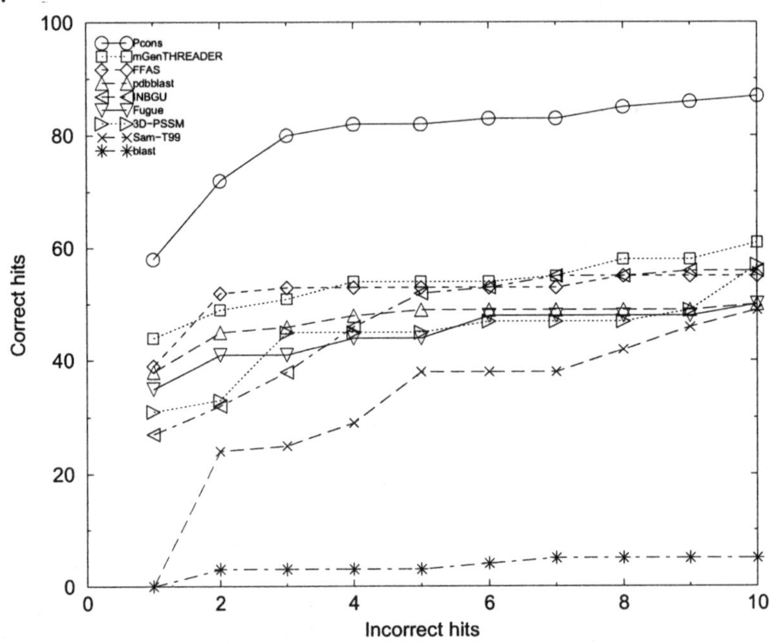

Figure 16.3. Cumulative plot of correct versus incorrect models in LiveBench-2. Correct and incorrect models are defined by using a combination of several different measures. The X-axis reports the number of incorrect models while the y-axis indicated number of correct models.

It is important not only to predict the correct fold, but also to produce a high quality model for a query sequence. One method of measuring the quality of the models is to use the sum of the MaxSub scores for the best model according to each server. In Table 16.2 it can be seen that, for the hard targets, Pcons produce better models than any single server (18.7 *vs.* 16.1 for 3D-PSSM); while there is not a significant improvement for the easy targets (14.4 *vs.* 14.4 for 3D-PSSM).

Table 16.2. Sum of MaxSub scores for individual servers.

Method	Easy	Hard	All
FFAS	14.0	12.4	26.4
pdbblast	14.5	5.7	20.2
inbgu	13.3	14.1	27.3
mGenTHREADER	13.0	13.4	26.4
3D-PSSM	14.4	16.1	30.6
Sam-T99	12.4	8.6	21.0
fugue	13.6	12.6	26.1
Pcons	14.4	18.7	33.1
Pcons-II	16.5	20.1	36.5
Pcons-II-MLR	15.6	22.6	38.2

16.5.2. Why Does Pcons Perform Better?

It is not obvious that Pcons should perform better than the best single protein structure predictor, however it does. In the introduction of this chapter we claimed that the main improvement was due to the "consensus analysis" used in Pcons, but not used by the other servers. However, it is also possible that the improvement is obtained from some other type of information.

During the development of Pcons, we developed several versions of Pcons. These include: NN-score that only used the scores from the severs as its input, NN-noscore that only used the "consensus analysis" as its input, and NN-model which was the final version of Pcons using both scores and network.[32] In Table 16.3 we can see that for the easy targets the three different Pcons implementations perform equally well; while for the hard targets, the performance of NN-score performs worse than the other two. NN-score is the only Pcons predictor that does not use the "consensus analysis." In fact NN-score does not perform significantly better than the best single method. This indicates that the consensus analysis is the most important contribution to the improvements obtained by Pcons. Once this type of information is included into the individual servers, it is very likely that a consensus predictor does not perform significantly better than the individual servers do. Interestingly, even completely ignoring the scores from the different methods, NN-noscore performs quite well. In the development of an MLR-based Pcons, we also observed that the weights for the scores were very small, indicating again that the consensus analysis is the most important part of Pcons.

Table 16.3. Number of correct first ranked models by each server and network for the LiveBench-1.

Method	Easy (44)	Hard (155)	All (199)
GenTHREADER	23	13	36
Sam-T98	22	16	38
FFAS	28	14	42
inbgu	23	21	44
3D-PSSM	22	21	43
pdbblast	28	10	38
NN-score	28	20	48
NN-noscore	28	30	58
NN-model	28	29	57

16.6. Pcons-II

The first version of Pcons was developed using data from LiveBench2 and benchmarked in the LiveBench2 process. In LiveBench2 it was concluded that several servers had increased their performance and that some new servers performed very well. Therefore, we decided to create a new version of Pcons using the data from LiveBench2.

16.6.1. Improvements Using More Servers

The first generation of the Pcons server used predictions from six different servers. After some minor benchmarking, we decided to use a set of seven servers for Pcons2. One completely new server was included (FUGUE,[28] while two new servers were updated (GenTHREADER to mGenTHREADER and SAM-T98 to SAM-T99). In Tables 16.1 and 16.2 these results are presented as Pcons-II. It can be seen that a small, but significant improvement is obtained using these additional servers. About 5% more correct structures are identified. The improvement is noticeable both for easy and hard targets.

16.6.2. Speed-Up of Structural Comparisons

The time for running Pcons increases with the square of the number of models. This makes it roughly twice as time-consuming to use seven servers (4900 comparisons) instead of five (2500 comparisons). This made the response time from Pcons too long to be acceptable. We have made a significant speed-up of the structural comparison algorithm used to calculate the LGscore. This was obtained by ending the structural comparison if a fragment had an RMSD larger than (N+225)/45, where N is the number of residues in a fragment. Look at http://www.sbc.su.se/~arne/lgscore/ for more details.

16.6.3. Using Better Statistics

During the evaluation of CASP4 we noted that the LGscore did not give significant scores to short proteins. To deal with this problem, we have re-calculated the measure of statistical significance, which the base for the LGscore measures. The original statistical measure was obtained from a study by Levitt and Gerstein.[36] However, the statistics were only calculated for proteins larger than 120 residues, while the fragments used by LGscore often are much shorter. Using these new measures gave no significant improvements, but as the statistics were better we decided to use it anyway. As mentioned above, we have also tried to use other measures of the model quality.

16.6.4. Improvements Using Linear Regression

Although neural networks are powerful tools to detect patterns, they are not always ideal. One such problem is that they might easily be over-trained and thereby not perform ideally for unseen data. To avoid these potential problems, we examined the possibility of using Multiple Linear Regressions instead of the neural networks. The results are shown in Tables 16.1 and 16.2 as Pcons-II-MLR. It can be seen that about 15% more correct hard targets are found using Pcons-II-MLR than the standard Pcons-II method. When studying the terms obtained from the MLR, it was shown that the terms for the different scores were small in comparison with the terms for the "consensus analysis." The model quality for the hard targets has also been improved. In LiveBench3 the Pcons-II-MLR method will be used.

16.7. Summary

In this article we show that by using a "consensus analysis," fold recognition methods can be improved significantly. Using this type of analysis we think that automatic fold recognition methods can challenge the performance of manual experts.

References

1. Smith TF, Waterman MS: **Identification of common molecular subsequences.** *J Mol Biol* 1981, **147**:195–197.
2. Needleman SB, Wunsch CD: **A general method applicable to the search for similarities in the amino acid sequence of two proteins.** *J Mol Biol*1970, **48**:443–453.
3. Gribskov M, McLachlan AD, Eisenberg D: **Profile analysis: Detection of distantly related proteins.** *Proc Natl Acad Sci USA* 1987, **84**:4355–4358.
4. Altschul SF, Madden TL, Schaffer AA, Zhang J, Zhang Z, Miller W, Lipman DJ: **Gapped blast and PSI-BLAST: A new generation of protein database search programs.** *Nucleic Acids Res* 1997, **25**:3389–3402.
5. Karplus K, Barrett C, Hughey R: **Hidden Markov models for detecting remote protein homologies.** *Bioinformatics* 1998, **14**:846–856.
6. Rychlewski L, Jaroszewski L, Li W, and Godzik A: **Comparison of sequence profiles. Strategies for structural predictions using sequence information.** *Protein Sci* 2000, **9**(2):232–241.
7. Fischer D, Eisenberg D: **Protein fold recognition using sequence-derived predictions.** *Protein Sci* 1996, **5**:947–955.
8. Rost B, Schneider R, and Sander C: **Protein fold recognition by prediction-based threading.** *J Mol Biol* 1997, **270**:471–480.
9. Rice D, Eisenberg D: **A 3D-1D substitution matrix for protein fold recognition that includes predicted secondary structure of the sequence.** *J Mol Biol* 1997, **267**:1026–1038.
10. Kelley L, MacCallum R, and Sternberg M: **Enhanced genome annotation using structural profiles in the program 3D-PSSM.** *J Mol Biol* 2000, **299**(2):523–544.
11. Jones DT, Taylor WR, Thornton JM: **A new appoach to protein fold recognition.** *Nature* 1992, **358**:86–89.

12. Abagyan RA, Batalov S: **Do aligned sequences share the same fold?** *J Mol Biol* 1997, **273**:355–368.

13. Park J, Teichmann SA, Hubbard T, Chothia C. **Intermediate sequences increase the detection of homology between sequences.** *J Mol Biol* 1997, **273**:249–254.

14. Park J, Karplus K, Barrett C, Hughey R, Haussler D, Hubbard T, Clothia C: **Sequence comparisons using multiple sequences detect three times as many remote homologues as pairwise methods.** *J Mol Biol* 1998, **284**:1201–1210.

15. Lindahl E, Elofsson A: **Identification of related proteins on family, superfamily, and fold level.** *J Mol Biol* 2000, **295**:613–625.

16. Moult J, Hubbard T, Bryant SH, Fidelis K, Pedersen JT: **Critical assesment of methods of proteins structure predictions (CASP): Round II.** *Proteins* 1997, **Suppl 1**:2–6.

17. Fischer D, Barret C, Bryson K, Elofsson A, Godzik A, Jones D, Karplus K, Kelley L, MacCallum R, Pawowski K, Rost B, Rychlewski L, Sternberg M: **Critical assessment of fully automated protein structure prediction methods.** *Proteins* 1999, **Suppl 3**:209–217.

18. Fischer D, Elofsson A, Rychlewski L, Pazos F, Valencia A, Rost B, Ortiz A, Dunbrack RL: **Cafasp2: The critical assessment of fully automated structure prediction methods.** *Proteins* 2001, **45 Suppl 5**:171-183.

19. Bujnicki J, Elofsson A, Fischer D, and Rychlewski L: **LiveBench: Continous benchmarking of protein structure prediction servers.** *Protein Sci* 2001, **10**(2):352–361.

20. Bujnicki M, Elofsson A, Fischer D, Rychlewski L: **LiveBench-2: Large-scale automated evaluation of protein structure prediction servers.** Proteins 2001, **45 Suppl 5**:184–191.

21. Domingues F, Lackner P, Andreeva A, Sippl MJ: **Structure based evaluation of sequence comparison and fold recognition alignment accuracy.** *J Mol Biol* 2000, **297**(4):1003–1013.

22. Sauder J, Arthur J, and Dunbrack JRL: **Large-scale comparison of protein sequence alignment algorithms with structure alignments.** *Proteins* 2000, **40**:6–22.

23. Elofsson A: **A study on how to best align protein sequences.** *Proteins* 2002, **46**(3):300–309.

24. Murzin AG, Brenner SE, Hubbard T, Chothia C: **Scop: A structural classification of proteins database for the investigation of sequences and structures.** *J Mol Biol* 1995, **247**:536–540.

25. Orengo CA, Michi AD, Jones S, Jones DT, Swindels MB, Thornton JM: **Cath—A hierarchical classification of protein domain structures.** *Structure* 1997, **5**:1093–1108.

26. Pearson WR, Lipman DJ: **Improved tools for biological sequence analysis.** *Proc Natl Acad Sci USA* 1988, **85**:2444–2448.

27. Fischer D: **Hybrid fold recognition: Combining sequence derived properties with evolutionary information.** In: *Pacific Symposium on Biocomputing,* vol. 5. Altman R, Dunker A, Hunter L, Klien T, eds. World Scientific, 2000:116–127.

28. Shi J, Blundell T, Mizuguchi K: **Fugue: Sequence-structure homology recognition using environment-specific substitution tables and structure-dependent gap penalties.** *J Mol Biol* 2001, **310**(1):243–257.

29. MacCallum R, Kelley LA, Sternberg M: Sawted: **Structure assignment with text description-enhanced detection of remote homologues with automated swiss-prot annotation comparisons.** *Bioinformatics* 2000, **16**(3):125–129.

30. Bairoch A, Apweiler R: **The swiss-prot protein sequence data bank and its new supplement trembl.** *Nucleic Acids Res* 1996, **24**:17–21.

31. Jones D: **Genthreader: An efficient and reliable protein fold recognition method for genomic sequences.** *J Mol Biol* 1999, **287**(4):797–815.

32. Lundström J, Rychlewski L, Bujnicki J, Elofsson A: **Pcons: A neural network based consensus predictor that improves fold recognition.** *Protein Sci* 2001, **10**(11):2354–2362.

33. Hadley C, Jones DT: **A systematic comparison of protein structure classifications: Scop, cath, and fssp.** *Structure* 1999, **7**(8):1099–1112.

34. Cristóbal S, Zemla A, Fischer D, Rychlewski L, Elofsson A: **How can the accuracy of a protein model be measured?** BMC: *Bioinformatics* 2001, 2:**5**.

35. Siew N, Elofsson A, Rychlewski L, and Fischer D: **Maxsub: An automated measure to assess the quality of protein structure predictions.** *Bionformatics* 2000, **16**(9):776–785.

36. Levitt M, Gerstein M: **A unified statistical framework for sequence comparison and structure comparison.** *Proc Natl Acad Sci USA* 1998, **95**(11):5913–5920.

17

New Insights into Protein Fold Space and Sequence-Structure Relationships

Ilya N. Shindyalov and Philip E. Bourne

Abstract

With the growth of genome and proteome data the need in quality annotation of genes and proteins dramatically increases. Comparative analysis of protein structures is essential part of the process of annotation, particularly in annotation done by structure homology and as an accessory in annotation based on sequence homology. This is in turn increasing need to understand the domain of structure comparison in big scale—at the level of fold space. Structural similarities in particular areas of fold spaceare commonly detected by all approaches while approximately 50-60% of similarities are unique to each method. To date, limited efforts have been spent to interpret the observed differences and to understand the underlying evolutionary mechanisms behind the similarities revealed. In this work we introduce several new views of protein fold space which will help us to further understand protein evolution and interpret structural similarities. The following closely related studies are reported here: (i) Analysis of sequence-structure space. (ii) Comparison between approaches to the structural classification problem based on human expertise (SCOP) and automated (CE). (iii) Understanding redundancy in structure data and how structure and sequence redundancies interrelated.

17

New Insights into Protein Fold Space and Sequence-Structure Relationships

Ilya N. Shindyalov and Philip E. Bourne

University of California, San Diego; San Diego Supercomputer Center, La Jolla, California

17.1. Introduction

Comparative analysis of protein structures has been addressed in many studies[1-3] with a variety of computer algorithms and through human annotation.[4] Structural similarities in particular areas of fold space are commonly detected by all approaches while approximately 50-60% of similarities are unique to each method.[2,5] To date, limited efforts have been spent[2] to interpret the observed differences and to understand the underlying evolutionary mechanisms behind the similarities revealed. In this work we introduce several new views of protein fold space which will help us to further understand protein evolution and interpret structural similarities. The following closely related studies are reported here:

(i) Analysis of sequence-structure space as determined by the Combinatorial Extension (CE) algorithm.[5]

(ii) SCOP (Structural Classification of Proteins)[4] *vs.* CE fold space comparison, based on a random set of 1000 SCOP domains with non-identical sequences (with sequence identity of 50% or less.)[6]

(iii) A sequence and structure redundancy study of the current PDB[7] using CE and BLAST[8] algorithms.

Areas in fold space are related to modes of protein evolution and degree of divergence. Observed differences between different structural classifications schemes allow us to propose new directions in the comparative analysis of structure.

17.2. Overview of CE Sequence-Structure Space

Plate 17.1 provides a distribution of significant structure (RMSD) *versus* sequence (Sequence Identity) similarity. As expected a peak can be seen in the area of sequence identity between 90% and 100% and RMSD between 0 Å and 1 Å (shown in magenta). These are redundant protein chains resulting from point mutations and alternative determinations of identical or near identical structures. The region extends towards 80%–90% and 1 Å–2 Å, respectively

The largest area on the plot has sequence identity between 0%–20% and RMSD between 3 Å–6 Å (shown in yellow). This area of sequence identity referred[9] as 'midnight zone' in protein similarity, where structural similarity cannot be reliably inferred from sequence similarity hence random sequence matches cannot be discriminated from those corresponding to proteins with similar structures. There are plenty of reliable structure similarities in this area, many of them can be thought as *analogous* similarities which did not have any common ancestor in the course of evolution and cannot be reliably detected based on similarity in sequence. This area presumably contains partial domain and subdomain similarities including the so-called recurrent structures.[10,11]

Another large area of significant structure similarity has a sequence identity between 20% and 50% and an RMSD between 1 Å and 3 Å, respectively (shown in green), represents *homologous* similarities. This area presumably contains superfamily and family level similarities.[4]

Area with sequence identity between 50% and 80% and RMSD between 0 Å and 2 Å, respectively (shown in orange), represents *homologous* similarities as well. This area contains close family members including identical proteins from various species (orthologs) and members of the multi-gene families (paralogs).

17.3. SCOP *vs.* CE Fold Space Comparison

Table 17.1 provides results of comparison between protein similarities detected by CE and SCOP for a random set of 1000 domains taken from SCOP (Astral) with <50% sequence identity. The number of structural similarities between protein domains based on the SCOP classification and different ranges of z-score evaluated using the CE algorithm is given.

Table 17.1. Comparison of two structure similarity methods—CE and SCOP. Numbers given are for similar domains. Comparisons are made for all combinations from a random set of 1000 SCOP domains below 50% sequence identity. Classes with proteins not similar by one or both criteria are shaded in gray.

		z-score (CE)					
		<3.5	3.5-3.75	3.75-4.0	4.0-4.25	4.25-4.5	>4.5
hierarchy (scop)	not similar	468572	14340	2583	2634	626	699
	fold	1507	1087	542	973	324	874
	superfamily	621	625	406	603	241	600
	family	154	126	64	182	84	914
	protein/species	20	14	5	10	3	67

One can see that there are many structural similarities detected by both methods. On the other hand there are similarities that are detected by only one method.

Table 17.2. Comparison of two structure similarity methods—CE and SCOP. Numbers given are for average RMSD (standard deviation is given in parenthesis). Comparisons are made for all combinations taken from a random set of 1000 SCOP domains below 50% sequence identity.

		z-score (CE)					
		<3.5	3.5-3.75	3.75-4.0	4.0-4.25	4.25-4.5	>4.5
hierarchy (scop)	not similar		4.9(1.3)	4.7(1.2)	4.6(1.3)	4.4(1.2)	4.4(1.4)
	fold		3.6(0.6)	3.6(0.5)	3.5(0.6)	3.7(0.6)	3.8(0.6)
	superfamily		3.5(0.6)	3.4(0.6)	3.2(0.7)	3.0(0.8)	3.3(0.7)
	family		3.4(1.0)	3.3(1.0)	3.0(1.2)	2.9(0.8)	2.3(0.8)
	protein/species		3.4(0.8)	2.8(0.9)	2.8(0.7)	3.1(1.0)	2.1(0.8)

To understand better the discrepancy between the two methods consider RMSD (Table 17.2) and relative alignment length (number of aligned residues related to the length of the smaller protein, Table 17.3). Similarities detected by CE tend to have a higher RMSD and lower coverage of protein length. This leads to common subdomains detected by CE belonging to different SCOP folds.

Table 17.3. Comparison of two structure similarity methods—CE and SCOP. Numbers are for the relative length of alignment (standard deviation is given in parenthesis). Comparisons are made for all combinations taken from a random set of 1000 SCOP domains below 50% sequence identity.

		z-score (CE)					
		<3.5	3.5-3.75	3.75-4.0	4.0-4.25	4.25-4.5	>4.5
hierarchy	not similar		51.9(20.4)	52.8(19.9)	53.6(20.5)	54.3(19.4)	59.6(19.4)
(scop)	fold		73.5(15.9)	75.0(15.7)	75.4(15.6)	72.2(16.3)	69.6(14.3)
	superfamily		71.5(14.6)	72.1(13.3)	73.1(13.9)	73.4(15.3)	71.6(15.6)
	family		68.6(22.2)	70.4(16.8)	76.3(17.4)	78.6(17.0)	86.3(11.7)
	protein/species		37.5(29.5)	38.2(26.6)	54.2(27.6)	74.7(13.3)	88.5(12.2)

Conversely there are a number of proteins with relatively low CE z-scores yet have similar folds as assigned by SCOP. This is attributed to manual similarity assignment, detecting similarities beyond the automated alignment based upon a rigid body model used by CE. These discrepancies call for alignment scoring schemes beyond those for two rigid bodies or alternatively a search for local similarities within the global alignment produced between two rigid bodies.

On the other hand, there are some cases where SCOP similarity is indicated in error. For example, chains C and D from PDB entry 1BCC are both annotated in scop as fold: *Membrane all-alpha*, protein: *Cytochrome Bc1 transmembrane subunit.* While chain C is all-alpha as classified, the D chain has a quite different fold with beta structures involved.

17.4. Analysis of Structure Redundancy

Earlier efforts to analyze redundancy in the PDB were undertaken based on both sequence[12-14] and structure criteria.[6] Here we compare the two criteria and introduce a combined criterion.

17.4.1. Size of NR Set as a Function of Criteria Used

Figs. 17.1 and 17.2 depict size of the non-redundant (NR) set as a function of selection criteria.

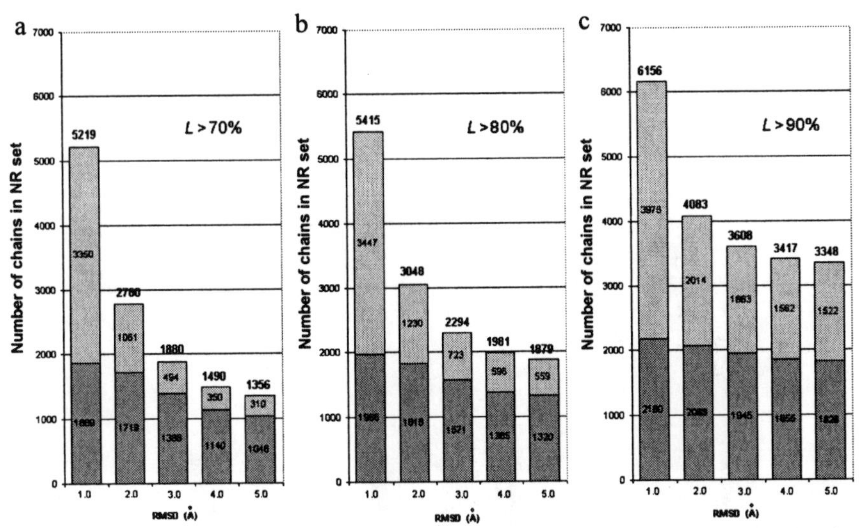

Figure 17.1. Size of structure NR set for various selection criteria: RMSD threshold and length coverage (L) threshold. **(a)** $L > 70\%$. **(b)** $L > 80\%$. **(c)** $L > 90\%$. The total number of chains used in the analysis is 20276. *Dark gray* indicates chains in structure NR set which are also non-redundant by sequence 25% identity criterion. *Light gray* indicates chains in structure NR set that are redundant by sequence criteria. Numbers on the top and inside the bars indicate total number of chains in structure NR set and those non-redundant and redundant by sequence criterion, respectively.

The structure NR set is presented on Fig. 17.1 as a function of RMSD and length coverage threshold (L). The size of the set varies from 1356 to 6156, i.e., from 6.7% to 30.4% from a total of 20276 chains.

Sequence with a length coverage at 80% provides for 1984 to 4815 chains in NR set with sequence identity changing from 25% to 95%, respectively (Fig. 17.2). There is remarkably linear growth of NR set size with sequence identity threshold.

17.4.2. Characterization of Chains Excluded from the Set

It is important to characterize chains excluded from the NR set. Fig. 17.3 provides distributions of structurally redundant chains by sequence identity and BLAST score. Similarity is measured between a given redundant chain and a chain in NR set that represents this chain i.e., was used as a basis for its exclusion. As one would expect there is a substantial number of chains 7159 (40%) with above 95% sequence identity, a result of structure determinations of point mutants and alternative determinations for the same structure. Most of the structurally redundant chains exhibit substantial sequence similarity, 16789 (93.6%) with sequence identity above 25%. There are as well some chains with low sequence similarity, 1193 (6.6%) with sequence identity below 25% or 2176 (12.1%) with a BLAST score below 30.

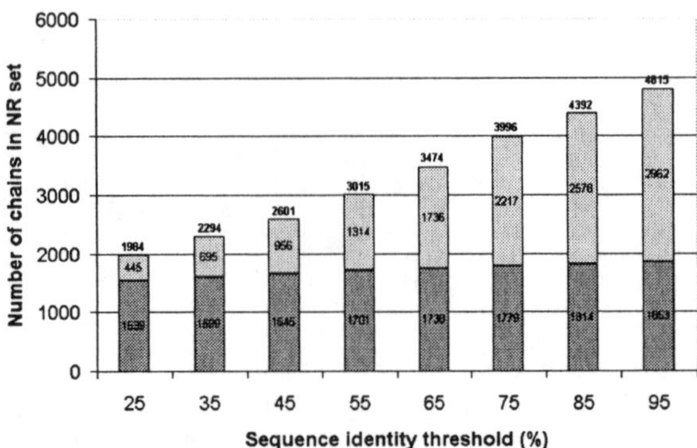

Figure 17.2. Size of sequence NR set for various sequence identity thresholds and a length coverage of 80%. The total number of chains used in analysis is 20276. *Dark gray* indicates chains in sequence NR set which are also non-redundant by 3 Å RMSD, 70% coverage structure similarity criterion *Light gray* indicates chains in sequence NR that are redundant by structure criteria. Numbers on the top and inside the bars indicate total number of chains in sequence NR set and those non-redundant and redundant by structure criterion, respectively.

Fig. 17.4 provides a distribution of sequence redundant chains against structure similarity. As expected the majority have similar structures 16944 (96.6%). However, there are 613 (3.4%) which do not have any significant structural similarity.

17.4.3. Characterization of Similarity Between Chains in the Set

Characterization of similarity between chains in structure and sequence NR sets was analyzed and presented in Figs. 17.5 and 17.6 for the structure NR set and in Figs. 17.7 and 17.8 for the sequence NR set.

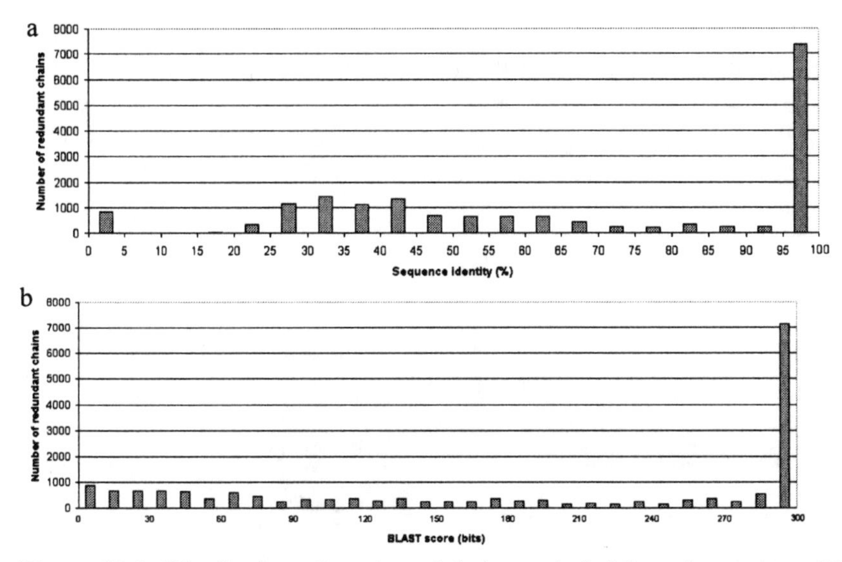

Figure 17.3. Distribution of number of chains excluded from the structure NR set derived with an RMSD threshold of 3.0 Å and a length coverage threshold of 80%. (**a**) By sequence identity. (**b**) By BLAST score (in bits).

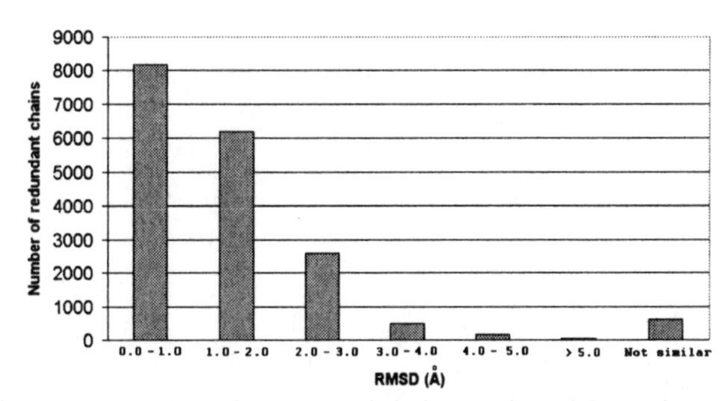

Figure 17.4. Distribution of number of chains excluded from the sequence NR set derived with sequence identity threshold of 25% and length coverage threshold of 80%.

Fig. 17.5 depicts distribution of structural similarities in structure NR set based on length coverage threshold of 80% and RMSD threshold of 3 Å. For each chain in NR set the best similarity based on z-score to a chain from the NR set was considered. Only similarities with z-score above 4.0 were used. The area with structural similarities excluded from NR set by selection criteria is shown as a bold line box. From 2294 chains in the set 1740 (75.9%) have significant structure similarity to another chain in the set. Certainly many of those similarities are partial. Considering similarities with a length coverage $L \geq 50\%$ and an RMSD above 4 Å then the number of similar chains goes down to 1023 (44.6%). Nevertheless there remain many structure similarities in the NR.

RMSD (A)

Length coverage (%)	0.0–1.0	1.0–2.0	2.0–3.0	3.0–4.0	4.0–5.0	5.0–
0–10	0	2	1	5	4	3
10–20	3	5	17	35	27	22
20–30	4	10	30	41	21	20
30–40	2	11	26	41	29	26
40–50	9	26	55	38	28	22
50–60	6	34	47	49	26	7
60–70	34	51	59	74	16	8
70–80	141	106	81	105	32	11
80–90				133	19	6
90–100				103	17	12

Figure 17.5. Distribution of structure similarity for chains in the structure NR set derived with RMSD threshold of 3.0 Å and length coverage threshold of 80%. Similarities not seen because of exclusion by derivation criteria are marked with bold line box. Grades of gray highlight populated areas in the distribution: *dark gray*—$n \geq 100$; *medium gray*—$100 > n \geq 50$; *light gray*—$50 > n \geq 30$. Where n is the number of chains from NR set with a given level of similarity to another structure in the set (see text for details).

Length of alignment

Sequence identity (%)	0	20	40	60	80	100	120	140	160	180	200	220	240	260	280	300	320	340	360	380 ...
0–10	0	0	0	0	0	0	0	0	0	0	0	0	0	0	0	0	0	0	0	0
10–20	0	0	0	1	0	1	1	0	0	0	0	0	0	0	0	0	0	0	0	0
20–30	0	0	4	10	26	21	14	13	9	6	4	4	10	2	0	4	3	4	6	11
30–40	0	9	30	29	28	27	18	9	11	9	4	3	12	2	2	4	2	1	2	9
40–50	0	17	22	8	14	6	2	8	0	4	1	0	1	3	1	0	0	0	0	6
50–60	1	14	5	8	1	3	1	4	2	0	0	0	2	0	0	0	1	3	0	4
60–70	0	3	6	3	6	0	6	1	0	2	0	0	3	4	1	0	0	0	2	0
70–80	2	6	5	4	4	0	0	2	0	0	3	2	0	0	1	1	0	0	1	1
80–90	1	11	0	6	4	4	2	1	6	5	3	3	0	2	1	1	0	0	0	4
90–100	2	76	68	73	79	43	26	51	31	18	17	21	6	22	48	29	23	2	30	99

Figure 17.6. Distribution of sequence similarity for chains in structure NR set derived with RMSD threshold of 3.0 Å and length coverage threshold 80%. Grades of gray highlight populated areas in the distribution: dark gray—$n \geq 50$; medium gray—$50 > n \geq 20$; light gray—$20 > n \geq 10$. Where n is the number of chains from NR set with a given level of similarity to another chain in the set (see text for details).

Fig. 17.6 depicts the distribution of sequence similarity for the same NR set as used in Fig. 17.5. For this analysis again the best similarity to a chain from the NR set was used. BLAST score (in bits) was used to select the best similarity. Only similarities with a BLAST score above 30 bits were used. From 2294 chains in the NR set 1393 (60.7%) have demonstrated significant sequence similarity to another chain in the set. There are clearly a number of chains with high sequence identity (from 90% to 100%), but are kept in the set because they do not satisfy the redundancy criteria. Also a number of chains with sequence identity in 20% to 40% range and length of alignment below 160 are present. With the decrease in alignment length there are more chains with higher sequence identity (up to 60%), similar to HSSP curve.[15]

Figs. 17.7 and 17.8 show the distribution of structure and sequence similarity, respectively, for the sequence NR set based on a length coverage threshold of 80% and sequence identity threshold of 25%. An analysis was performed similar to that for the structure NR set (see above, Figs. 17.5 and 17.6). From 1984 chains in the NR set 1638 (82.6%) and 864 (43.5%) have demonstrated structure and sequence similarity, respectively. Chains showing structure similarity cluster around the area with RMSD from 1 Å to 3 Å and length coverage from 50% to 100%. Chains showing sequence similarity closely follow HSSP curve pattern.

RMSD (A)

Length coverage (%)	0.0–1.0	1.0–2.0	2.0–3.0	3.0–4.0	4.0–5.0	5.0–
0–10	0	4	4	2	3	3
10–20	3	7	20	34	27	19
20–30	4	5	36	37	27	22
30–40	2	12	34	27	24	20
40–50	6	20	59	29	19	17
50–60	2	21	55	43	22	6
60–70	7	32	63	56	11	7
70–80	8	41	107	68	22	7
80–90	8	77	151	90	17	6
90–100	13	71	71	26	4	0

Figure 17.7. Distribution of structure similarity for chains in sequence NR set derived with sequence identity threshold of 25% and length coverage threshold of 80%. Grades of gray highlight populated areas in the distribution: *dark gray*—$n \geq 100$; *medium gray*—$100 > n \geq 50$; *light gray*—$50 > n \geq 30$. Where n is the number of chains from NR set with a given level of similarity to another chain in the set (see text for details).

Length of alignment

Sequence identity (%)	0–20	20–40	40–60	60–80	80–100	100–120	120–140	140–160	160–180	180–200	200–220	220–240	240–260	260–280	280–300	300–320	320–340	340–360	360–380	380 ...
0–20																				
20–30	0	0	13	29	34	30	16	21	17	8	11	4	14	1	1	2	2	2	3	9
30–40	0	10	63	90	39	30	33	11	16	18	5	8	9	3	2	6	7	3	0	7
40–50	0	35	56	15	16	6	4	2	0	0	0	0	2	5	0	0	0	0	0	2
50–60	1	31	16	7	4	0	0	0	0	0	0	0	0	2	0	0	0	0	0	2
60–70	0	23	2	0	0	4	0	0	2	0	0	0	0	0	0	0	0	0	0	0
70–80	1	15	0	0	0	0	0	0	0	0	0	0	0	0	0	0	0	0	0	0
80–90	1	3	0	0	0	0	0	0	0	0	2	0	0	0	0	0	0	0	0	0
90–100	3	8	5	2	4	0	0	2	0	0	0	0	2	0	0	0	0	2	0	0

Figure 17.8. Distribution of sequence similarity for chains in sequence NR set derived with a sequence identity threshold of 25% and length coverage threshold of 80%. Similarities not seen because of exclusion by derivation criteria are marked with a bold line box. Grades of gray highlight populated areas in distribution: *dark gray*—$n \geq 50$; *medium gray*—$50 > n \geq 20$; *light gray*—$20 > n \geq 10$. Where n is the number of chains from NR set with a given level of similarity to another chain in the set (see text for details).

17.4.4. Complementary Sequence and Structure NR Sets

Consider NR set built either by sequence or structure criteria and then used again to search for a NR set of a different kind, i.e., sequence in structure NR set and structure in sequence NR set. Results of such experiments are provided in Figs. 17.1 and 17.2 with light gray portions of the bars indicating redundant chains revealed in the search for a complementary NR set. Consider the following choice of criteria: length coverage threshold of 80%, sequence identity threshold of 25% and RMSD of 3 Å. Then 31.5% of sequence redundant chains are revealed in the structure NR set and 22.4% of structure redundant chains are revealed in the sequence NR set.

17.4.5. Combined NR Set

We attempted to build a NR set combining sequence and structure criteria. The following criteria were used: an RMSD threshold of 3 Å, sequence identity threshold of 25% and length coverage threshold of 80%. The size of combined NR set was 2772 chains. The size of closest structure and sequence NR sets was 2294 and 1984, respectively. The increase in set size was 21% relative to the corresponding structure NR set and 40% relative to the corresponding sequence NR set.

References

1. Holm L, Sander C: **Searching protein structure databases has come of age.** *Proteins* 1994, **19**:165–173.

2. Godzik A: **The structural alignment between two proteins: is there a unique answer?** *Protein Sci* 1996, **5**:1325–1338.

3. Gibrat JF, Madej T, Bryant SH: **Surprising similarities in structure comparison.** *Curr Opin Struct Biol* 1996, **6**:377–385.

4. Murzin AG, Brenner SE, Hubbard T, Chothia C: SCOP: **A structural classification of proteins database for the investigation of sequences and structures.** *J Mol Biol* 1995, **247**:536–540.

5. Shindyalov IN, Bourne PE: **Protein structure alignment by incremental combinatorial extension (CE) of the optimal path.** *Protein Eng* 1998, **9**:739–747.

6. Brenner SE, Koehl P, Levitt M: **The ASTRAL compendium for sequence and structure analysis.** *Nucleic Acids Res* 2000, **28**:254–256.

7. Berman HM, Westbrook J, Feng Z, Gilliland G, Bhat TN, Weissig H, Shindyalov IN, Bourne PE: **The Protein Data Bank.** *Nucleic Acids Res* 2000, **28**:235–242.

8. Altschul SF, Madden TL, Schaffer AA, Zhang J, Zhang Z, Miller W, Lipman DJ: **Gapped BLAST and PSI-BLAST: A new generation of protein database search programs.** *Nucleic Acids Res* 1997, **25**:3389–3402.

9. Rost B: **Twilight zone of protein sequence alignment.** *Protein Eng* 1999, **12**:85–94.

10. Holm L, Sander C: **Dictionary of recurrent domains in protein structures.** *Proteins* 1998, **33**:88–96.

11. Shindyalov IN, Bourne PE:. **An alternative view of protein fold space.** *Proteins* 2000, **38**:247–260.

12. Heringa J, Sommerfeldt H, Higgins D, Argos P: **OBSTRUCT: A program to obtain largest cliques from a protein sequence set according to structural resolution and sequence similarity.** *Comput Appl Biosci* 1992, **8**:599–600.

13. Hobohm U, Scharf M, Schneider R, Sander C: **Selection representative set of protein structures.** *Protein Sci* 1992, **1**:409–417.

14. Hobohm U, Sander C: **Enlarged representative protein data set.** *Protein Sci* 1994, **3**:522–524.

15. Sander C, Schneider R: **Database of homology-derived protein structures and the structural meaning of sequence alignment.** *Proteins* 1991, **9**(1):56–68.

18

A Flexible Method for Structural Alignment: Applications to Structure Prediction Assessments

Vladimir Kotlovyi, Igor Tsigelny, and Lynn F. Ten Eyck

Abstract

In this chapter we discuss the design principles of a structure alignment system that can be used for structure prediction assessments. This system is based on a hierarchical representation of a protein shape. Such a representation makes the system suitable for effective alignments of structures with low similarity. We outline the mathematical background of the methods used for the structural alignment, with a special attention to angular measures determining the shape of a protein. A concise example of the suggested data representation scheme using XML forms is given along with some illustrative results obtained from our structure alignment and structure prediction assessment web-servers.

18

A Flexible Method for Structural Alignment: Applications to Structure Prediction Assessments

Vladimir Kotlovyi, Igor Tsigelny and Lynn F. Ten Eyck

University of California, San Diego; San Diego Supercomputer Center, La Jolla, California

18.1. Introduction

Many approaches to automated structure prediction are organized as an iterative process consisting of the following stages:

(i) generation of primary sequence or various characteristic patterns (pattern libraries), along with their mappings onto a selected set of secondary, supersecondary, tertiary, etc. structures,

(ii) development and tuning (training) of special structure prediction or fold recognition models based, for example, on Hidden Markov Models, or Neural Networks,

(iii) testing and assessment of the developed and tuned models.

Since structural alignment methods are used in all the stages of such a process, it is evident that the predictive properties of the models depend on the consistency (soundness) and accuracy of the alignment methods applied throughout the whole process. It is much easier to assess the accuracy of predictions and tune up the models, if the same methods of structural alignment and similarity criteria are used in all the stages. An actual process is frequently more convoluted, because the methods and criteria of the structural alignment itself may be tuned in the process. In our opinion, the following inconsistencies and drawbacks of structural alignment methods may reduce the effectiveness of structure prediction methods:

- The pattern libraries are built with one structural alignment method, but the accuracy of prediction is assessed with another alignment method and/or using a different set of similarity criteria;
- Many structural alignment methods depend on "hidden," uncontrollable by the user, parameters which may affect the process of tuning the prediction models;
- Some alignment methods give only one alignment in situations where several competitive alignments are possible; even more harmful are the methods that produce at least one alignment for *any* pair of proteins (for example, methods based on unrestricted least-square algorithms);
- Some structural alignment methods are applied beyond the range of their well-founded applicability (e.g., rigid distance-based methods when used to align structures with a low, but possibly recognizable by an expert eye, similarity);
- Most of the methods were developed without having specific needs of the structure prediction in view.

We believe that many of these inconsistencies and drawbacks can be eliminated in an integrated structure alignment and prediction assessment system. Such a system should be specifically oriented towards the structure prediction problems and have, as an integral part, a simple, complete, easily tunable, and fast method of structural alignment that is uniformly applied in all the stages of the structure prediction process. Our special attention is devoted to creating an integrated structural alignment subsystem that satisfies the following requirements:

- It should be based on a simple method with a clear and straightforward interpretation of the results;
- The algorithms used must be fast enough to process thousands of protein structures in acceptable time;
- The subsystem must find all possible alignments (the requirement

of completeness, e.g., see the paper of Lackner et al.);[1]

- It should be specifically oriented toward the needs of building structure prediction models (e.g., use the same similarity criteria, input/output format, database structure, etc.);
- The tunable parameters must have clear interpretation and must be under full control of the user (no hidden parameters).

We have designed such a system and are now at the final stage of its implementation.

Our goal in this chapter is to give a short description of the integrated structure prediction system, as well as some background theoretical material.

18.2. Theoretical Background

To reach the specified goals, we use a hierarchically organized set of alignment methods and strategies, which may be roughly classified into the following categories:

- topological, i.e., based on the order and type of secondary/tertiary structures or folds/motifs along the protein chain, as well as incidence relations (links),
- shape-based, i.e., analyzing the shape of a backbone and ignoring some unimportant differences in shape along the back-bones of the aligned proteins,
- distance-based, i.e., aligning proteins (or backbones only, or any set of supposedly important atoms) by applying any method using interatomic distances and measures derived from them, such as root mean square deviation (RMSD), or other Euclidean invariants (e.g., angles, volumes).

Shifting of interest from relatively simple topological and distance-based methods towards more complicated, hierarchical methods seems to be a common trend.

In the present implementation of our system, the topological alignment strategy is reduced to matching the secondary structure signatures (defined below) of proteins, taken directly from PDB files or calculated with the algorithm due to W. Kabsch and C. Sander.[2] The shape-based methods are the foundation of our approach. Their importance can be seen from the fact that, using the characterization of a protein's backbone with a set of angular curvatures and torsions (see below), we completely eliminate unacceptable alignments such as helix-to-strand or strand-to-coil; these, and similar false alignments, would have been quite numerous had we restricted ourselves to purely distance-based (e.g., aiming to reach the minimum RMSD only) methods.

All the distance-based, and some shape-based, alignment methods have this mathematical result in their background: All the metric invariants in Euclidean space can be built up from one bilinear form—the scalar product of two vectors.[3] Examples of such invariants include Euclidean distance, norm of a vector, angle between vectors, volume and area of the convex hull of a chain, etc. All these invariants are used in one or another alignment method, the most popular being the Euclidean (interatomic) distance and its variations (like RMSD).

Although the pure distance-based methods are unsurpassed in situations where an alignment of an arbitrarily selected set of atoms is needed, a more shape-oriented approach is preferable when we have in view applications to the structure prediction. We can define the "shape" of a protein (or its fragment) as the shape of its backbone or, more precisely, as a space curve which goes along (or through) the C_α atoms of the backbone. We say that this curve approximates the protein chain if it goes along the backbone and minimizes a given distance measure; if the approximating curve goes through all the C_α atoms, we say that it interpolates the protein chain. The simplest example of the latter curve is the broken line connecting all the C_α atoms along a backbone. It is evident that an interpolating curve closely follows the shape of a protein chain. From this it follows that when matching two interpolating curves, we shall obtain a very precise alignment. On the other hand, an approximating curve is more suitable for the analysis of the overall shape of a protein chain, which is important for the alignment of structures with low similarity.

Any smooth space curve can be uniquely represented by its curvature and torsion at every point on the curve.[4,5] In other words, the curvature and torsion are metric invariants characterizing smooth space curves. For the purposes of alignment we can use simpler, finite "analogues" of these infinitesimal invariants. These analogues, being also metric invariants, uniquely characterize protein chains represented as ordered sets of C_α atoms. Actually, the infinitesimal invariants historically followed the finite invariants; so, it is not fair to call the latter "analogues" of the former. In some older books on differential geometry[6] the precedence is clear, i.e., the curvature and torsion are derived as infinitesimal analogues of the finite invariants. S. Rackovsky and H. A. Scheraga introduced the finite curvature and torsion for the purposes of structural alignment in an important early paper.[7] We follow their approach with some simplifying modifications.

A protein's backbone is represented as an ordered set (sequence) of C_α atoms. The virtual bonds between the consecutive C_α atoms are represented by the vectors connecting two adjacent atoms. The lengths of the virtual bonds are equal to the norms of the corresponding vectors, by definition. The angular curvature is defined as the angle between two consecutive vectors. Any two consecutive vectors determine a bivector, which can be represented as the oriented plane or, equivalently, as their cross-product. The angular torsion can then be defined as the angle between two adjacent bivectors (planes). Thus, a protein chain with N C_α atoms is represented by N–1 virtual bonds, N–2 angular curvatures, and N–3 angular torsions. The set of these three parameters is invariant under translations and rotations, and uniquely determines the shape of the protein chain. Moreover, if we fix the initial coordinates of a C_α atom, and choose an initial direction (or plane) in the space to calculate the missing values of curvature and torsion, then the original coordinates of all the C_α atoms in the chain can be determined from the set of invariants. A proof of this statement is sketched in the paper by Rackovsky and Scheraga;[7] a detailed proof can be found in the book by Akimov.[4]

It is clear from the foregoing definitions that curvature and torsion are determined by the coordinates of three or four consecutive C_α atoms, respectively. Therefore, the smallest chain fragment that can possibly be aligned with another fragment by these criteria must consist of four consecutive C_α atoms.

According to the hierarchy of alignment strategies and methods, an adequate representation of alignments must have a hierarchical structure, comprising topological features at the top and strict distance-based (metric) characteristics at the bottom. In our system, we used a simplified approach to such a representation. Thus, the topological features are represented by the secondary structure signature, i.e., a linear sequence of secondary structure types which are present (and can be unambiguously recognized) in the given protein segment. On the lower level, the overall shape of the chain is represented as a sequence of the approximating or interpolating curve segments which fit into the given protein backbone. At the lowest level, we have a set of metric invariants (virtual bond length, angular curvature, angular torsion), which uniquely determines the given chain. All the further developments are based on this hierarchical representation and on the similarity measures defined for each level of the hierarchy.

In our simplified model, the topological and shape-based characteristics are represented as sequences of basic types (symbols), which we call signatures. This representation makes possible the application of some well-known sequence similarity (homology) measures, the simplest of which is the number of matching symbols. Thus, we can speak about the homology of secondary structure signatures. To compare primary sequences of the structurally aligned fragments, we use the number of matching residues normalized by the length (i.e., the number of residues) of the fragments.

As to the metric similarity measures, we are inclined to give preference to the maximum interatomic distance (MAXD) rather than to the commonly accepted RMSD. RMSD poorly reflects the distribution of the interatomic distances in the aligned segments and is not informative enough to be taken as the sole or the main similarity criterion. On the other hand, MAXD immediately gives the upper bound of the interval comprising all the interatomic distances, including the RMSD itself. To combine the best features of RMSD and MAXD, we use RMSD-based criteria and algorithms for the primary (raw) structural alignment, and the MAXD-based criteria for the refinement of raw alignments and as the basic structural similarity criterion.

18.3. Algorithms and Their Implementation

The designed software system comprises various interconnected components, the most important of which are:

- the structural alignment module,
- the structural alignment web server,
- the structure prediction models assessment web server,
- the database (repository) of aligned fragments/domains.

In this section we describe the structural alignment module; the implementation of the web servers is outlined in the next section.

A protein chain is represented as the sequence of *x*, *y*, and *z* coordinates of its C_α atoms. For further shape comparison, this sequence is transformed into the naturally ordered set of angular curvature, angular torsion, and virtual bond length values as described in the previous section. The curvature/torsion pairs are discretized; that is, they are rounded to integer values with a tunable threshold from 2 to 12 degrees. This discretization allows us to use a fast comparison algorithm based on sequence matching, which finds all the possible matches of the fragments longer than three consecutive residues. Therefore, the requirement of completeness is satisfied. The user sets the (local) matching criteria as the maximum angular (in degrees) or interatomic (in Ångstrøms) distance, and the minimum number of residues in the matching fragments.

In the second stage, the matched fragments are combined into possibly longer fragments of two different types: continuous and broken. The latter type we shall call "domain." The continuous fragments consist of a number of contiguous residues in their natural order, i.e., without gaps. The domains comprise two or more separate continuous fragments. This distinction is conventional and relative, since the types and the lengths of aligned fragments depend on the similarity criteria.

Matched fragments are combined if they meet two similarity criteria. The first is that the composition of the rotation and the translation that superposes the two components of the first pair also superposes the two components of the second pair. The second criterion is that the MAXD be below a threshold. We compute the similarity of the two rotations by first computing the two 3×3 rotation matrices $\boldsymbol{R_1}$ and $\boldsymbol{R_2}$ that superpose the corresponding components.[10,11] The angular distance D_a between $\boldsymbol{R_1}$ and $\boldsymbol{R_2}$ is defined as

$$D_a = (trace(\boldsymbol{R_1}^{-1}\boldsymbol{R_2})) - 1)/2.$$

Since the trace is invariant under an invertible linear transformation of the coordinate system ($trace(\boldsymbol{B}^{-1}\boldsymbol{AB}) = trace(\boldsymbol{A})$ for any matrix $\boldsymbol{A}$ and an invertible matrix ***B***), the angular distance does not depend on the initial orientations of the two fragments.

The estimate of the MAXD is based on the following data:

- the angular distance,
- the distance between the centroids of the two matched fragments,
- the norm of the longest of vectors connecting the atoms in a fragment with its centroid.

We use a well-known algorithm to divide up all the matched fragments into equivalence classes, which form continuous fragments or domains according to the above equivalence (similarity) criteria. This algorithm is described in section 2.3.3 (algorithm E.) of D. Knuth's book,[8] but our implementation is closer to that presented in section 8.6 of the "Numerical Recipes in C."[9] All the fragments from an equivalence class, when combined, form a unique continuous fragment or domain.

In the third stage, we superpose the fragments or domains, using the method developed by W. Kabsch[10] (and taking into account the corrections from the subsequent paper.)[11] As a result, we have the list of all possible alignments (satisfying the requirement of completeness) with the associated RMSD and MAXD values. This list can be sorted by different values, e.g., the aligned fragment length, RMSD or MAXD, the number of fragments in a domain, etc. Some alignments may be removed from the list if they are considered by the user as uninteresting (e.g., a helix-to-helix alignment, RMSD or MAXD too large, fragment length too short, etc.) The alignments from this list can be further refined using stricter similarity criteria.

As the final stage of the process, we have a simple algorithm, which builds primary sequence alignments based on the list of structural alignments. These primary sequence alignments are used to build pattern libraries.

18.4. Representation of Data in XML Forms

An issue of great importance is the representation of the results. Since we aimed at the development of an integrated system comprising the alignment and the structure prediction assessment modules, we must have a common form of representation for those modules. The main idea is to have a form that needs only minor pre- and post-processing when transferred from one module to another, and can be kept in the same database serving both modules. In our opinion, it would be convenient to base such a form on XML (eXtendable Markup Language) specifications,[12] because this format is easily extendable to include new elements and their parameters, it is a free format (i.e., the interpretation of data is not position-dependent.) It is well-structured and, if properly specified, is easy to understand by a non-technical user. Moreover, powerful tools have been developed in many popular scripting/programming languages which facilitate fast implementation of interfaces between XML forms, web servers, and databases. To illustrate the convenience of XML specifications, we give a commented example of XML form below. This form represents a predicted model in terms of the CASP4 AL format, though slightly modified.

An example of XML form:

```
<?xml version="1.0"?>
<model id="example" format="CASP4_AL">
	<target id="1G29"></target>
	<probe id="1BOU"></probe>
	<method type="HMM" id="SPECTRM">
		<parameter P1="0.2" P2="A" P3="32.4></parameter>
	</method>
	<alignment>
		<score type="size" value="68"></score>
		<fragment id="1" target="A140-A164" probe="A154-A178">
			<score type="RMSD" value="1.264"></score>
			<score type="MAXD" value="2.955"></score>
			<score type="size" value="25"></score>
		</fragment>
		<fragment id="2" target="A190-A203" probe="A203-A216">
			<score type="RMSD" value="1.603"></score>
			<score type="MAXD" value="2.972"></score>
			<score type="size" value="14"></score>
		</fragment>
		<fragment id="3" target="A157-A169" probe="A171-A183">
			<score type="RMSD" value="1.580"></score>
			<score type="MAXD" value="2.915"></score>
			<score type="size" value="13"></score>
		</fragment>
	</alignment>
</model>
```

This form corresponds to the results of an actual prediction model assessment represented by the table from our web-server (see three upper rows in Table 18.3.1, Section 18.7.) The form contains several sections, which may be nested, each section beginning and ending with a recognizable keyword. Within a section there is a set of subsections, specifying such entities as the range of the aligned residues and different scores assigned to each aligned fragment. The nested structure of the XML form allows us to use the same keywords in different sections and with different interpretations. Thus, the "score" in the "alignment" section means the total size of the aligned fragments, but the same keyword with different "type" modifiers in the "fragment" subsections, specifies the values of RMSD, MAXD, and the size of a separate fragment. The main advantage of the XML specification is its ability to represent the results of structural alignment, structure prediction, and prediction accuracy assessment in a uniform, well-structured manner. On our web-server, the XML forms are used for the input data specifications, for the representation of results, and for the exchange of the data between different web-server modules.

18.5. Timing

The timing tests show that our alignment software is one of the fastest among other similar programs, without losing precision or convenience. For example, for medium-size proteins (300–500 residues), alignments take less than 0.3 sec on 600 MHz Pentium-III machines.

18.6. Web-Servers

The structural alignment web-server tries to find all possible structural alignments between given proteins or their fragments specified by the user. The user also sets up the similarity criteria for an alignment and, if needed, selects the fragments and chains to be aligned. The protein structures are either taken directly from the Protein Data Bank repository at SDSC or uploaded to the server by the user. The similarity criteria determine the upper bounds for acceptable interatomic distances—MAXD, or for the "average" interatomic distance—RMSD. As the result of an alignment, if there is any, the user obtains one or more tables comprising the following data:

- the list of the aligned fragments or domains, specified by their residue numbers, along with their lengths (sizes),
- MAXD and RMSD for the aligned fragments,
- the profile of interatomic distances along the aligned protein chains,
- the primary sequence alignment based on the structural alignment,
- the secondary structure signatures of the aligned fragments,
- the superposed proteins in the PDB format,
- pictures of the superposed proteins in the VRML format.

The web-server address: http://www.npaci.edu/CCMS/sa

18.7. Illustrative Examples

Table 18.1 shows the six longest alignments for the 1G29/1B0U pair, as an illustration. 1G29[13] is the ATPase subunit of the maltose transport protein from *Thermococcus litoralis* and 1B0U[14] is the histidine permease from *Salmonella typhimurium*; both are ABC transporters. The table contains the list of the aligned fragments sorted by fragment size, i.e., the number of residues in the fragment. Each fragment is described by its number, size, the number of continuous subfragments ("Pieces" column), the first and the last residues of the segment comprising the fragment ("Segments" column), and the alignment scores for the fragment ("RMSD" and "MAXD" columns). The primary sequence alignment, based on the structural alignment for the first domain from Table 18.1, is shown in Table 18.2. In this table, the secondary structure signature is represented with symbols "h" and "s"; "h" stands for atoms in a helix, "s"—for atoms in a strand. The primary sequence similarity is measured as the number of identical residues in the aligned fragments.

The profile plot (Fig. 18.1) shows the distribution of the interatomic distances (in Ångstrøms) along the aligned protein chains (for the first domain). Such plots may be instrumental in selecting fragments for further, refined alignments, i.e., the alignments with stricter similarity criteria.

Table 18.1. The alignment table for 1G29 (chain A) and 1B0U (chain A); only the six biggest domains are shown. The alignment images for the first and second domains are shown in Plates 18.1 and 18.2.

SUMMARY ALIGNMENT TABLE FOR 1G29:A/1B0U:A						
Domains			Segments		Distance	
Domain	Size	Pieces	1G29:A	1B0U:A	MAXD	RMSD
1	92	6	A 3 - A 227	A 6 - A 240	1.987	1.124
2	87	6	A 4 - A 227	A 7 - A 240	1.978	1.106
3	83	7	A 4 - A 224	A 7 - A 237	1.949	1.114
4	64	3	A 91 - A 161	A 103 - A 175	1.967	0.908
5	43	3	A 98 - A 162	A 110 - A 176	1.998	1.169
6	22	1	A 108 - A 129	A 121 - A 142	1.660	0.594

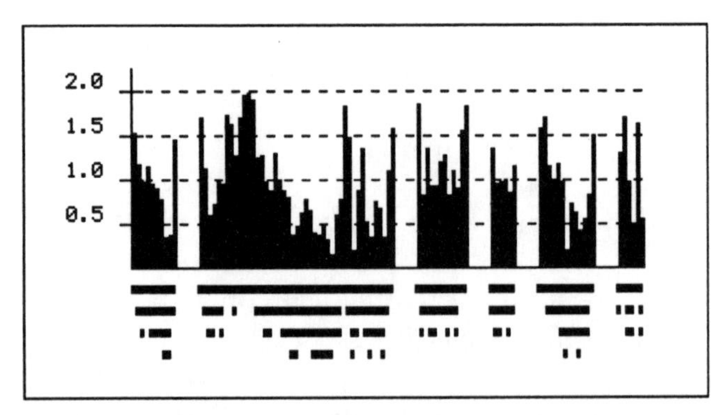

Figure 18.1. The interatomic distances profile for the domain No. 1 in the Table 18.1. The distances are shown at four cut-off levels from 0.5 to 2.0 Å. The set of bars under the plot represents the same profile at the mentioned levels. The continuous fragments are separated with blank spaces. See also the corresponding primary sequence alignment in Table 18.2.

The prediction assessment web-server is based on the same structural alignment code, but represents the results in a different form. The results are shown in two tables (see Tables 18.3.1 and 18.3.2 and comments to Table 18.1). In Table 18.3.1, the predicted superposition of the probe onto the target is compared with the superposition obtained from the structural alignment server for all the fragments specified in the input model. Table 18.3.2 shows all the alignments found by the structural alignment server without this restriction, i.e., ignoring the intervals of residue numbers specified in the model file, but operating in the same range of residues determined by the first and the last residue in the model file.

Table 18.2. The primary sequence alignment based upon the structural alignment of the first domain in Table 18.1. The recognized secondary structure elements are marked as "h" (helix) or "s" (β-strand). The alignment image for this domain is shown in Plate 18.1.

Domain	Size	MAXD	RMSD
1	92	1.987	1.124

Piece	1G29:A	1B0U:A	MAXD	RMSD	Size	Identical AA	
1	A 3 - A 12	A 6 - A 15	1.545	1.043	10	2	20.00 %

```
GVRLVDVWKV
KLHVIDLHKR
 ssss sss
```

Piece	1G29:A	1B0U:A	MAXD	RMSD	Size	Identical AA	
2	A 17 - A 61	A 20 - A 64	1.987	1.108	45	18	40.00 %

```
TAVREMSLEVKDGEFMILLGPSGCGKTTTLRMIAGLEEPSRGQIY
EVLKGVSLQARAGDVISIIGSSGSGKSTFLRCINFLEKPSEGAII
  ssss sss    sssss      hhhhhhhh        sss
```

Piece	1G29:A	1B0U:A	MAXD	RMSD	Size	Identical AA	
3	A 157 - A 168	A 171 - A 182	1.864	1.276	12	7	58.33 %

```
KPQVFLMDEPLS
EPDVLLFDEPTS
   sssss
```

Piece	1G29:A	1B0U:A	MAXD	RMSD	Size	Identical AA	
4	A 192 - A 197	A 205 - A 210	1.368	1.066	6	3	50.00 %

```
TTIYVT
TMVVVT
 ssss
```

Piece	1G29:A	1B0U:A	MAXD	RMSD	Size	Identical AA	
5	A 207 - A 219	A 220 - A 232	1.701	1.064	13	1	7.69 %

```
GDRIAVMNRGVLQ
SSHVIFLHQGKIE
  ssssss sss
```

Piece	1G29:A	1B0U:A	MAXD	RMSD	Size	Identical AA	
6	A 222 - A 227	A 235 - A 240	1.710	1.218	6	3	50.00 %

```
GSPDEV
GDPEQV
s hhhh
```

Table 18.3.1. The comparison of the alignments made by the server and the alignments from a structure prediction model. The server performed alignment only for the residues specified in the model.

MODEL/SERVER MATCHED ALIGNMENT FOR 1G29:A/1B0U:A							
Model			Server			Distance	
Size	1G29:A	1B0U:A	1G29:A	1B0U:A	Size	MAXD	RMSD
25	140 - 164	154 - 178	A 140 - A 164	A 154 - A 178	25	2.955	1.264
14	190 - 203	203 - 216	A 190 - A 203	A 203 - A 216	14	2.972	1.603
13	157 - 169	171 - 183	A 157 - A 169	A 171 - A 183	13	2.915	1.580
10	175 - 184	189 - 198	A 175 - A 184	A 189 - A 198	10	2.704	2.261
6	190 - 195	203 - 208	A 190 - A 195	A 203 - A 208	6	2.078	1.367
Total matched size: 68							

Table 18.3.2. The alignment made by the server without relying on information from a structure prediction model. (See Table 18.3.1, Plate 18.3, and comments in the text.)

SERVER ALIGNMENT FOR 1G29:A/1B0U:A						
Domains			Segments		Distance	
Domain	Size	Pieces	1G29:A	1B0U:A	MAXD	RMSD
1	37	3	A 157 - A 203	A 171 - A 216	2.972	1.798
2	31	2	A 140 - A 195	A 154 - A 208	2.955	1.284
3	16	1	A 141 - A 156	A 188 - A 203	1.001	0.435
4	7	1	A 129 - A 135	A 185 - A 191	1.110	0.640
5	7	1	A 192 - A 198	A 222 - A 228	2.739	1.217
6	6	1	A 191 - A 196	A 94 - A 99	2.342	1.139

Acknowledgements

We would like to thank William L. Nichols for his important contributions to the design and implementation of the alignment subsystem at the early stages of its development in 1994–2000. This research was supported by grants from the National Science Foundation (DBI 9911196) and the National Institutes of Health (5P41RR008605).

References

1. Lackner P, Koppensteiner WA, Sippl MJ, Domingues FS: **ProSup: A refined tool for protein structure alignment.** *Protein Eng* 2000, **13**(11):745–752.

2. Kabsch W, Sander C: **Dictionary of protein secondary structure: Pattern recognition of hydrogen-bonded and geometrical features.** *Biopolymers* 1983, **22**:2577–2637.

3. Weyl H: The *Classical Groups: Their Invariants and Representations.* Princeton NJ: Princeton University Press, 1939.

4. Aminov Y: *Differentsial'naya g'eometriya i topologiya krivykh.* (*Differential Geometry and Topology of Curves*; in Russian). Moscow: Nauka, 1987.

5. O'Neill B: *Elementary Differential Geometry.* San Diego: Academic Press, 1997.

6. Aoust X: *Analyse infinitésimal des courbes dans l'espace.* Paris: Gauthier-Villar, 1876.

7. Rackovsky S, Scheraga HA: **Differential geometry and polymer conformation I: Comparison of protein conformations.** *Macromolecule* 1978, **11**(6):1168–1174.

8. Knuth D: *The Art of Computer Programming.* Vol. 1. Boston, MA:Addison-Wesley Publishing Company, Inc., 1968.

9. Press WH, Teukolsky SA, Vetterling WT, Flannery BP: *Numerical Recipes in C.* Cambridge, U.K.: Cambridge University Press, 1992.

10. Kabsch W: **A solution for the best rotation to relate two sets of vectors.** *Acta Cryst* 1976, **A32**:922–923.

11. Kabsch W: **A discussion of the solution for the best rotation to relate two sets of vectors.** *Acta Cryst* 1978, **A34**:827–828.

12. Harold ER, Means WS: *XML in a Nutshell.* Cambridge, MA: O'Reilly & Associates, Inc., 2001.

13. Diederichs K, Diez J, Greller G, Mueller C, Breed J, Schnell C, Vonrhein C, Boos W, Welte W: **Crystal structure of MalK, the ATPase subunit of the trehalose/maltose ABC transporter of the archaeon *Thermococcus Litoralis*.** *EMBO J* 2000, **19**:5951–5961.

14. Hung L-W, Wang IX, Nikaido K, Liu P-Q, Ames GF-L, Kim S-H: **Crystal structure of the ATP-binding subunit of an ABC transporter.** *Nature* 1998, **396**:703–707.

19

Comparative Analysis of Protein Structure: New Concepts and Approaches for Multiple Structure Alignment

Chittibabu Guda, Eric D. Scheeff, Philip E. Bourne, and Ilya N. Shindyalov

Abstract

Multiple structure alignment is of increasing importance as we move into the era of structural genomics that will bring forth a large number of unannotated structures, for which automated functional assignments will be needed. This paper presents evidence that the Monte Carlo (MC) algorithm proposed here can contribute in this regard and discusses some new concepts useful in further development of structure alignments. The overall effect of the MC optimization is the compression of many of the gapped regions generated between and at the ends of secondary structural elements within the multiple alignments, resulting in increased alignment length. Two specific families, protein kinases and aspartic proteinases were tested and compared against curated alignments from HOMSTRAD and manual alignments. This algorithm has improved the overall number of aligned residues by preserving key catalytic residues and excluding residues from aligned regions which do not fit well, but were present in the initial alignment. Further refinement of the method and its application to generate multiple alignments for all protein families in the PDB, is currently in progress.

19

Comparative Analysis of Protein Structure: New Concepts and Approaches for Multiple Structure Alignment

**Chittibabu Guda, Eric D. Scheeff,
Philip E. Bourne, and Ilya N. Shindyalov**

*University of California, San Diego; San Diego
Supercomputer Center, La Jolla, California*

19.1. Introduction

Rapid growth of protein structure data over the past decade has prompted the development of sophisticated methods for the alignment of protein structures. Many algorithms have been developed for the pair-wise alignment of protein structures.[1-4] However, very few approaches are available for the alignment of multiple structures.[5-7] A global and comprehensive study of protein structures is possible only by comparison of multiple structures and investigation of their folding similarities and evolutionary relationships. With the availability of vast amounts of structural information, accurate and fully automated structural alignment algorithms are needed for a better understanding of sequence-structure-function relationships in proteins. Recently, we have developed a new algorithm for the alignment of multiple protein structures using Monte Carlo (MC) optimization method.[8] It was shown that the quality of the alignment optimized by MC is far exceeding the quality of alignments obtained with traditional strategy. Here we discuss some new concepts useful in further development of structural alignment using Monte Carlo algorithm.

19.2. Algorithm for Aligning Multiple Protein Structures Using Monte Carlo Optimization

A new algorithm (MC-CE) has been developed for the alignment of multiple protein structures based on a Monte Carlo optimization technique. The algorithm uses pair-wise structural alignments as a starting point. Four different types of moves were designed (Fig. 19.1) to generate random changes in the alignment.[8] A distance-based score is calculated for each trial move; and moves are accepted or rejected based on the improvement in the alignment score until the alignment is converged.

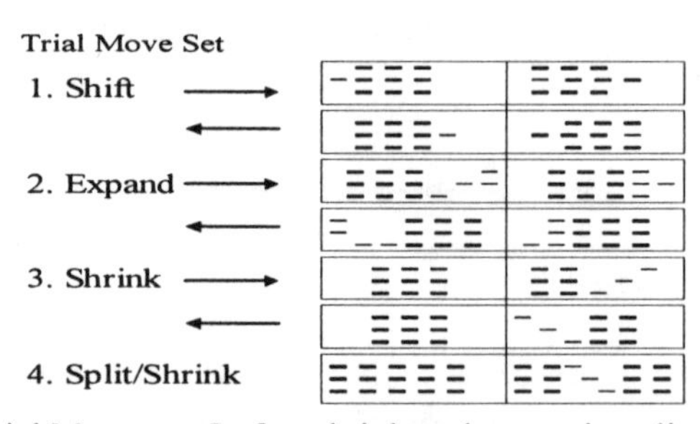

Figure 19.1. Trial Move set. Left and right columns give alignment before and after move, respectively. Solid and split strips represent residues in the aligned (Blocks) and unaligned (Free pool) regions, respectively. Location of alignment blocks shown in boxes. 1—shift, 2—expand, 3—shrink, 4—split and shrink.

19.2.1 Scoring Function

A distance-based score was calculated for each column in the alignment block. Geometric distances were calculated from the 3-D coordinates of

C_α atoms for each pair of residues in a column for $R(R-1)/2$ combinations, where R is the number of residues in a column. Column distances were defined as average geometric distances calculated for each column. The alignment score S was calculated (similar to Gernstein and Levitt, 1998)[9] from the column distances in aligned blocks, using the following scoring function:

$$S = \sum_{i=0}^{l} \left[\frac{M}{1+(d_i/d_0)^2} - A \right] - G \ , \qquad (19.1)$$

where l is the total number of aligned columns, $M = 20$ is the maximum score of a match, d_i is the average distance for column i, d_0 is the maximum distance that is not penalized, $A = \begin{cases} 0, \text{if } d_i \leq d_0 \\ 10, \text{if } d_i > d_0 \end{cases}$, and G is the affine gap penalty term with gap initiation and gap extension penalties of 15 and 7, respectively.

19.3. Approaches for Optimization of Multiple Structure Alignment

In the original algorithm, the value of d_0 is chosen from the initial distance distribution for all columns in the multiple alignments at the boundary for the top 10% of column distances. However, by adjusting d_0 values together with other parameters, multiple alignments can be obtained at different precisions. We have investigated the combined effect of different d_0 values along with weights based on the number of residues in the columns, on the alignment length and alignment distance.

19.3.1. Effect of Weights Based on Number of Residues on Alignment Length and Alignment Distance

Individual column scores were weighed based on the number residues present in the column. Columns with no gaps contribute full score to the total score while the ones with gaps contribute partial score, depending upon the number of gaps. To reflect this idea, the original scoring function (19.1) was modified as follows:

$$S = \sum_{i=0}^{l} \left[\left(\left(\frac{M}{1 + (d_i / d_0)^2} \right) * (c / n) \right) - A \right] - G, \qquad (19.2)$$

where c is the number of residues in the column and n is the total number of structures being aligned.

Monte Carlo runs were obtained for a set of 7 families at different d_0 values with the new scoring function (Fig. 19.2) and average results were compared with those obtained from the original scoring function (Fig. 19.3). When no weights were used in the scoring function, the alignment length increased sharply at lower d_0 values (Fig. 19.2). However, when weights were used, similar increase in the alignment length was observed at higher d_0 values (Fig.19.3). This is because, in the former case, new columns contribute full score to the total score resulting in higher acceptance of trial moves; while only partial score is added in the latter case, which slows down the rate of acceptance of trial moves. However, at higher d_0 values, more residues are added to the new columns resulting in near complete column scores added to the total score that increases the alignment length. As expected, at higher d_0 values residues with higher pair-wise distances are incorporated into the alignment resulting in rapid increase in the alignment distance (Fig. 19.3) compared to the scoring scheme where no weights were used (Fig. 19.2).

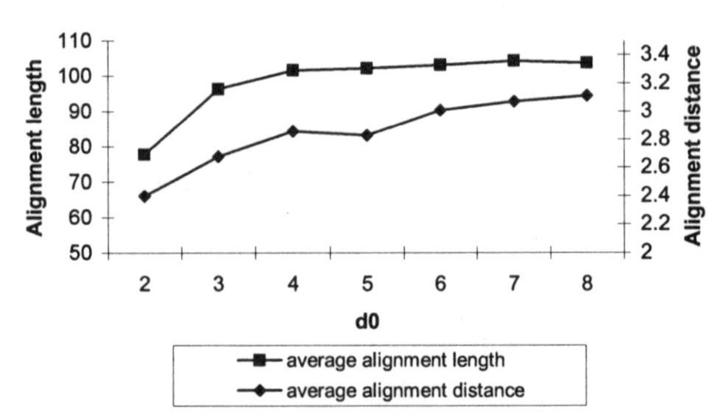

Figure 19.2. Effect of d_0 on the alignment length and alignment distance using scoring function (19.1) where, full column scores are used irrespective of the number of residues in the column.

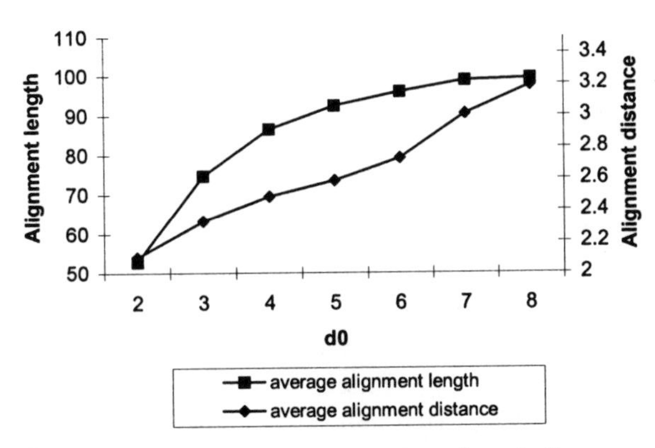

Figure 19.3. Effect of d_0 on the alignment length and alignment distance using scoring function (19.2) where, column scores are proportional to the number of residues in the column.

19.4. Analysis of Specific Protein Families

Two specific protein families have been selected for further evaluation of MC algorithm: protein kinases and aspartic proteinases. MC optimized alignments have been compared to assembled pair wise CE alignments and to curated (manually optimized) alignments. Curated alignments for aspartic proteinases have been obtained from HOMSTRAD.[10] Curated alignment for protein kinases have been built in a separate research effort, briefly described below.

19.4.1. Analysis of an Alignment of Protein Kinases

An assessment of the MC algorithm was performed by aligning a set of 17 divergent protein kinase catalytic cores taken from the PDB. The representatives were chosen such that the sequence identity upon structural alignment with CE was <50% between any two structures. The protein kinases are composed of a small, mostly beta sheet N-terminal domain and a larger, mostly alpha-helical C-terminal domain joined by a flexible hinge region. Residues important for ATP binding and phosphotransfer line the active site cleft between the domains.[11] The MC algorithm has improved the alignment score by 153%, which is accompanied by a 17% increase in number of aligned columns, a 19% increase in average alignment distance, and a 29% reduction in the total alignment length.

A hand-curated alignment was established by careful examination and adjustment of the automated CE structural alignments, curated alignments provided in the HOMSTRAD database,[10] and two reviews.[11,12] Much of the MC alignment largely agreed with the curated alignments, particularly in the C-terminal subunit, a less flexible portion of the kinase structure. Two alignment sections of the N-terminal subunit are highlighted in Figs. 19.4 and 19.5. These sections represent the most challenging sections of the protein to align because of the positional variability seen in this region of the molecule.[13] They also illustrate the improvements the MC optimization provides over the standard CE protocol.

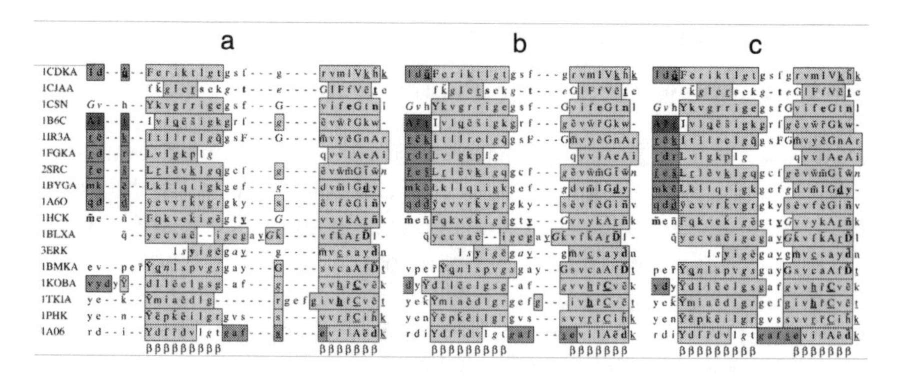

Figure 19.4. Alignment of strand 1, the glycine rich loop, and strand 2 of the protein kinases, by standard CE, CE + MC, and hand curated alignment. Alignment is shown in the JOY format[14] which annotates the sequence alignment for structural features. Shaded boxes: *light gray*: β-strand, *medium gray*: 3-10 helix, *dark gray*: α-helix. Residue (letter) characteristics: *uppercase*: solvent inaccessible, *lowercase*: solvent accessible, *italic*: positive Φ, *breve(˘)*: cis-peptide, *tilde(~)*: hydrogen bond to other sidechain, **bold**: hydrogen bond to mainchain amide, underline: hydrogen bond to mainchain carbonyl. Blank regions signify portions of the sequence which lack atomic coordinates in the structure.[8] Courtesy of World Scientific Publishing Co. Pte. Ltd. (Singapore).

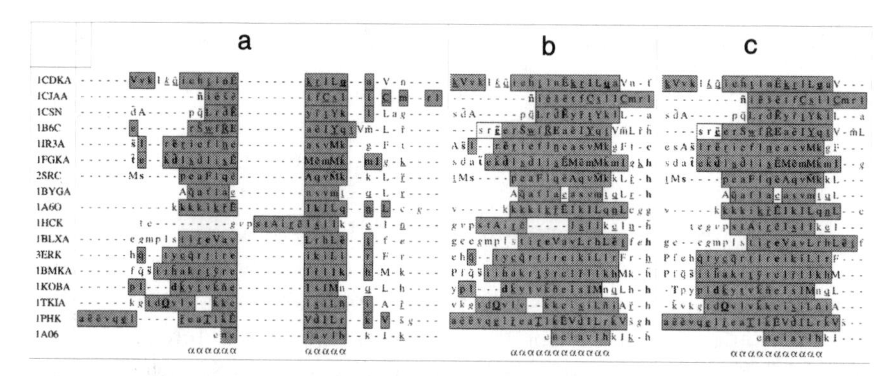

Figure 19.5. Alignment of helix C of the protein kinases by standard CE, CE + MC, and hand curated alignment. Alignment is shown in the JOY format. Shaded boxes: *medium gray*: α-helix, *white*: 3-10 helix.[8] Courtesy of World Scientific Publishing Co. Pte. Ltd. (Singapore).

Consider the examples:

Example 1: Strand 1 and strand 2 of the glycine-rich loop (Fig. 19.4). This region of the kinase domain is flexible and often in different conformations.[12] However, it contains the well-conserved GxGxxG motif, which is important for the binding of ATP in the active site.[11] With the exception of one structure (1CJA:A), it should be aligned without gaps to properly align this motif. Standard CE alignment splits off some of the sequence leading up to strand 1, and unnecessarily separates off a row of conserved glycines in the loop between strands 1 and 2 (Fig. 19.4a). The MC alignment compresses the sequence leading up to strand 1, and closes the gap, which causes the glycine displacement in CE (Fig. 19.4b). However, MC does not correct the misaligned glycine residues seen in some structures in the original CE alignment (Fig. 19.4b-c).

Example 2: Helix C (Fig. 19.5). This helix is found at different angles in the various protein kinase structures[12] making it difficult to align. However, it should be aligned without gaps, and a highly conserved Glu residue should be lined up in all structures. The standard CE alignment produces multiple small gaps in the alignment at the ends of the helix, as well as one large gap based on 1HCK (Fig. 19.5a). Inspection of 1HCK reveals that helix C is displaced and rotated to a particularly large degree in this structure. The MC alignment compresses most of the gaps at the ends of the helix and realigns the improperly gapped section (Fig. 19.5b). However, it does not correct the misaligned Glu residues seen in some structures in the original CE alignment (Figs. 19.5a-c).

19.4.2. Analysis of an Alignment of Aspartic Proteinases

We have selected aspartic proteinases, which are composed of a high proportion of beta sheets and relatively few alpha helices, as a second family for testing the MC algorithm. Important members of this family are renins (1BBS:_, 1SMR:A), pepsins (5PEP:_, 1PSN:_, 1JXR:A, 1MPP:_, 1AM5:_) and proteinases (3APP:_, 4APPE:_, 2APR:_, 2ASI:_) that are associated with several pathological conditions in humans. We have used the same 12 structures as classified by HOMSTRAD under this family for ease of comparison and reference. These structures have an average sequence similarity of 37%. Seed alignments were constructed from CE pair-wise data as explained previously, using 3APP:_ as the master.

The MC algorithm has improved the alignment score by 19%, which is accompanied by a 10% increase in number of aligned columns, a 13% increase in average alignment distance, and a 35% reduction in the total alignment length. Many improvements were observed in the overall alignment especially in the areas that CE failed to align properly. Due to space limitations, examples of only two major improvements are presented in Figs. 19.6 and 19.7 (For an explanation on the structural features of residues in JOY format refer to Fig. 19.4).

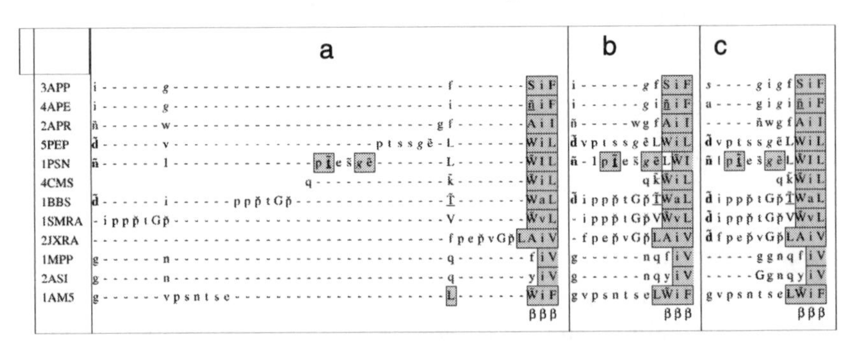

Figure 19.6. Comparison of multiple alignments generated by CE, MC and HOMSTRAD, in the poly-proline segment of aspartic proteinases family. Shaded boxes: *light gray*: β-strand.[8] Courtesy of World Scientific Publishing Co. Pte. Ltd. (Singapore).

Consider the examples:

Example 1: Aspartic proteinases exhibit a bilobal structure with an active site cleft in the middle of two lobes. On the opposite side of the active site cleft, there is a "poly-proline" loop contributed by the C-terminal domain. In the case of renins, the sequence contains ~P-P-P-T-G-P~ (although the analogous structure in some aspartic proteinases contain fewer or even no proline residues) that influences binding of the S2' and S3' pockets in the active site cleft.[15] In the CE alignment this region is widely spread out with no alignment of the poly-proline residues (Fig. 17.6a). However, this region is well aligned by the MC algorithm in all the three chains (1BBS:_, 1SMR:A, 2JXRA) that contain this region (Fig. 17.6b) and these results compare well with the HOM-STRAD alignments (Fig. 17.6c).

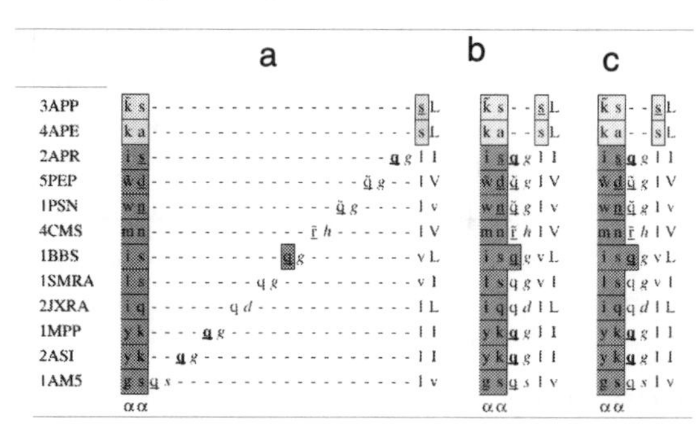

Figure 19.7. Comparison of multiple alignments generated by CE, MC and HOMSTRAD, in the N-terminal lobe of aspartic proteinases family. Shaded boxes: *medium gray*: α-helix, *white*: 3-10 helix.[8] Courtesy of World Scientific Publishing Co. Pte. Ltd. (Singapore).

Example 2: Another improvement is seen in the loop region between an α-helix and a β-sheet in the N-terminal lobe. As seen in Fig. 19.7, the CE alignment has spread out the residues that are well conserved in all but the first two structures (Fig. 19.7a), whereas the MC algorithm has realigned these residues (Fig, 19.7b) making it comparable to that of the HOMSTRAD alignment (Fig. 19.7c).

19.5. Summary

Multiple structure alignment is of increasing importance as we move into the era of structural genomics that will bring forth a large number of unannotated structures, for which automated functional assignments will be needed. This chapter presents evidence that the MC algorithm proposed here can contribute in this regard. The application of MC improves some aspects of CE pair-wise alignments considerably, while other aspects are minimally affected. Most importantly, the MC optimization does not introduce any significant new errors into the alignment. The effects of the optimization are nearly always positive. The overall effect of the optimization is the compression of many of the gapped regions generated between, and at the ends of, secondary structural elements within the multiple alignments. The MC optimization has little effect within regions of the alignments where a small misalignment error is present well within a large block of aligned structure/sequence. These types of errors do not seem to benefit from MC optimization. Current work seeks to: (i) further explore the behavior of MC by empirical means; (ii) compare MC to other automated multiple alignment techniques; (iii) provide a public database of aligned protein families.

Acknowledgment

This work was supported through NSF grant DBI 9808706.

References

1. Taylor W, Orengo C: **Protein structure alignment.** *J Mol Biol* 1989, **208**:1–22.
2. Holm L, Sander C: **Protein structure comparison by alignment of distance matrices.** *J Mol Biol* 1993, **233**:123–138.
3. Shindyalov IN, Bourne PE: **Protein structure alignment by incremental combinatorial extension (CE) of the optimal path.** *Protein Eng* 1998, **11**:739–747.
4. Mirny LA, Shakhnovich EI: **Protein structure prediction by threading. Why it works and why it does not.** *J Mol Biol* 1998, **283**:507–526.
5. Leibowitz N, Fligelman ZY, Nussinov R, Wolfson HJ: **Multiple structural alignment and core detection by geometric hashing.** *ISMB* 1999:169–177.

6. Feng DF, Doolittle RF: **Progressive sequence alignment as a prerequisite to correct phylogenetic trees.** *J Mol Evol* 1987, **25**:351–369.

7. Thompson JD, Plewniak F, Poch O: **A comprehensive comparison of multiple sequence alignment programs.** *Nucleic Acids Res* 1999, **13**:2682–2690.

8. Guda C, Scheeff ED, Bourne PE, Shindyalov IN: **A new algorithm for the alignment of multiple protein structures using Monte Carlo optimization.** *Proc Pacific Symp Biocomput 2001,* Altman RB, Dunker AK, Hunter L, Lauderdale K, Klein TE, eds. Singapore: World Scientific Publishing Co., 2001:275–286.

9. Gerstein M, Levitt M: **Comprehensive assessment of automatic structural alignment against a manual standard, the SCOP classification of proteins.** *Protein Sci* 1998, **7**:445–456.

10. Mizuguchi K, Deane CM, Blundell TL, Overington JP: **HOMSTRAD: A database of protein structure alignments for homologous families.** *Protein Sci* 1998, **7**:2469–2471.

11. Taylor SS, Radzio-Andzelm E: **Three protein kinase structures define a common motif.** *Structure* 1994, **2**:345–355.

12. Hanks SK, Hunter T: **Protein kinases 6. The eukaryotic protein kinase superfamily: Kinase (catalytic) domain structure and classification.** *FASEB J* 1995, **9**:576–596.

13. Sowadski JM, Epstein LF, Lankiewicz L, Karlsson R: **Conformational diversity of catalytic cores of protein kinases.** *Pharmacol Ther* 1999, 82:157–164.

14. Mizuguchi K, Deane CM, Blundell TL, Johnson MS, Overington JP: **JOY: Protein sequence-structure representation and analysis.** *Bioinformatics* 1998, **14**:617–623.

15. Humphreys MJ, Berry C: **Aspartic proteinases from the animal parasites *Plasmodium berghei* and *Eimeria tenella*.** In: *Structure and Function of Aspartic Proteinases: Retroviral and Cellular Enzymes*, James MNG, ed. NewYork: PlenumPress, 1998:416.

20

Comparative Analysis of Protein Structure: Automated vs. Manual Alignment of the Protein Kinase Family

Eric D. Scheeff, Philip E. Bourne, and Ilya N. Shindyalov

Abstract

Alignment of similar protein structures can provide extremely useful data. Therefore, a variety of methods have been developed to automatically produce high-quality protein structure alignments. However, these methods face a range of difficulties in producing alignments that are satisfactory at a fine level of detail. These challenges will be discussed from the standpoint of specific case study, where an alignment of eukaryotic protein kinases generated using the combinatorial extension algorithm (CE) is compared with a manually derived alignment. Implications for CE will be discussed, as well as implications for automated structural alignment in general.

20

Comparative Analysis of Protein Structure: Automated vs. Manual Alignment of the Protein Kinase Family

Eric D. Scheeff, Philip E. Bourne, and Ilya N. Shindyalov

University of California, San Diego; San Diego Supercomputer Center, La Jolla, California

20.1. Introduction

The most effective way to compare two protein structures is a spatially derived (structural) alignment. A structural alignment makes it possible to directly compare protein structures, and determine the presence or absence of secondary structural elements, motifs, and critical residues. In the case of closely related proteins, structural alignments can be used to compare molecules in different conformations, such as an enzymes bound and unbound to substrate. These comparisons can illuminate critical aspects of protein activation or catalytic mechanism. Additionally, in the case of distantly related proteins, structural alignment can be used to generate an alignment of sequences that could never be aligned accurately using sequence information alone. Because of the degeneracy of protein sequence,[1,2] distantly related proteins can only be compared accurately with the inclusion of structural information. Indeed, alignments of protein structures can be used to help determine protein ancestry when two proteins are so distantly related that their sequences have no detectable similarity.[3,4]

20.2. The Challenge of Automated Protein Structure Alignment

Because the manual alignment of protein structures is often arduous, a variety of methods have been developed to provide an automated alignment of protein structures.[5-9] However, the automated alignment of protein structures presents special challenges beyond those encountered in the automated alignment of protein sequences. In protein sequence alignments, the "optimal" alignment between two sequences can be directly and quantitatively defined. Protein sequence alignment is based on the concept of residue ancestry, and, by extension, DNA codon ancestry.[10,11] Thus, the problem is one of producing the correct one-to-one correlation of residues based on the presumed ancestry of each position in the protein chains. The scoring systems for such alignments are generally based on straightforward, empirically derived models where particular residue pairings provide particular scores, with an additional score penalty for gaps (insertions and deletions). An optimal sequence alignment may not be the *correct* alignment when considered in the context of known features of the proteins being aligned, such as active site residues, functional motifs, and structural features. However, for a given scoring system, the dynamic programming method will generate an alignment (or alignments) that can be considered without question optimal, since this method considers all possible alignment solutions before determining the highest-scoring one.[12,13] Further, because of the simplicity of protein sequence alignment, the dynamic programming method is computationally tractable for all but large database searches, where heuristics are employed.[14,15]

Protein structures present a much richer, complex form of data, and as a result, automated protein structure alignments present challenges not experienced in sequence alignment. First, defining the "optimum" alignment of protein structures is not as straightforward as in sequence alignments. Thus, the definition of an ideal structure alignment is left to the interpretation of the researchers developing the method.[16] There is, at present, no agreed upon method to measure optimality in structure alignment. Root mean squared deviation (RMSD) is commonly used to provide an estimate of protein structure similarity, but it does not provide a true measure of alignment quality. Even as a measure of structure similarity, RMSD can be shown to be limited,[17] and for this reason Levitt and Gerstein have proposed a new measure which attempts to converge metrics between sequence and structure comparisons. However, it is clear that the issue of alignment *optimality* measures is far from being answered and further research is needed.

Though it will not be described in detail in this chapter, the second difficulty of protein structure alignment is that the complexity of protein structures means that the problem of aligning them is NP-hard, and thus a truly rigorous alignment method cannot be created.[18] Therefore, heuristics must be used in the alignment process. Even if an optimal alignment could be defined in a universal manner, these heuristics would inevitably produce some degradation in the output quality of the method.

20.3. A Case Study: Alignment of the Eukaryotic Protein Kinases and Their Relatives

Some of the challenges inherent in protein structural alignment will be detailed here in the context of a specific set of alignments. We have generated a hand-curated multiple structure alignment of 18 divergent eukaryotic protein kinase catalytic cores and their homologues for purposes of evaluation of the Combinatorial Extension (CE) protein structure alignment algorithm.[9] This alignment was carefully prepared through direct visual observation of superpositions of the aligned structures, and adjustment to yield an "optimal" result, based on criteria that will be described below. All available information was used in the preparation of the alignment, such as peptide backbone configuration, residue position, and secondary structure types. Information on conserved residues in the family and alignments from two expert reviews were also taken into account.[19,20] The aim of the alignment was to produce a result that had maximal biological meaning, while still being structurally reasonable (in effect, an "expert" alignment.)

The protein kinases are phosphotransferases that possess a distinctive structure composed of two domains. A tyrosine kinase, c-Src (PDB id 2SRC), is shown in Plate 20.1. Conserved secondary structural elements are numbered (strands) and lettered (helices). The N-terminal domain is primarily anti-parallel β-sheet, with one or more helical insertions between the 3rd and 4th strands. The C-terminal domain is primarily helical, with some small B-sheet structure facing the N-terminal domain. The ATP binding region sits in the cleft between the two domains, and the active site residues line the opening of the cleft. The two domains are joined by a flexible hinge region, which allows the N-terminal domain to flex into an open configuration to receive ATP.[21,22] In Plate 20.1, an ATP analogue is shown in the binding site in green. Three conserved residues important in the phosphotransfer reaction are shown in yellow, D386, N391, and D404 (numbers are based on the PDB file for 2SRC.)

The structures selected for the alignment discussed here are a non-redundant representative set, selected such that no structure has a sequence identity of greater than 40% to any other. This ensures that the alignment will provide a substantial challenge to the alignment method used. Further, some of the kinases in the alignment are distant relatives, such as the Actin-Fragmin kinase (PDB id 1CJA:A).[23] This structure is distinguished primarily by a differently folded C-terminal region (though the essential elements of the active site and ATP binding regions are well preserved). It is also, because of its divergence, one of the most challenging structures to align to the standard protein kinase fold, and will form the basis for some of the discussions to follow.

20.4. An Example of an Automated Alignment: The Combinatorial Extension Algorithm

The Combinatorial Extension (CE) algorithm is a method of automatically aligning pairs or groups of structures. The method has been previously described,[9] and will be only briefly reviewed here. The CE algorithm compiles an alignment of a given pair of protein chains by considering the chains sectioned into all possible octapeptide fragments, as defined by the backbone α-carbons. Those octapeptide pairs that have a high distance-based similarity score are deemed Aligned Fragment Pairs (AFPs) and retained for the next step in the analysis. CE then tries to join each AFP to a maximal number of other AFPs in order to create the longest possible alignment path through the two proteins in consideration (with an allowance for gaps of up to 30 residues in either protein chain). In effect this "stitches together" a set of AFPs covering contiguous regions in each protein that represent a possible alignment path.

Once the possible paths through the two proteins are determined, CE uses additional heuristics in an attempt to improve the final alignment. The 20 best scoring paths are compiled and the proteins are directly compared based upon the superposition of the aligned residues. The path which yields the lowest RMSD is then retained as the "optimal path." This path is then subjected to a dynamic programming structural alignment directly between the two structures, which tests all possible residue equivalences and the resulting RMSD from their superposition. During this phase of the alignment, gaps are allowed in the initial AFPs, which cause most of the octapeptide signature to disappear in the final alignments. Note that while a dynamic programming optimization is used in this step, it does not generate a rigorous alignment, because the dynamic programming performs the optimization based on a superposition that was derived via a heuristic method.

In addition to the standard CE process described above, we have created an additional optimization step to further improve fine details of the alignment. We have tested parameters for this optimization step against manual alignments of the kinases, and from these results selected the most successful heuristic parameter set. Essentially, this step utilizes an additional dynamic programming step that explicitly incorporates coordinates of, not only α-carbons, but also backbone oxygen atoms, and β-carbons when they are present in the side chains of both residues. We have found that inclusion of these atoms in the optimization helps CE to achieve alignment results closer to our optimality criteria, probably because it provides a more complete picture of the conformation of the polypeptide than α-carbons alone.

This extended variant of CE was used to align the same 18 structure kinase set as was aligned in the manual alignment. Further, CE was directed to align only the specific sections of the structures that were aligned manually. Because CE can automatically select the sections of the proteins to align, it was important to limit the comparison to the issue of specific residue alignments by starting with an alignment of the same regions of the protein chains. Thus, the aim of this alignment was to provide a "best case" CE alignment for comparison with the "optimal" manually derived results.

20.5. Parameters for the Determination of an "Optimal" Structure Alignment

In order to create an optimal structure alignment, it is important to first define what is meant by "optimal." Consider the extreme example of a researcher aligning analogous structures (structures that do not have a common genetic ancestry). Here, residue ancestry does not have any meaning, so the researcher has no interest in attempting to align residues with likely common ancestry, or in using the structural alignment to generate an evolutionary meaningful alignment of the sequences. The focus of the alignment is on spatial positioning of residues, and the similarity that may exist between the structures despite their differing evolutionary origins. Alternately, in a case where two proteins with very similar sequences but different conformations are being aligned, the researcher may be interested in obtaining an alignment that is meaningful with respect to residue pairings, and in particular, critical residues known to be important to function. Here, correct residue pairings take precedence over spatial similarity (and the researcher may be more satisfied with a sequence-based alignment in this case!). Thus, optimality in structure alignment is somewhat dependent on the aims of the researcher aligning the structures.

Presuming that multiple implementations of an algorithm tuned to specific needs of researchers are not a possibility, a compromise must be made between these competing requirements. Because no such compromise is generally agreed upon, the general criteria used to produce our "optimal" alignment will be explicitly described. As mentioned previously, the overall goal of the alignment was to produce an "expert" alignment of the protein kinase set, which made maximal biological sense while still respecting spatial similarity in the proteins. All current structure alignment programs (including CE) seek to generate specific residue equivalences, much as are produced in protein sequence alignments. Therefore, the criteria used in the manual alignment are based around what constitutes equivalent residues:

- *Equivalent positions should be spatially similar.*
- *Residue side chains and backbone atoms of aligned positions should be doing similar things in the structure.*
- *Residues should be aligned based on evolutionary ancestry where it applies.*
- *Positions with conserved sequence should be aligned, provided they do not conflict severely with structure.*

These criteria are arbitrary, but seem reasonable based on the requirements in most structural alignments, and the way in which proteins are believed to evolve.[1] The application of these criteria, and the difficulty in achieving automated alignments that reliably meet them, will be explored in the examples that follow.

20.6. Comparison of CE Alignments with Manual Alignments

We have selected several interesting cases that are particularly illustrative of the challenges inherent in automated structural alignment. In all of these cases but one, the manual alignment displays a distinct disagreement with the results from CE. It should be noted that large sections of the structure alignment are identical between the manual and CE results. However, the sections where CE "fails" to meet our criteria are most informative and so will be primarily described here. We will use the standard helix and strand numbering conventions used for the protein kinases.[19]

***Example #1*: The difficulty caused by displacement of domains—alignment of helix C of two tyrosine kinases.**

Helix C of the protein kinases contributes residues to the ATP binding pocket, and is the only helix universally present in the n-terminal domain in every structure in the extended kinase family.[4,19] Here, an alignment of the C helix will be examined in two tyrosine kinases,

Insulin Receptor Tyrosine Kinase (PDB id 1IR3:A)[24] and c-Src (PDB id 2SRC.)[25] These two structures are reasonably similar, and can be aligned with an RMSD of 3.0Å and a sequence identity of 40% based on our manual alignment. Thus, they are two of the most closely related proteins in our selected set.

However, when these proteins are aligned, CE places a gap within the n-terminal end of the C helix in 1IR3:A relative to 2SRC. This gap occurs because of a displacement of the n-terminal domain of 2SRC relative to 1IR3:A. A superposition of the n-terminal regions of 1IR3:A (orange, PDB file residue numbers 993-1080) and 2SRC (light blue, PDB file residue numbers 264-342) is shown in Plate 20.2a. The C-terminal regions of the molecules are not shown for the sake of clarity. The view of the structure is upward from the ATP binding pocket, based on the rotation used for the kinase shown in Plate 20.1. The n-terminal domain of the kinases is connected to the c-terminal domain by a flexible hinge region, and the n-terminal domain can flex within a range of "open" or "closed" positions.[21,22] In Plate 20.2a, Helix C can be seen to begin in a highly displaced position between the two structures, and then gradually become more spatially similar as it progresses.

In the context of the CE algorithm, the gap made is reasonable because the displacement is such that the backbone atoms of the two proteins are rather distant from each other in space. CE therefore chooses to align the n-terminal section of helix C in 1IR3:A with atoms in the loop region prior to helix C in 2SRC, as shown in Plate 20.2b. The residues displayed are 1036-1056 for 1IR3:A (orange) and 299-319 for 2SRC (light blue), and constitute all of Helix C and a small amount of the connecting loops on both ends. C_α atoms are shown as small spheres, and the helix backbone is shown as a C_α trace. Green C_α atoms are paired between the two structures; yellow C_α atoms (and the yellow backbone connecting them) represent atoms that are gapped by CE in the alignment, and thus have not partner in the other structure. The two lower unpaired atoms in the image (in 2SRC, residues 304-305), are within the helix but are not paired with their corresponding atoms in the 1IR3:A helix, because the helical displacement is too great. This same gap is shown as a yellow section in Plate 20.2a.

While these residues are spatially similar, according to our other criteria their alignment is erroneous. The residues aligned by CE are not involved in similar interactions in the two structures, and they are not participating in the same secondary structural element. Further, it is more likely that the residues in the two helices have the same evolutionary origin. Therefore, our manual alignment aligns the helical residues together despite their spatial displacement, as shown in Plate 20.2c. Here, residues 304-305 in 2SRC are paired with their helical counterparts in 1IR3:A (1041-1042). These paired residues are highlighted in white. Though this alignment makes less sense in terms of pure spatial distances, it makes more biological sense; the aligned residues are both parts of corresponding helices, and they are displaced largely because of domain movements in the structures.

***Example #2*: The limitations of secondary structure as guide to structure alignment—alignment of strand 4 of Actin-Fragmin Kinase and cAMP Dependent Protein Kinase.**

The above example might suggest that our criteria call for all secondary structure to be aligned without gaps. However, ungapped secondary structure alignments can, in some cases, lead to results that are not satisfactory. An alignment of the Actin-Fragmin Kinase (PDB id 1CJA:A)[23] and cAMP-Dependent Protein Kinase (PDB id 1CDK:A.)[26] will be explored here. These two structures can be aligned with an RMSD of 3.3Å, and a sequence identity of 13%, based on our manual alignment. Thus, they are far more divergent than the tyrosine kinase pair explored earlier, but they do share a reasonably similar N-terminal region.

Strand 4 of the kinases lines the outer edge of the N-terminus, on the opposite side of the protein from the ATP binding pocket. Despite this dislocation from the active site, this element is well conserved across all members of the extended kinase family. Most of the protein kinases have a slight bend in the strand near its N-terminus, but in 1CJA, the strand is almost perfectly straight. An alignment of the n-terminal regions of the Actin-Fragmin kinase (orange, PDB id 1CJA:A, residues 40-123) and cAMP-Dependent Protein Kinase (light blue, PDB id 1CDK:A, residues 72-165) is shown in Plate 20.3a. The n-terminal domain is shown rotated 180° about the vertical axis relative to the kinase in Plate 20.1. Strand 4 is shown in yellow in 1CDK:A (residues 106-111). The bend in this strand can be seen relative to strand 4 in the 1CJA:A.

As a result of the bend in strand 4 of 1CDK:A, an alignment of the strand 4 of 1CJA:A and 1CDK:A without gaps will produce an alignment where residues in part of strand 4 of 1CJA:A are aligned with residues on the opposite side of the strand in 1CDK:A (Plate 20.3b). Plate 20.3b shows a hypothetical alignment of strand 4 (residues 103-111 for 1CDK:A and 120-127 for 1CJA:A). C_α atoms are shown as small spheres, color coded based on their alignment status and membership in secondary structure. Purple atoms represent positions aligned between the two structures that are members of the loop structure leading to strand 4. Green and white atoms represent aligned positions within the strand. The green and white positions are alternated so that pairing can be deciphered. Yellow positions represent unpaired (gapped) positions. As can be seen, paired positions in this ungapped secondary structural alignment are quite spatially distant and often on different sides of the sheet.

This alignment makes little sense, since it means that positions are being made equivalent that have side chains projecting in opposite directions in space. This directly contradicts our criteria that side chains of aligned positions must be doing similar things in the structure. Therefore, we align these strands with a mid-strand gap in 1CJA:A that re-establishes the correct pairing between the residues, as shown in Plate 20.3c. It should be noted that in this case, CE aligns these strands in exactly the same way as the manual alignment. This example is given simply to illustrate the difficulties that would be posed by an alignment scheme based only on making secondary structure positions equivalent without gaps.

***Example #3*: Disturbance in alignment of critical catalytic residues caused by a sequence insertion—Actin-Fragmin Kinase aligned to cAMP Dependent Protein Kinase.**

The examples above concern sections of the alignment that do not include functional or known conserved residues. However, alignment errors involving important residues can also occur, for example in active site loops. In fact, these can be more difficult to deal with from a structural perspective, because loop regions will tend to be more flexible, and show up in slightly different conformations in different structures.

All of the kinase structures in our alignment conserve a specific Asp residue in their c-terminal regions that is believed to constitute the catalytic base in the phosphotransfer reaction.[19,20] These structures also conserve a Asn residue that is important for coordination to a metal atom in the active site complex (with one exception, the lipid kinase phosphatidylinositol phosphate kinase IIβ (PDB id 1BO1:A.)[27] Most of the structures that have both residues display the characteristic motif DxxxxN, where x represents reasonably well conserved residues but no absolute residue type. The Asp residue is located in a loop region, while the Asn residue is located in a 3-10 helix that is followed immediately by a β-strand. The Actin-Fragmin Kinase (PDB id 1CJA:A) is the only exception, with a 9 residue loop insertion that occurs in the middle of the motif, as shown in Plate 20.4a. The two highly conserved residues, D204 and N218 (numbering from 1CJA:A), are shown in yellow. The insertion in 1CJA:A does not severely disrupt the position of these functional residues, but it shifts them enough that CE misaligns them (Plate 20.4a).

An alignment of 1CJA:A with the cAMP-Dependent Protein Kinase (PDB id 1CDK:A) will be explored once again here. The overall similarity in the catalytic regions of these proteins is reasonably high. Still, the spatial translation of the loops in 1CJA is enough to place the α-carbons of residues in the structure in close proximity to non-equivalent α-carbons in 1CDK:A, resulting in the misalignment of the critical Asp residue. In addition, this shift in the alignment also translates through to the conserved Asn residue and the 3-10 helix / β-strand that it participates in. In Plate 20.4b, the conserved Asp and Asn positions shared between the two structures are shown in yellow for 1CJA:A (D204, N218) and green for 1CDK:A (D166, N171.) Apparently, CE attempts to make an extensive alignment between the insertion in 1CJA:A and the standard connection between these residues in 1CDK:A, leading to a faulty result for the two most important positons. This alignment violates almost all of our criteria for optimality. Therefore, in our manual alignment we shift the residues back so that the similar positions in the two structures (and similar sequence in the structures) are aligned.

***Example #4*: Flexibility in functional motifs—alignment of the DFG motif in two tyrosine kinases.**

In the previous example, enough structural information is present that an ideal structural alignment algorithm should be able to determine the correct alignment, even without knowledge of the conserved sequence features. However, this is not always the case with conserved motifs. The DFG motif is a highly conserved motif that follows the DxxxxN motif in the sequences of almost all of the kinases. The Asp residue in this motif is involved in an important coordination with a metal atom in the active site complex.[19,20] Though the DFG motif is well conserved, this portion of the structure is in a poorly stabilized loop that is often in different conformations.

Alignment of the two tyrosine kinases discussed above, Insulin Receptor Tyrosine Kinase (PDB id 1IR3:A) and c-Src (PDB id 2SRC), will be explored here once again. Though these two structures are closely related, the protein chain in the area of the DFG motif is extremely displaced, as can be seen in Plate 20.5a. The effect of the displacement is that the protein backbone of the Gly residue in 2SRC aligns better with the backbone of the Phe residue of 1IR3:A. Plate 20.5b presents the CE alignment of the DFG motif and the surrounding sequence. The two chains are colored orange for 1IR3:A and light blue for 2SRC, with the DFG motif in each in a darker color (red for 1IR3:A, blue for 2SRC). Residues shown are 1148-1153 for 1IR3:A and 402-408 for 2SRC. C_α atoms are shown as small spheres, color coded based on status. Green and white atoms represent aligned positions within the structure. These colors alternate so as to make the pairing between residues in the alignment clear. Yellow atoms represent unpaired (gapped) positions in the alignment. As can be seen, the CE alignment makes a great deal of spatial sense, with positions whose C_α atoms are closest in space being paired. However, this alignment does not respect the stringent conservation of this motif, because it does not align the Phe residue of 2SRC with the Phe residue of 1IR3:A, and indeed shifts the alignment from this point on.

Because there are no structural cues in the protein backbone, it is difficult to imagine how a structure alignment program would be able to correctly align this motif. Interestingly, the Phe residue side chains of the two structures do occupy similar locations in the structure, despite the backbone displacement. This suggests that an enhanced emphasis on side chain positioning in the alignment might be helpful. Based on sequence conservation and this positioning, our alignment of the DFG motif is made without gaps, as shown in Plate 20.5c. While this makes for a more biologically meaningful alignment, it has consequences for the spatial correctness, since some of the paired atoms between the structures are distant from each other in space.

20.7. Conclusion

The examples presented here illustrate the difficulties inherent in producing a high-quality automatic alignment. However, the ability to produce expert-quality alignments automatically would be extremely valuable, and therefore we are continuing to modify our heuristics to further improve the results. Several potential modifications are suggested by our explorations of the CE alignments.

While it is clear that secondary structure is not always a perfect guide to alignments, it also often provides spatially repetitive regions that can be extremely helpful to the alignment process. However, as seen in example #1, considering these elements by simply superimposing the structures can sometimes lead to incorrect residue equivalences because of the displacement they undergo. This effect can also be seen in well conserved looping regions such as described in example #3. For this reason, we are exploring methods for a local sub-alignment optimization that is guided only loosely by the overall initial CE alignment. Such a method would be able to compare secondary structures based on local geometry, despite their rotation or translation in space. Additionally, it is clear that this kind of local alignment must consider as many atoms of the protein chain as can be reasonably compared, as this guards against alignments of positions that have side chains in completely different places in space (as was discussed in example #2). In example #4, for example, some of the residues CE fails to align do have side chains reasonably close in space despite backbone dislocation. Thus, an expansion of the use of the whole of the peptide chain in optimizations will likely be useful.

Specifically, our analysis suggests that a context-dependent scoring function would be helpful in achieving the aims above, as uniform affine gap penalties alone are clearly not sufficient for the satisfactory alignment of all parts of the proteins. It appears that particular features of the manual alignment can only be captured within a specific affine gap function. For example, gap penalties could be set differently based on context such as secondary structure type, with strands having a lower gap penalty than helices. Further, many alignment difficulties might be corrected by using a scale-dependent scoring function that treats local and global levels of protein structure similarity differently, particularly by applying a different stringency of similarity criteria at different levels. Finally, there are certain features in manual alignments that cannot be captured with any affine gap function, for example large domain movements. Our analysis suggests that a multiple rigid body approach could be used to deal with this kind of alignment difficulty. In this approach, the structures would be able to undergo multiple superpositions in order to achieve the best superposition for each specific subsection of the alignment under consideration. The scores for these other possible alignments would then be added to the primary score, based on the overall structural superposition, to provide additional guidance in aligning any displaced regions. We anticipate that such an enhanced scoring function will be developed and tuned using manual structure alignments as a "gold standard."

Ultimately, an important aspect of any structure alignment program is outputs that can be easily viewed, adjusted, and transformed by the researcher using the software. Because of the limitations inherent in automated structure alignments, and the different requirements different researchers may have for their alignments, it does not seem likely that the structure alignment problem will be completely solved. Therefore, enabling expert input into the final alignment result would be expected to produce the best possible outcomes.

References

1. Lesk AM, Chothia C: **How different amino acid sequences determine similar protein structures: The structure and evolutionary dynamics of the globins.** *J Mol Biol* 1980, **136**(3):225–270.
2. Chothia C, Lesk AM: **The relation between the divergence of sequence and structure in proteins.** *EMBO J* 1986, **5**(4):823–826.
3. Pastore A, Lesk AM: **Comparison of the structures of globins and phycocyanins: Evidence for evolutionary relationship.** *Proteins* 1990, **8**(2):133–155.
4. Grishin NV: **Phosphatidylinositol phosphate kinase: A link between protein kinase and glutathione synthase folds.** *J Mol Biol* 1999, **291**(2):239–247.
5. Orengo CA, Brown NP, Taylor WR: **Fast structure alignment for protein databank searching.** *Proteins* 1992, **14**(2):139–167.

6. Holm L, Sander C: **Protein structure comparison by alignment of distance matrices.** *J Mol Biol* 1993, **233**(1):123–138.

7. Madej T, Gibrat JF, Bryant SH: **Threading a database of protein cores.** *Proteins* 1995, **23**(3):356–369.

8. Gibrat JF, Madej T, Bryant SH: **Surprising similarities in structure comparison.** *Curr Opin Struct Biol* 1996, **6**(3):377–385.

9. Shindyalov IN,. Bourne PE: **Protein structure alignment by incremental combinatorial extension (CE) of the optimal path.** *Protein Eng* 1998, **11**(9):739–747.

10. Dayhoff MO, Schwartz RM, Orcutt BC: **A model of evolutionary change in proteins.** In: *Atlas of Protein Sequence and Structure,* vol.5, suppl. 3. Dayhoff MO, ed. Washington, DC: National Biomedical Research Foundation. 1978:345-352.

11. Henikoff S, Henikoff JG: **Amino acid substitution matrices from protein blocks.** *Proc Natl Acad Sci USA* 1992, **89**(22):10915–10919.

12. Needleman SB, Wunsch CD: **A general method applicable to the search for similarities in the amino acid sequence of two proteins.** *J Mol Biol* 1970, **48**(3):443–453.

13. Smith TF, Waterman MS: **Identification of common molecular subsequences.** *J Mol Biol* 1981, **147**(1):195–197.

14. Pearson WR, Lipman DJ: **Improved tools for biological sequence comparison.** *Proc Natl Acad Sci USA* 1988, **85**(8):2444–2448.

15. Altschul SF, Madden TL, Shaffer AA, Zhang J, Zhang Z, Miller W, Lipman DJ: **Gapped BLAST and PSI-BLAST: A new generation of protein database search programs.** *Nucleic Acids Res* 1997, **25**(17):3389–3402.

16. Godzik A: **The structural alignment between two proteins: is there a unique answer?** *Protein Sci* 1996, **5**(7):1325–1338.

17. Levitt M, Gerstein M: **A unified statistical framework for sequence comparison and structure comparison.** *Proc Natl Acad Sci USA* 1998, **95**(11):5913–5920.

18. Lathrop RH: **The protein threading problem with sequence amino acid interaction preferences is NP-complete.** *Protein Eng* 1994, **7**(9):1059–1068.

19. Taylor SS, Radzio-Andzelm E: **Three protein kinase structures define a common motif.** *Structure* 1994, **2**(5):345–355.

20. Hanks SK, Hunter T: **Protein kinases 6. The eukaryotic protein kinase superfamily: Kinase (catalytic) domain structure and classification.** *FASEB J* 1995, **9**(8):576–596.

21. Cox S, Radzio-Andzelm E, Taylor SS: **Domain movements in protein kinases.** *Curr Opin Struct Biol* 1994, **4**(6):893–901.

22. Sowadski JM, Epstein LF, Lankiewicz L, Karlsson R: **Conformational diversity of catalytic cores of protein kinases.** *Pharmacol Ther* 1999, **82**(2-3):157–164.

23. Steinbacher S, Hof P, Eichinger L, Schleicher M, Gettemans J, Vandekerckhove J, Huber R, Benz J: **The crystal structure of the Physarum polycephalum actin-fragmin kinase: an atypical protein kinase with a specialized substrate-binding domain.** *EMBO J* 1999, **18**(11):2923–2929.

24. Hubbard SR: **Crystal structure of the activated insulin receptor tyrosine kinase in complex with peptide substrate and ATP analog.** *EMBO J* 1997, **16**(18):5572–5581.

25. Xu W, Doshi A, Lei M, Eck MJ, Harrisson SC: **Crystal structures of c-Src reveal features of its autoinhibitory mechanism.** *Mol Cell* 1999, **3**(5):629–638.

26. Bossemeyer D, Engh RA, Kinzel V, Ponstingl H, Huber R: **Phosphotransferase and substrate binding mechanism of the cAMP—Dependent protein kinase catalytic subunit from porcine heart as deduced from the 2.0 A structure of the complex with Mn2+ adenylyl imidodiphosphate and inhibitor peptide PKI(5-24).** *EMBO J* 1993, **12**(3):849–859.

27. Rao VD, Misra S, Boronenkov IV, Anderson RA, Hurley JH: **Structure of type IIbeta phosphatidylinositol phosphate kinase: A protein kinase fold flattened for interfacial phosphorylation.** *Cell* 1998, **94**(6):829–839.

Index